Passive Vibration Isolation

by
Eugene I. Rivin

NEW YORK   ASME PRESS   2003

© 2003 by The American Society of Mechanical Engineers
Three Park Avenue, New York, NY 10016
ISBN: 0-7918-0187-X

Co-published in the UK by Professional Engineering Publishing Limited,
Northgate Avenue, Bury St. Edmunds, Suffolk, IP32 6BW, UK
ISBN: 1-86058-400-4

All rights reserved. Printed in the United States of America. Except as permitted under the United States Copyright Act of 1976, no part of this publication may be reproduced or distributed in any form or by any means, or stored in a database or retrieval system, without the prior written permission of the publisher.

INFORMATION CONTAINED IN THIS WORK HAS BEEN OBTAINED BY THE AMERICAN SOCIETY OF MECHANICAL ENGINEERS FROM SOURCES BELIEVED TO BE RELIABLE. HOWEVER, NEITHER ASME NOR ITS AUTHORS OR EDITORS GUARANTEE THE ACCURACY OR COMPLETENESS OF ANY INFORMATION PUBLISHED IN THIS WORK. NEITHER ASME NOR ITS AUTHORS AND EDITORS SHALL BE RESPONSIBLE FOR ANY ERRORS, OMISSIONS, OR DAMAGES ARISING OUT OF THE USE OF THIS INFORMATION. THE WORK IS PUBLISHED WITH THE UNDERSTANDING THAT ASME AND ITS AUTHORS AND EDITORS ARE SUPPLYING INFORMATION BUT ARE NOT ATTEMPTING TO RENDER ENGINEERING OR OTHER PROFESSIONAL SERVICES. IF SUCH ENGINEERING OR PROFESSIONAL SERVICES ARE REQUIRED, THE ASSISTANCE OF AN APPROPRIATE PROFESSIONAL SHOULD BE SOUGHT.

ASME *shall not be responsible for statements or opinions advanced in papers or . . . printed in its publications* (B7.1.3). Statement from the Bylaws.

For authorization to photocopy material for internal or personal use under those circumstances not falling within the fair use provisions of the Copyright Act, contact the Copyright Clearance Center (CCC), 222 Rosewood Drive, Danvers, MA 01923, tel: 978-750-8400, www.copyright.com.

Library of Congress Cataloging-in-Publication Data

Rivin, Eugene I.
Passive vibration isolation / Eugene I. Rivin.
    p. cm.
  Includes bibliographical references.
  ISBN 079810187X
1. Vibration.
2. Damping (Mechanics).
I. Title
TA355.R52 2003
620.3'7—dc21   2002034287

CONTENTS

Preface ix

Chapter 1 Dynamic Properties of Vibration Isolation Systems 1

1.0 General Comments 3
1.1 Inertia and Geometric Properties of Typical Machines and Other Mechanical Devices 5
1.2 Basic Characteristics of Elastic Mounts (Vibration Isolators) 8
1.3 Elastically Supported Solid Object Dynamic Coupling 12
1.4 Dynamics of Single-Degree-of-Freedom Vibration Isolation System 25
   1.4.1 Vibration Isolator with Viscous Damper 28
   1.4.2 Vibration Isolator with Internal (Hysteresis) Damping 31
   1.4.3 "Relaxation" Isolation System with Viscous Damper 36
   1.4.4 Single-Degree-of-Freedom System with Motion Transformation 41
1.5 Dynamics of Two-Degrees-of-Freedom Vibration Isolation System 48
   1.5.1 Generic Two-Degrees-of-Freedom System with Damping 48
   1.5.2 Vibration Isolator with Intermediate Mass (Two-Stage Isolator) 56

1.6	Three-Degrees-of-Freedom (Planar) Vibration Isolation System		61
	1.6.1	Dynamics of Planar Isolation System	62
	1.6.2	Natural Frequencies of Planar Vibration Isolation System	65
	1.6.3	Planar Vibration Isolation System with Inclined Mounts	70
		1.6.3A Focal Equi-Frequency System	74
1.7	Vibration Isolation System under Random Excitation		77
	1.7.1	Definitions	77
	1.7.2	Travel of Random Vibration Through Dynamic System	79
1.8	Vibration Isolation System under Pulse Excitation		83
1.9	Nonlinearity in Vibration Isolation Systems		89
1.10	Wave Effects in Vibration Isolators		96

References — 100

Chapter 2 Principles and Criteria of Vibration Isolation — 103

2.1	General Comments		105
	2.1.1	Damping vs. Heat Generation in Vibration Isolators	107
2.2	Isolation of Vibration-Sensitive Objects		109
	2.2.1	Ambient Vibrations	110
	2.2.2	Detrimental Effects of Vibration	119
	2.2.3	Model of Vibration Transmission	123
	2.2.4	Principles and Criteria of Vibration Isolation	129
		2.2.4A Isolation from Steady-State Vibration	130
		2.2.4B Practical Selection of Vibration Isolation Parameters for Precision Objects	139
		2.2.4C Isolation from Impulsive Vibration	142

	2.2.5		Vibration Isolation Systems	144
	2.2.6		Side Issues for Vibration Isolated Precision Equipment	145
		2.2.6A	Reduction of Mobility of Isolated Objects Caused by Internal Dynamic Forces	146
		2.2.6B	Influence of Vibration Isolation on Effective Stiffness of Isolated Object	149
	2.2.7		Vibration Protection of Civil Engineering Structures	156
2.3	Isolation Requirements for Vibration-Producing Objects			160
	2.3.1		Objects Producing Single Frequency Excitations	160
		2.3.1A	Objects Producing Unidirectional Excitation	160
		2.3.1B	Objects Producing Multidirectional Excitation	162
	2.3.2		Objects Producing Polyharmonic Excitations	166
	2.3.3		Objects Producing Conservative Impact Excitations	169
	2.3.4		Objects Generating Pulses of Inertial Nature	172
2.4	General Purpose Machinery and Equipment			180
	2.4.1		Influence of Mounting Conditions on Dynamic Stability (Chatter)	180
	2.4.2		Vibration Levels of General Purpose Machines	183
	2.4.3		Influence of Vibration Isolation on Bearing Loads	185
	2.4.4		Influence of Vibration Isolation on Noise	188
2.5	Vibration Isolated Objects Installed on Non-Rigid Supporting Structures			188

2.6	Engine and Machinery Mounting in Vehicles		192
2.7	Experimental Selection of Isolators		202
2.8	Role of Damping in Vibration Isolation		203

References 207

Chapter 3   Realization of Elasticity and Damping in Vibration Isolators   211

3.0	General Comments		213
3.1	Metal Elastic Elements (Springs)		213
	3.1.1	Coil Springs	214
		3.1.1A Statically Nonlinear Coil Springs	218
	3.1.2	Slotted Springs	220
	3.1.3	Belleville Springs	222
3.2	Dampers		224
	3.2.1	Viscous Dampers	225
	3.2.2	Dry (Coulomb) Friction Dampers	227
	3.2.3	Electromagnetic Dampers	229
3.3	Elasto-Damping Materials (EDM)		229
	3.3.1	Meshed Fibrous Materials	230
		3.3.1A Wire-Mesh Materials and Cables	230
		3.3.1B Felt	233
	3.3.2	Elastomeric (Rubberlike) Materials	235
		3.3.2A Static Deformation Characteristics of Rubberlike Materials	238
		3.3.2B Streamlined ("Ideal Shape") Rubber Elements	243
		3.3.2C Thin-Layered Rubber-Metal Laminates	250
		3.3.2D Elastic Stability of Laminated Rubber Parts	255

		3.3.2E	Dynamic Characteristics of Rubberlike Materials	259
		3.3.2F	Creep of Rubberlike Materials	270
	3.3.3	Plastic Materials		271
3.4	High Damping Metals			272
3.5	Pneumatic Flexible Elements			275
3.6	Hydromounts (Basics)			284

References — 292

Chapter 4 Passive Vibration Isolation Means — 297

- 4.1 General Comments — 299
- 4.2 Material Selection for Vibration Isolators — 301
- 4.3 Isolating Mats and Pads — 304
  - 4.3.1 Rubber Mats — 305
  - 4.3.2 Plastic and Fibrous Pads — 310
- 4.4 Vibration Isolators with Rubber Flexible Elements — 314
  - 4.4.1 Machinery and Engine Mounts — 315
- 4.5 Constant Natural Frequency (CNF) Vibration Isolators — 337
  - 4.5.1 General Comments; Variable Stiffness Isolators — 337
  - 4.5.2 Designs of CNF Isolators — 346
- 4.6 Vibration Isolators with Metal Flexible Elements — 362
  - 4.6.1 General Comments — 362
  - 4.6.2 Vibration Isolators with Coil Springs — 363
  - 4.6.3 Wire Mesh and Cable Isolators — 368
- 4.7 High Angular Stiffness and Anisotropic Vibration Isolation Systems — 372
- 4.8 Low Stiffness Isolators — 377
  - 4.8.1 Buckling and Stiffness — 378

4.9	Pneumatic Isolators		386
	4.9.1	General Comments	386
	4.9.2	General Purpose Pneumatic Isolators	388
	4.9.3	Self-Leveling Isolators	392
4.10	Vibration Isolated Foundations and Installation Systems		395
	4.10.1	Installation Systems with Inertia Blocks	395
	4.10.2	Installation System with Reference Frame	399
4.11	Isolation of Non-Supporting Connections of Isolated Objects		402
4.12	Power Transmission Couplings		407

References	415
Index	421

PREFACE

On the one hand, the intensity of vibration sources around us is increasing. Some of the reasons for this phenomenon are: increasing speeds of machinery, proliferation of off-road vehicles and four-cylinder internal combustion engines, continuous improvements of cutting inserts allowing heavier cuts on machine tools, etc. On the other hand, tolerances on allowable vibration levels are becoming more and more stringent. This is due to higher and higher required precision, continuously improving sensitivity of measuring instruments, proliferation of light (but strong) structures that are more prone to vibration, competitive pressure on improving human comfort both in stationary structures and in vehicles, etc. To accommodate these contradictory phenomena, vibration protection means are continuously improving.

Vibration isolation is one of the vibration control techniques whereby the *source* of vibration excitation and the *object* to be protected are separated by an auxiliary system comprising special devices called *vibration isolators* or *vibration isolating mounts*. The effect of vibration isolation amounts to weakening of *dynamic coupling* between the source and the object, thus reducing transmission of unwanted vibration excitations to or from the object. The weakening of dynamic coupling may also result in "side" effects such as increasing static displacements between the source and the object; increasing low frequency and transient relative displacements between the source and the object; and in increases in size, weight, and cost of the installation. Installation of several connected units on compliant vibration isolators increases misalignment between these units, which is usually undesirable. Thus, in many cases a multi-parameter optimization is required to provide acceptable vibration reduction effects while satisfying other constraints.

Servo-controlled ("active") vibration isolators are increasingly used in cases when low frequency vibration isolation is required while large static deformations of the low frequency passive vibration isolators are undesirable or unacceptable. However, "passive" isolators are still used in an overwhelming majority of applications.

Vibration isolation is possibly the most widely used approach for vibration protection. It breaks the vibration transmission path from the source to the vibration-sensitive unit of equipment or the civil engineering structure and, if properly designed, improves the performance of the connected objects. The other vibration protection approaches are 1) reduction of vibration excitations at the source by design, changes in mass and/or stiffness and/or damping of the structural components, use of dynamic vibration absorbers, etc., and 2) increase in robustness of the object so it is less sensitive to transmitted vibrations.

In today's environment, not only is better protection from vibration required from the vibration isolation means, they also have to comply with additional constraints that have become more and more difficult to accommodate. Some examples include a need to install highly vibration-sensitive equipment in light and low-stiffness structures, exposure of elastomeric passive vibration control elements or electronic active vibration control elements to high temperatures (e.g., in engine compartments of vehicles), as well as a need for very space-restrictive packaging of vibration control devices. So, where a set of spacers made up of some off-the-shelf rubber would have been sufficient 30 to 50 years ago, much more sophistication and optimization are required now. However, with few exceptions, commercially available passive vibration isolation means have not changed much during the last 30 to 50 years. The new developments, such as hydromounts, are not as reliable and are much more expensive than the vibration isolators with elastomeric or metal flexible elements. They often require a tedious tuning for a specific application. Even less reliable and more expensive are active isolators. Accordingly, such "high tech" isolators have to be used in cases where their unique performance characteristics justify living with their shortcomings. However, their shortcomings are often overlooked and their advantages are often overblown by clever promotion, and they are used in the systems where much less expensive, more reliable, and more compact passive isolators could be used, provided that they are properly selected. But this requires a better understanding of principles and basic concepts of vibration isolation of real-life objects, which is frequently lacking.

One reason for this is a lack of literature dedicated to a comprehensive understanding of vibration isolation techniques. Vibration isolation is presented in a very basic way in vibration textbooks, in somewhat more details in handbooks. But in real life, much more information is required for an in-depth understanding of vibration isolation systems. Without such in-depth understanding, optimization of real-life systems is impossible. A wealth of information is available in papers published in transactions, proceedings, and patent specifications but their collection and assessment is very difficult for a practicing designer and even more so for a user. The last (and, possibly, also the first) comprehensive book on vibration isolation was the classic book "Vibration and Shock Isolation" authored by Ch.E. Crede and published in 1951.

This book attempts to present a more-or-less comprehensive treatment of passive vibration isolation, covering analysis of basic isolation systems and dynamic processes in them, requirements for isolation of various objects, as well as materials used in and designs of passive vibration isolators. The theory and designs of active vibration isolators are not addressed, with two exceptions of "static" active systems, one of self-leveling isolators (Section 4.9.3), and another of an installation system with a reference frame (Section 4.10.2).

Vibration isolation is a very broad area, including isolation of stationary equipment and measuring instruments, both vibration-sensitive and vibration-producing; isolation of vibration- and shock-sensitive electronic boards; isolation of engines and other parts of surface, underwater, and flying vehicles; etc. This book mainly addresses vibration isolation of stationary equipment, although a large amount of information can be used for the other cases of vibration isolation.

The importance of vibration isolating mounting of the stationary equipment is increasing because of several reasons. As noted above, increasing working speeds and installed power—frequently achieved together with reduction in weight of machinery—lead to increasing dynamic loads and more intense vibration generation. At the same time, the degree of precision of both production machines and measuring instruments is increasing at a fast pace. In some cases, increasing precision is accompanied by increasing productivity, e.g., in coordinate-measuring machines (CMM). The high-performance CMMs, as well as such ultra-precision devices as photolithography tools for semiconductor production, have fast-moving heavy structural parts whose acceleration/deceleration rates often must be reduced in order to reduce influence of the transient vibrations on measuring or production accuracy. The popular "flexible production" lines often require placing precision finishing machines and measuring apparatuses next to basic production machines, sometimes intensive vibration producers. Multistory production facilities are used more and more frequently overseas; their advent to the United States is a matter of time due to increasing cost of land and, in some cases, due to a potential for savings in material handling. Installation of production and measuring equipment on the upper floors (or over the basement) may create problems because of reduced stiffness of the floor structures. Recently, vibration isolation became a critical issue for military equipment transported on vehicles, frequently off-road, and exposed to the harsh battleground environment. Previously, military optical and electronic equipment were designed with special ruggedization techniques, making them less susceptible to damage in harsh environments. The prevailing trend now is "commercial off the shelf" (COTS) electronic, optical, etc., components and units that do not have the built-in protection and require external vibration and shock isolation means.

As was stated above, the "classic" principles of vibration isolation described in textbooks and handbooks are often not applicable to real-life situations because of dynamic complexity of the isolated objects, supporting structures, excitations, and vibration isolators. Detailed dynamic analysis/modeling of each practical case using available instrumentation and/or software packages would result in huge expenses in both time and money. Even more important, often the representative modeling is impossible because the accurate input data on structures and excitations are not available.

Currently, there are two most widely used approaches to vibration isolation problems. The "down-to-Earth" approach considers the isolated object as a single-degree-of-freedom system following the basic treatments in vibration textbooks. Selection of isolators is performed by the classical "trial-and-error" approach, which is very time-consuming and results in vibration isolation systems that are far from optimal. Another, "modern" or "modernistic" approach treats the vibration isolation system as a complex six-degrees-of-freedom system (or even as a more complex system considering dynamics of the supporting structure and/or of the isolated object). This approach requires complex and time-consuming computer modeling using big, usually proprietary, software packages. However, due to the above mentioned lack of reliable input data, the result is largely the same—an extensive and lengthy trial-and-error process at the final stages of the design process. The latter approach is typical, for example, for automotive industry (engine mounting).

However, it has been practically demonstrated that an in-depth understanding of typical groups of vibration isolation problems allows formulation of general approaches to typical cases of vibration isolation, resulting in near-optimal performance without the need for complex individual studies.

This book presents such generalized approaches to typical cases of vibration isolation. In its four chapters, four basic areas important for realization and optimization of vibration isolation are presented and/or discussed.

Chapter 1 discusses dynamics of vibration isolation systems with the aim of addressing the core issues for practical applications while using only the minimal mathematical apparatus. The treatment is performed largely with the assumptions of the solid body isolated object and the solid body supporting structure. The classical six-degrees-of-freedom system is initially considered, but it is shown that for many practical applications the properly designed vibration isolation system can be reduced to a set of single- and two-degrees-of-freedom systems. Areas such as isolation from random excitations and nonlinear dynamics of vibration isolation systems each deserve a separate book. Here, only special topics that are important for real-life vibration isolation systems are briefly addressed. Some "obscure" systems, such as old and mostly forgotten focalized systems and system with motion transformation, are described since they can be quite useful in some niche applications. The performed analysis clearly demonstrated advantages of so-called constant natural frequency (CNF) vibration isolators.

Chapter 2 formulates practical principles of vibration isolation of vibration sensitive objects, of vibration-producing objects, and of widely used, important, but not "very sexy" and thus attracting little attention "general purpose" group of machinery and equipment. It is shown that in the overwhelming majority of cases wherein vibration isolation is required, damping in isolators plays as important a role as stiffness or natural frequencies of the isolation system. In many cases, interrelation of stiffness/natural frequencies and damping can be expressed as criteria. These criteria make the tasks of designer/consultant easier and allow optimization of material selection for flexible elements of vibration isolators, depending on intended applications of these isolators. Since in many cases an increase in damping is very beneficial to the isolation system performance, Chapter 2 starts with a proof that the increased damping in vibration isolators wouldn't cause overheating of the flexible element, but just the opposite. Naturally, the most attention in this chapter is given to vibration isolation of precision/vibration sensitive equipment and machinery. It is demonstrated that such vibration isolation systems often develop the resonance condition, which becomes a legitimate working regime, statements in vibration textbooks notwithstanding. The importance of both damping in the vibration isolators and of the CNF characteristic of the isolators for vibration sensitive objects are clarified. Special attention is given to so-called side issues, such as influence of vibration isolating installation on effective stiffness of the isolated precision object, and reduction of negative effects of the dynamic loads generated during the working process inside the protected precision object itself. Many examples are given to illustrate the procedure for specifying the isolation means for typical precision objects. These examples clearly indicate that vibration isolation systems and/or civil engineering structures for many high precision objects are often grossly overdesigned.

Design of vibration isolators for various applications requires knowledge and understanding of both static and dynamic characteristics of materials used for flexible elements of the isolators. Chapter 3 summarizes some important static and dynamic characteristics of the most widely used materials, such as elastomers, wire mesh/cable elements, etc., as well as provides basic information on pneumatic flexible elements and on hydromounts. Special attention is given to such important (or potentially important) issues as elastomeric elements of streamlined shapes and thin-layered rubber-metal laminates. Since such condensed information on properties of materials used in passive vibration control devices is not available in other sources, this brief survey can be useful for readers interested not only in vibration isolation, but in other vibration control techniques.

Chapter 4 presents an analytical survey of vibration isolator designs. There are hundreds of such designs on the market, thus only typical ones are surveyed. It seems that such a survey provides much of the needed information both for users of vibration isolation means and for designers of vibration isolators. There is a special emphasis on CNF isolators having substantial advantages in many applications. Besides the designs of vibration isolators, some "side issues" are addressed, such as installation systems for stiffness-critical machinery and connections between the isolated objects and their environment.

While the Chapters were intended to be self-contained, all the issues addressed in the book are closely intertwined. So, even though some readers might be especially interested in one Chapter/Section, some tables or figures from other Chapters/Sections, even appearing far later in the book page count, might be quoted. This is explained by the author's conviction that these later Chapters/Sections are more logical locations for placement of these tables/figures. Some statements are repeated if it makes the respective Sections more complete and understandable.

The book aims to summarize the author's many years of experience in analysis and design of vibration isolation and other vibration control systems, in R&D on vibration isolator designs, and in practical "shop floor" applications. The need for such a book became obvious from many encounters with engineers both designing and using vibration isolation techniques and means, and especially from internationally attended tutorials on vibration isolation, which the author ran (together with Prof. Daniel B. DeBra, Stanford University) for many years at the Annual Meetings of the American Society for Precision Engineering (ASPE). The author is grateful and indebted to Dan DeBra, who read the manuscript and made numerous comments, which significantly improved the book.

The book is quite condensed and may not cover some issues in adequate detail. Many more details are available in periodicals and proceedings. The rather extensive reference database in the book may be useful for the readers requiring more detailed information. The mathematical apparatus used in the book is rather elementary, but even these simple derivations can be skipped by the readers more interested in practical conclusions and advice.

The book may be useful both for vibration consultants, designers and users of vibration isolated equipment, and for designers, users, and manufacturers of vibration control devices. The author hopes that instructors in both basic and advanced vibration courses would find some sections worthy of presenting in the classroom.

CHAPTER 1

Dynamic Properties of Vibration Isolation Systems

This Chapter provides the basic information necessary for designing well functioning vibration isolation systems for various groups of objects. The textbook-like analysis of single-degree-of-freedom vibration isolation systems illustrates influence of damping on vibration isolation as well as importance of the damping mechanism involved in a particular application. Consideration of intentional or non-intentional nonlinearity of vibration isolators is shown to be important.

Vibration isolation systems for real-life objects are characterized by much more than one degree-of-freedom. This issue is considered in two aspects. On one hand, the basic model of a vibration isolation system, assuming a rigid object isolated by resilient mounts from a rigid supporting structure, has six degrees-of-freedom. Importance of dynamic coupling in this system and, in most of the cases, importance of decoupling between the vibratory modes in the "natural" coordinate frame are addressed together with guidelines for achieving such decoupling. Both desirable and undesirable nonlinear effects are illustrated.

Another consideration of the multi-degrees-of-freedom character of the system is briefly given in relation to high frequency (wave) resonances in vibration isolators.

Special design concepts for vibration isolators are described. These can be useful in many practical applications.

1.0 GENERAL COMMENTS

A vibration isolation system comprises the object being isolated, the supporting surface (floor, foundation), and the vibration isolators (mounts) placed between them. Fig. 1.0.1 illustrates the simplest "uni-axial" vibration isolation system. Depending on circumstances, the object must be protected from the objectionable vibratory motions of the supporting surface, or the supporting surface must be protected from the dynamic loads generated within the object. In some cases both tasks have to be addressed simultaneously (see Chapter 2). Each of the three subsystems (object, isolator, supporting surface) shown in Fig. 1.0.1 generally represents a distributed parameter dynamic system with many or an infinite number of degrees of freedom.

The object being isolated (e.g., a machine) is usually made up of numerous connected (joined) components. If each component is approximated as an absolutely rigid massive body, and the connections (joints) are approximated as lumped-parameter springs, then each component has six degrees of freedom. At higher frequencies each component behaves as a distributed parameter dynamic system with an

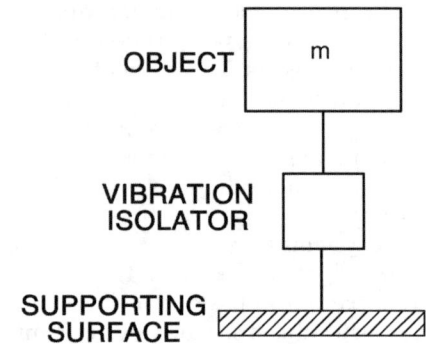

Fig. 1.0.1. Uni-Axial Vibration Isolation System.

infinite number of degrees of freedom. The floor/foundation structure is also a distributed parameter dynamic system with an infinite number of degrees of freedom. The same statement can be made about vibration isolators. Keeping in mind that many connections/joints inside and between the sub-systems in Fig. 1.0.1 are nonlinear, one can imagine the dynamic complexity of the vibration isolation system. Fortunately, in the overwhelming majority of practical vibration isolation cases, only a low-frequency range is of interest. Also, in many cases the lowest (the most important for vibration isolation) natural frequencies of the object as well as of the supporting structure are significantly higher than the lowest natural frequencies of the vibration isolation system "object-isolators-supporting structure/floor." Usually (but not always) the mass of the flexible element of each isolator is a small fraction of the mass of the object being isolated. It can be neglected at low frequencies and the isolator can be considered as a massless spring. At higher frequencies this assumption cannot be used and wave resonances in the flexible element have to be considered (e.g., see Section 1.10). In many practical cases, the effectiveness of vibration isolation is especially critical in the low-frequency range, about 5 to 100 Hz. Usually, only the lowest structural modes of the object, of the isolators, and of the supporting structure are positioned in this frequency range, and the higher modes can be neglected. Many useful conclusions can be made when both the object and the supporting structure are considered as ideal rigid bodies and the isolators are considered as massless.

Based on these assumptions, the vibration isolation system can be considered, at least in the first approximation, as having six degrees of freedom. Consideration of special inertia and geometrical properties, as well as design properties of typical isolated objects often allows to further simplify the system.

A correct design of the vibration isolation system results in its effective performance (adequate isolation) with the maximum stiffness of the isolators, i.e., the maximum resistance of the system to rocking motion.

Due to the above-described extreme complexity of the vibration isolation system, it would be very useful to have a general understanding of the system and its subsystems. The generic features of vibration isolation systems of stationary objects are addressed in this chapter.

1.1 INERTIA AND GEOMETRIC PROPERTIES OF TYPICAL MACHINES AND OTHER MECHANICAL DEVICES

Inertia and geometric properties of the isolated object [*mass, moments/ radii of inertia*, position of the *center of mass/gravity* (C.G.), directions of *principal axes of inertia*] significantly influence dynamic characteristics of the vibration isolation system. Determining these parameters for a real-life object such as a heavy machine or an engine is not easy. It requires a lengthy experimental and/or analytical effort [1.1], [1.2], which can be justified only in special cases. However, typical objects such as machines, engines, etc., exhibit some generic trends related to their inertia and geometric properties [1.3], [1.4].

For many pieces of industrial equipment, the *natural coordinate frame* is such a frame in which axis Z is vertical and axes X, Y are in a horizontal plane. For example, the rotational axes of rotors in rotary machines are usually either vertical (centrifuges, separators, machine tools with vertical spindles, etc.) or horizontal (fans, compressors, turbines, lathes, machining centers with horizontal spindles, etc.). Similarly located are massive reciprocating parts (rams of forging hammers, tables and carriages of machine tools and CMMs, etc.). In vibration-producing machines dynamic loads are vertical (crank presses, etc.) or horizontal (cold headers, reversing tables and gantries of surface grinders and CMMs, etc.). Vibration-sensitive machines most frequently have their axes of maximum sensitivity in a horizontal plane. At the same time, maximum amplitudes of floor vibration are usually in the vertical direction, with horizontal vibrations being much weaker (see Section 2.2.1).

Positioning of the motion axes, as described above, results in such mass distribution whereas the principal axes of inertia are parallel to the basic structural axes of the isolated object (e.g., machine or apparatus) if the object has axes of symmetry. However, even if there are no axes of symmetry, the principal axes of inertia are still quasi-parallel to the structural axes. Detailed calculations of mass distribution have been performed for a medium-size lathe not having an axis of symmetry. It was shown that the principal axes of inertia deviate from the axes of the natural coordinate frame (one axis is vertical, and one horizontal axis coincides with the spindle axis) by not more than 8°–10°. Similar calculations have been performed for a medium-size surface grinder having a vertical plane of symmetry when the table is in its central position. In this case, one principal horizontal axis of inertia (perpendicular to the plane of symmetry) is directed along the design axis (direction of longitudinal motion of the table). Two other principal axes of inertia deviate from the respective (vertical and transverse) design axes by 9° 15'. The degree of deviation is characterized by the cosine of the deviation angle, and cos 10° = 0.985 ≈ 1.0. For a planar system, the *centrifugal moment of inertia* ("*product of inertia*") in the coordinate frame X', Y' inclined at angle α to the principal coordinate frame X, Y is (e.g., [1.5])

$$I_{x'y'} = 1/2(I_x - I_y)\sin 2\alpha. \qquad \text{Eq. (1.1.1)}$$

For $\alpha = 10°$, $\sin 2\alpha = 0.34$. Thus, in the worst case when the magnitude of one of the principal moments of inertia I_x, I_y is significantly larger that the other, $I_{x'y'}$ is less than 0.17 of the larger principal moment of inertia. Since in most practically important cases magnitudes of the principal moments of inertia are relatively close to one another, magnitudes of the products of inertia become even less significant. Thus, in the first approximation it can be assumed that for the machine-like objects whose design axes are parallel to axes X, Y, Z of the natural coordinate frame, the centrifugal moments of inertia (products of inertia) are

$$I_{xy} \approx I_{xz} \approx I_{yz} \approx 0. \qquad \text{Eq. (1.1.2)}$$

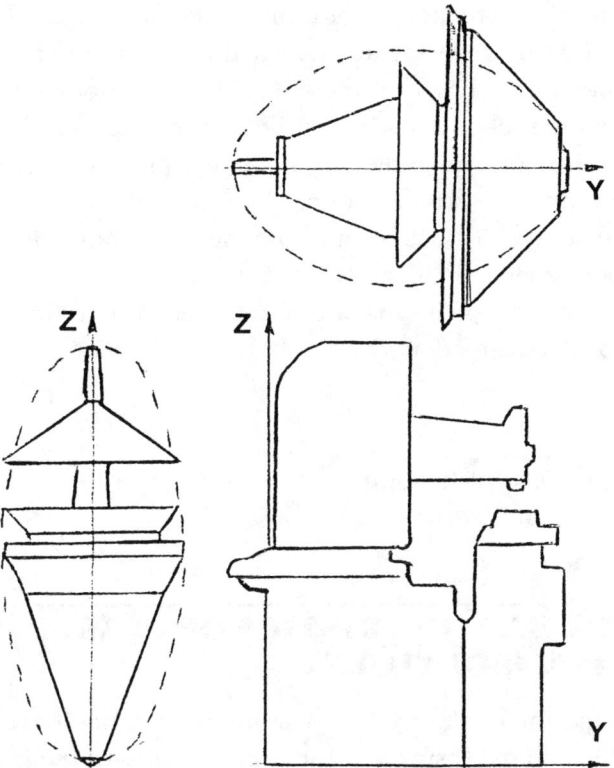

Fig. 1.1.1. Mass Distribution in Y-Z and X-Y Planes in a Surface Grinder.

Some general trends related to mass distribution in typical machine-like structures were also found. Usually, massive elements such as drives, gear boxes, and tables are packaged inside the machine. The external part is composed of relatively thin shells. A detailed analysis of mass distribution had been performed for several machine tools that were considered as typical representatives of the class "stationary machinery." This analysis has shown that the volumetric mass distribution is similar to a tri-axial ellipsoid. This is illustrated in Fig. 1.1.1, showing mass distribution along two axes for a medium-size surface grinder. Radii of inertia for a tri-axial ellipsoid are

$$\rho_i = \sqrt{\frac{1}{20}\left(H_j^2 + H_k^2\right)} \qquad \text{Eq. (1.1.3)}$$

where H are diameters of the ellipsoid, and i, j, k are cyclically assigned meanings of x, y, z designating directions of the axes.

Radii of inertia of a machine or a machine-like unit can be calculated from Eq. (1.1.3) if H designate overall dimensions of the unit in the respective directions. The use of Eq. (1.1.3) for various diverse objects, such as machine tools, IC engines, railway cars, etc., resulted in finding values of ρ with errors not exceeding 10–15%. Application of Eq. (1.1.3) to a non-symmetrical object—a vertical hammer crusher—gave values of radii of inertia only 3–10% different from the exact values determined by a very laborious experimental study.

Of course, if radii of inertia are known, corresponding moments of inertia can be determined as

$$I_i = M\rho_i^2 \qquad \text{Eq. (1.1.4)}$$

where M is the mass of the unit.

1.2 BASIC CHARACTERISTICS OF ELASTIC MOUNTS (VIBRATION ISOLATORS)

The elastic mount (isolator) is a volumetric elastic (flexible) element having two terminals or surfaces for connecting with the isolated object and with the supporting surface (see Chapter 4). Unless special restraints are employed, the flexible element develops resistance to motion between the terminals in all translational and angular directions. In the relatively low-frequency range, usually of the greatest importance for vibration isolation, the flexible element can be considered as massless. However, vibration transmissibility at high frequencies can be significantly affected by wave resonances in the body of the flexible element (see Section 1.10), and the mass of the flexible element has to be considered for analysis of such cases.

The flexible elements develop resistance to motion (forces) between the terminals. This resistance is dependent on the magnitude and direction of the relative displacement between the terminals, as characterized by the stiffness parameters, as well as the magnitude and direction of their relative velocity, as characterized by the damping parameters. In cases where the flexible elements are represented by metal (usually steel) springs, their damping is negligibly low and is frequently enhanced by specially designed dampers.

In general, application of a force to the flexible element results in its deformation in a direction different from that of the force. In an arbitrary coordinate frame, whose origin coincides with the *elastic center* of the flexible element (see below), elastic properties of a linear flexible element can be described by 36 *stiffness constants* k_{ij}, where the first subscript indicates direction of the forcing factor (force or moment) and the second subscript indicates direction of the displacement (linear or angular) caused by this forcing factor. Three "proper" translation stiffness constants k_{xx}, k_{yy}, k_{zz} describe deformations of the flexible elements in the directions of the coordinate axes from forces acting along the same axes. Six "cross" translational stiffness constants k_{xy}, k_{xz}, k_{yx}, k_{yz}, k_{zx}, k_{zy} describe translational deformations of the flexible element in the directions of the coordinate axes from forces acting along other axes; three "proper" angular stiffness constants $k_{\alpha\alpha}$, $k_{\beta\beta}$, $k_{\gamma\gamma}$ describe angular deformations of the flexible element about axes X, Y, Z, respectively, caused by the moments about the same axes. Six "cross" angular stiffness constants, $k_{\alpha\beta}$, $k_{\alpha\gamma}$, $k_{\beta\alpha}$, $k_{\beta\gamma}$, $k_{\gamma\alpha}$, $k_{\gamma\beta}$, describe angular deformations about one axis caused by moments about another axis. Another 18 "cross" stiffness coefficients represent translational-angular interactions, $k_{x\alpha}$, $k_{\alpha x}$, $k_{y\gamma}$, $k_{\gamma y}$, etc. The stiffness constants have reciprocity properties, $k_{ij} = k_{ji}$, or

$$k_{xy} = k_{yx}; k_{xz} = k_{zx}; \ldots; k_{\alpha\beta} = k_{\beta\alpha}; k_{\alpha\gamma} = k_{\gamma\alpha} \ldots; \quad \text{etc.} \qquad \text{Eq. (1.2.1)}$$

Directions of the deformation and of the forcing factor causing this deformation coincide only when the vector of the force or the moment is directed along (or about) one of the *principal elastic axes*. Any axis of symmetry is a principal elastic axis. The principal elastic axes are orthogonal to one another and intersect at the *elastic center* of the flexible element, which defines position of the element in the mathematical model of the system. This position is used in the equations of motion for the system. In a coordinate frame having its origin at the elastic center, translational/rotational cross-stiffness constants vanish. In the *principal coordinate frame* of the flexible element all "cross" stiffness constants vanish, $k_{ij} = 0$. If the principal elastic axes are designated as P, Q, R; the translational stiffness constants are k_p, k_q, k_r; the angular stiffness constants about axes P, Q, R are designated as k_π, k_θ, k_ρ; and the cosines of the angles between the principal elastic axes P, Q, R and axes X, Y, Z are

λ_{xp}, λ_{xq}..., then the following equations can be written for a coordinate frame having its origin at the elastic center of the mount:

$$k_{xx} = k_p \lambda_{xp}^2 + k_q \lambda_{xq}^2 + k_r \lambda_{xr}^2$$

$$k_{yy} = k_p \lambda_{yp}^2 + k_q \lambda_{yq}^2 + k_r \lambda_{yr}^2$$

$$k_{zz} = k_p \lambda_{zp}^2 + k_q \lambda_{zq}^2 + k_r \lambda_{zr}^2$$

$$k_{xy} = k_p \lambda_{xp} \lambda_{yp} + k_q \lambda_{xq} \lambda_{yq} + k_r \lambda_{xr} \lambda_{yr}$$

$$k_{xz} = k_p \lambda_{xp} \lambda_{zp} + k_q \lambda_{xq} \lambda_{zq} + k_r \lambda_{xr} \lambda_{zr}$$

$$k_{yz} = k_p \lambda_{yp} \lambda_{zp} + k_q \lambda_{yq} \lambda_{zq} + k_r \lambda_{yr} \lambda_{zr}$$

$$k_{\alpha\alpha} = k_\pi \lambda_{xp}^2 + k_\theta \lambda_{xq}^2 + k_\rho \lambda_{xr}^2$$

$$k_{\beta\beta} = k_\pi \lambda_{yp}^2 + k_\theta \lambda_{yq}^2 + k_\rho \lambda_{yr}^2$$

$$k_{\gamma\gamma} = k_\pi \lambda_{zp}^2 + k_\theta \lambda_{zq}^2 + k_\rho \lambda_{zr}^2$$

$$k_{\alpha\beta} = k_\pi \lambda_{xp} \lambda_{yp} + k_\theta \lambda_{xq} \lambda_{yq} + k_\rho \lambda_{xr} \lambda_{yr}$$

$$k_{\alpha\gamma} = k_\pi \lambda_{xp} \lambda_{zp} + k_\theta \lambda_{xq} \lambda_{zq} + k_\rho \lambda_{xr} \lambda_{zr}$$

$$k_{\beta\gamma} = k_\pi \lambda_{yp} \lambda_{zp} + k_\theta \lambda_{yq} \lambda_{zq} + k_\rho \lambda_{yr} \lambda_{zr} \quad \text{Eq. (1.2.2)}$$

Equation (1.2.2) allows computation of stiffness constants for any coordinate frame being used to describe the vibration isolation system if the principal axes and the principal stiffnesses are defined, or to find the principal stiffness constants and positions of the principal axes (keeping in mind that only six cosines λ are independent; e.g., see [1.5]).

While this may seem somewhat cumbersome, one has to remember that all of the above implicitly assume that all the considered coordinate frames have their origins in the elastic center whose position is not known for a given asymmetrical (or having less than three axes/planes of symmetry), flexible element. If the position of the elastic center is not known, the analysis becomes even more complicated, by at least an order of magnitude.

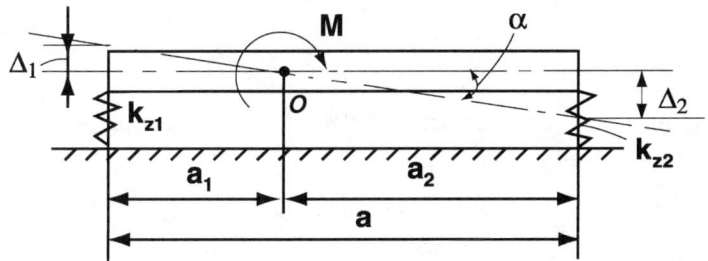

Fig. 1.2.1. Angular Stiffness of a Beam Supported by Translational Springs.

Fortunately, the overwhelming majority of real-life vibration isolators have flexible elements with at least one axis of symmetry or one—more frequently two or more—plane of symmetry. Also, their dimensions are usually much smaller than the dimensions of the isolated object and/or the distances between the isolators. As a result, it would be an acceptable approximation to assume that the elastic center of the flexible element is located in its geometrical or mass center.

An important issue is angular stiffness of vibration isolators and of the isolation system. For the most important case—when all isolators are located in one (e.g., horizontal) plane—the angular stiffness of the installation about a horizontal axis is determined by two factors: by vertical stiffness constants of the isolators combined with distances between them, and by angular stiffness constants of the isolators. The first factor is illustrated in Fig. 1.2.1, showing a beam installed on two isolators having vertical stiffnesses k_{z1}, k_{z2}, and acted upon by moment M causing vertical deformations of two isolators Δ_1, Δ_2, respectively. The beam is displaced by moment M for angle α, rotating about center O. Position of O along the beam can be found from the force equilibrium equation

$$k_{z_1}\Delta_1 - k_{z_2}\Delta_2 = 0, \quad \text{or} \quad \frac{\Delta_1}{\Delta_2} = \frac{a_1}{a_2} = \frac{k_{z_2}}{k_{z_1}} \qquad \text{Eq. (1.2.3)}$$

The moment equilibrium equation about point O is

$$(k_{z_1}\Delta_1)a_1 + (k_{z_2}\Delta_2)a_2 = M, \qquad \text{Eq. (1.2.4)}$$

and, since $\alpha = \Delta_1/a_1 = \Delta_2/a_2$,

$$\left(k_{z_1}a_1^2 + k_{z_2}a_2^2\right)\alpha = M \qquad \text{Eq. (1.2.5)}$$

12 ● PASSIVE VIBRATION ISOLATION

or

$$k_\alpha = \frac{M}{\alpha} = k_{z_1} a_1^2 + k_{z_2} a_2^2 \qquad \text{Eq. (1.2.6)}$$

If each isolator has some angular stiffness, $k_{\alpha 1}$ and $k_{\alpha 2}$, respectively, then total angular stiffness is

$$k_\alpha = \left(k_{\alpha_1} + k_{\alpha_2}\right) + \left(k_{z_1} a_1^2 + k_{z_2} a_2^2\right) \qquad \text{Eq. (1.2.7)}$$

Unless a high angular stiffness is specially designed into the isolators (e.g., see Section 4.7), the first parenthesis in Eq. (1.2.7) is usually negligible compared to the second parenthesis for the overwhelming majority of practical vibration isolation systems. Because of this, the angular stiffness of isolators is usually not considered in equations of motion for isolation systems.

Usually, it is assumed that damping constants of isolators are proportional to their stiffness constants, e.g., see [1.1]. However, while this is very convenient analytically, it cannot be considered correct. First of all, this statement can be used only if the isolators have viscous (linear) damping. While the overwhelming majority of isolators have elastomeric, wire mesh, and other elements with nonlinear damping characteristics (see Chapter 3), a few designs using metal springs have only insignificant damping. If they are equipped with viscous dampers, the latter are acting usually only in one principal (axial) direction, thus no proportionality can be assumed. Elastomeric and other mass-produced isolators not only have nonlinear damping characteristics (e.g., dependent on vibration amplitude, which is different for different isolators under the same object and for different directions for the same isolator), their damping constants in various directions can vary in ways totally different from stiffness. For example, deformation of the elastomeric flexible elements in compression is associated with a different damping than deformation of the same element in shear. In pneumatic isolators, see Section 3.5, the damping is high (by design) in the direction of the piston motion, but is negligible in the perpendicular direction.

1.3 ELASTICALLY SUPPORTED SOLID OBJECT DYNAMIC COUPLING

In the general case, the object is supported by an arbitrary number of arbitrarily located elastic mounts, only one of which is shown in Fig. 1.3.1.

Dynamic Properties of Vibration Isolation Systems • 13

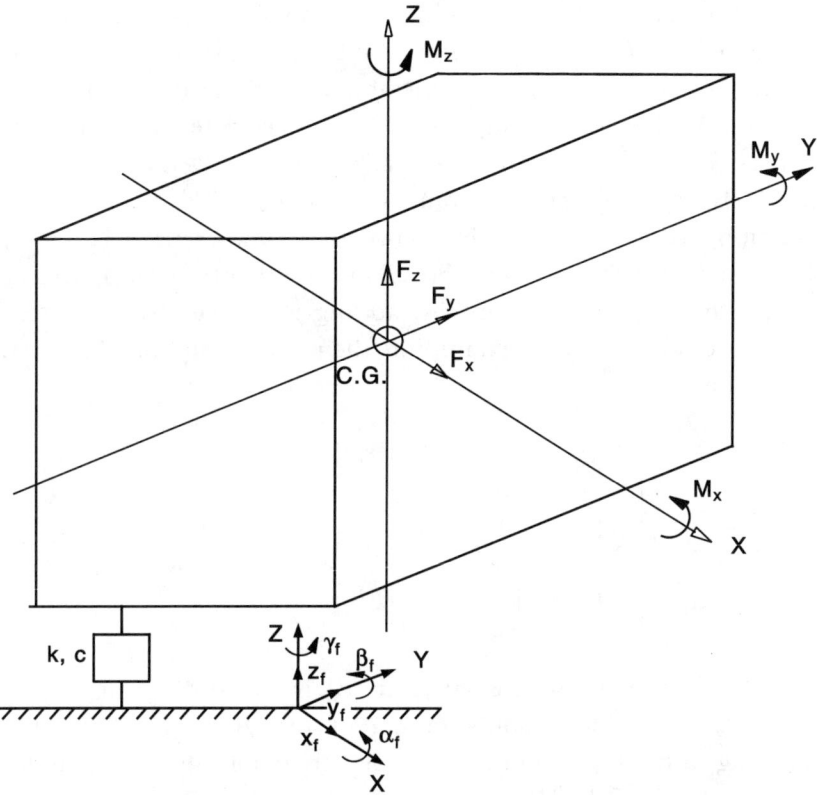

Fig. 1.3.1. A Six-Degrees-of-Freedom Vibration Isolation System.

The elastic mount shown is characterized by all of the 12 stiffness coefficients listed in Eq. (1.2.2), as well as by the corresponding damping coefficients. The system is excited by multiple forcing factors whose set can be resolved into three force components F_x, F_y, F_z along the axes X, Y, Z and three moment components M_x, M_y, M_z, about these axes, as well as by foundation motions acting on elastic mounts (translational motion components x_f, y_f, z_f and angular motion components α_f, β_f, γ_f). All components of the foundation motion are shown, for convenience, as defined in the same coordinate frame as the object itself, namely X, Y, Z. For the most important practical cases, the preferred coordinate frame is the "natural frame" as described in Section 1.1. As was shown in Section 1.2, angular stiffness of the isolators can be neglected. If the damping of elastic mounts is not considered, a translational displacement x_c of the C.G. of the object would result in the force along the X-direction in each elastic mount equal to $-k_{xx}(x_c - x_f)$, a moment about Y-axis (β-direction) equal to $-k_{xx}(x_c - x_f) a_z$, and a moment about Z-axis (γ-direction) equal to $-k_{xx}(x_c - x_f)a_y$. The cross-coupling stiffness parameters k_{xy} and k_{xz} result in forces $k_{xy}(x_c - x_f)$ and $-k_{xz}(x_c - x_f)$ in Y- and Z-directions, respectively, as well as in moments $(-k_{xz}a_y + k_{xy}a_x)(x_c - x_f)$ in α-direction, $k_{xz}a_x(x_c - x_f)$ in β-direction, and $-k_{xy}a_x(x_c - x_f)$ in γ-direction. Compiling similar expressions for other translational (Y and Z) and angular (α, β, γ) coordinates, adding responses from all resilient mounts, and not considering damping, the matrix equation of motion can be written as

$$[m]\begin{bmatrix}\ddot{x}\\ \ddot{y}\\ \ddot{z}\\ \ddot{\alpha}\\ \ddot{\beta}\\ \ddot{\gamma}\end{bmatrix} + [k]\begin{bmatrix}x\\ y\\ z\\ \alpha\\ \beta\\ \gamma\end{bmatrix} = [k]\begin{bmatrix}x_f\\ y_f\\ z_f\\ \alpha_f\\ \beta_f\\ \gamma_f\end{bmatrix} + [F], \qquad \text{Eq. (1.3.1)}$$

where $[m]$ is the inertia matrix and $[F]$ is the force/moment matrix. However, for a better understanding of the directions for achieving decoupling in the isolation system, it is beneficial to write the following equations of motion [1.1]:

$$m\ddot{x}_c + \Sigma k_{xx}x_c + \Sigma k_{xy}y_c + \Sigma k_{xz}z_c + \Sigma(k_{xz}a_y - k_{xy}a_z)\alpha$$

$$+ \Sigma(k_{xx}a_z + k_{xz}a_x)\beta + \Sigma(k_{xy}a_x - k_{xx}a_y)\gamma$$

$$= F_x + \Sigma k_{xx}x_f + \Sigma k_{xy}y_f + \Sigma k_{xz}z_f + \Sigma(k_{xz}a_y - k_{xy}a_z)\alpha_f$$

$$+\Sigma(k_{xx}a_z + k_{xz})\beta_f + \Sigma(k_{xy}a_x - k_{xx}a_y)\gamma_f$$

$$I_{xx}\ddot{\alpha} - I_{xy}\ddot{\beta} - I_{xz}\ddot{\gamma} + \Sigma(k_{xz}a_y - k_{xy}a_z)x_c + \Sigma(k_{yz}a_y - k_{yy}a_z)y_c$$

$$+\Sigma(k_{zz}a_y - k_{yz}a_z)z_c + \Sigma(k_{yy}a_z^2 + k_{zz}a_y^2 - 2k_{yz}a_ya_z + k_{\alpha\alpha})\alpha$$

$$+\Sigma(k_{xz}a_ya_z + k_{yz}a_xa_z - k_{zz}a_xa_y - k_{xy}a_z^2 + k_{\alpha\beta})\beta$$

$$+\Sigma(k_{xy}a_ya_z + k_{yz}a_xa_y - k_{yy}a_xa_z - k_{xz}a_y^2 + k_{\alpha\gamma})\gamma$$

$$= M_x + \Sigma(k_{xz}a_y - k_{xy}a_z)x_f + \Sigma(k_{yz}a_y - k_{yy}a_z)y_f$$

$$+ \Sigma(k_{zz}a_y - k_{yz}a_z)z_f + \Sigma(k_{yy}a_z^2 + k_{zz}a_y^2 - 2k_{yz}a_ya_z + k_{\alpha\alpha})\alpha_f$$

$$+ \Sigma(k_{xz}a_ya_z + k_{yz}a_xa_z - k_{zz}a_xa_y - k_{xy}a_z^2 + k_{\alpha\beta})\beta_f$$

$$+ \Sigma(k_{xy}a_ya_z + k_{yz}a_xa_y - k_{yy}a_xa_z - k_{xz}a_y^2 + k_{\alpha\gamma})\gamma_f$$

$$m\ddot{y}_c + \Sigma k_{xy}x_c + \Sigma k_{yy}y_c + \Sigma k_{yz}z_c + \Sigma(k_{yz}a_y - k_{yy}a_z)\alpha$$

$$+ \Sigma(k_{xy}a_z - k_{yz}a_x)\beta + \Sigma(k_{yy}a_x - k_{xy}a_y)\gamma$$

$$= F_y + \Sigma k_{xy}x_f + \Sigma k_{yy}y_f + \Sigma k_{yz}z_f + \Sigma(k_{yz}a_y - k_{yy}a_z)\alpha_f$$

$$+ \Sigma(k_{xy}a_z - k_{yz}a_x)\beta_f + \Sigma(k_{yy}a_x - k_{xy}a_y)\gamma_f$$

$$I_{yy}\ddot{\beta} - I_{xy}\ddot{\alpha} - I_{yz}\ddot{\gamma} + \Sigma(k_{xx}a_z - k_{xz}a_x)x_c + \Sigma(k_{xy}a_z - k_{yz}a_x)y_c$$

$$+ \Sigma(k_{xz}a_z - k_{zz}a_x)z_c + \Sigma(k_{xz}a_ya_z + k_{yz}a_xa_z - k_{zz}a_xa_y - k_{xy}a_z^2$$

$$+ k_{\alpha\beta})\alpha + \Sigma(k_{xx}a_z^2 + k_{zz}a_x^2 - 2k_{xz}a_xa_z + k_{\beta\beta})\beta$$

$$+ \Sigma(k_{xy}a_xa_z + k_{xz}a_xa_y - k_{xx}a_ya_z - k_{yz}a_x^2 + k_{\beta\gamma})\gamma$$

$$= M_y + \Sigma(k_{xx}a_z - k_{xz}a_x)x_f + \Sigma(k_{xy}a_z - k_{yz}a_x)y_f$$

$$+ \Sigma(k_{xz}a_z - k_{zz}a_x)z_f + \Sigma(k_{xz}a_ya_z + k_{yz}a_xa_z - k_{zz}a_xa_y$$

$$- k_{xy}a_z^2 + k_{\alpha\beta})\alpha_f + \Sigma(k_{xx}a_z^2 + k_{zz}a_x^2 - 2k_{xz}a_xa_z + k_{\beta\beta})\beta_f$$

$$+ \Sigma(k_{xy}a_xa_z + k_{xz}a_xa_y - k_{xx}a_ya_z - k_{yz}a_x^2 + k_{\beta\gamma})\gamma_f$$

$$m\ddot{z}_c + \Sigma k_{xz}x_c + \Sigma k_{yz}y_c + \Sigma k_{zz}z_c + \Sigma(k_{zz}a_y - k_{yz}a_z)\alpha$$

$$+ \Sigma(k_{xz}a_z - k_{zz}a_x)\beta + \Sigma(k_{yz}a_x - k_{xz}a_y)\gamma$$

$$= F_z + \Sigma k_{xz}x_f + \Sigma k_{yz}y_f + \Sigma k_{zz}z_f + \Sigma(k_{zz}a_y - k_{yz}a_z)\alpha_f$$

$$+ \Sigma(k_{xz}a_z - k_{zz}a_x)\beta_f + \Sigma(k_{yz}a_x - k_{xz}a_y)\gamma_f$$

$$I_{zz}\ddot{\gamma} - I_{xz}\ddot{\alpha} - I_{yz}\ddot{\beta} + \Sigma(k_{xy}a_z - k_{xx}a_y)x_c + \Sigma(k_{yy}a_x - k_{xy}a_y)y_c$$

$$+\Sigma(k_{yz}a_x - k_{xz}a_y)z_c + \Sigma(k_{xy}a_y a_z + k_{yz}a_x a_y - k_{yy}a_x a_z - k_{xz}a_y^2 + k_{\alpha\gamma})\alpha$$

$$+\Sigma(k_{xy}a_x a_z - k_{xz}a_x a_y - k_{xx}a_y a_z - k_{yz}a_x^2 + k_{\beta\gamma})\beta$$

$$+\Sigma(k_{xx}a_y^2 + k_{yy}a_x^2 - 2k_{xy}a_x a_y + k_{\gamma\gamma})\gamma$$

$$= M_z + \Sigma(k_{xy}a_z - k_{xx}a_y)x_f + \Sigma(k_{yy}a_x - k_{xy}a_y)y_f$$

$$+\Sigma(k_{yz}a_x - k_{xz}a_y)z_f + \Sigma(k_{xy}a_y a_z + k_{yz}a_x a_y - k_{yy}a_x a_z - k_{xz}a_y^2$$

$$+ k_{\alpha\gamma})\alpha_f + \Sigma(k_{xy}a_x a_z - k_{xz}a_x a_y - k_{xx}a_y a_z - k_{yz}a_x^2 + k_{\beta\gamma})\beta_f$$

$$+ \Sigma(k_{xx}a_y^2 + k_{yy}a_x^2 - 2k_{xy}a_x a_y + k_{\gamma\gamma})\gamma_f \qquad \text{Eq. (1.3.1')}$$

Six equations of motion Eq. (1.3.1') fully describe this six-degrees-of-freedom system. This system, of course, has six natural frequencies. It is clear from Eq. (1.3.1') that all coordinates are dynamically coupled. This means that any X, Y, Z coordinate component of the external force or moment, as well as any coordinate component of translational or angular motion of the foundation is causing, in general, displacements of the object in all coordinate directions. Also, any coordinate component of translational or angular motion of the installed object is causing forces/moments to and displacements of the foundation in all coordinate directions. Such global coupling is usually not desirable for performance in many important cases—as shown in many instances below and in Chapter 2—or for computational synthesis/optimization of the system. The inter-coordinate coupling may be desirable in some cases if only some axes of the flexible elements (vibration isolators) have a significant damping. An increase of damping in other coordinate directions can be achieved if a strong coupling is present in the system.

One danger of the inter-coordinate coupling can be formulated as follows: In some cases where vibration isolation is used, it is required that the highest natural frequency should be below the range of frequencies of the objectionable external excitations, regardless of their directions. If there is a strong coupling between all six natural vibratory modes of the isolation system, it would lead to a need to shift all natural frequencies down, thus requiring very low stiffness isolators and potentially causing large rocking motions in the system. Such condition is sometimes conveniently (but incorrectly in a rigorous sense) called instability of the system. This is true for both vibration sensitive objects and for vibration-producing objects. The computational analysis of this type of globally cross-coupled system becomes very complex and involved, partly due to the fact that spectral characteristics of all excitations have to be identified and entered into the model.

One of the most effective ways of improving the performance of a vibration isolation system is by eliminating or, at least, reducing the dynamic inter-coordinate coupling in the system. This would simplify computing natural frequencies of the system and make the influences of various design parameters on the natural frequencies and modes more transparent. Also, it would make possible or, at least, would simplify bringing natural frequencies closer together (by some reduction of higher frequencies and some increase of lower frequencies) without a significant reduction of isolator stiffness constants. Finally, it may improve performance when the coupling between specific vibratory modes is undesirable. For example, coupling between vertical and horizontal translational vibrations is undesirable when there is a high intensity of excitation in the vertical direction and the horizontal direction is the direction of high sensitivity (e.g. see Section 2.2).

Means for alleviation/elimination of inter-coordinate coupling can be devised by analyzing Eq. (1.3.1'). These equations have been written for the natural coordinate frame but with arbitrary attitudes of the isolators. However, with the exception of the special case of inclined mounts analyzed in Section 1.6.3, the isolators, usually having planes of symmetry, are placed in such a way that their principal axes are parallel to the natural coordinate axes. Also, except for the special cases, angular stiffnesses of isolators can be neglected. These statements can be rewritten as

$$k_{xx} = k_p = k_x; \; k_{yy} = k_q = k_y; \; k_{zz} = k_r = k_z; \; k_{xy} = k_{xz} = k_{yz} = 0;$$

$$k_\alpha = k_\beta = k_\gamma = k_{\alpha\beta} = k_{\alpha\gamma} = k_{\beta\gamma} = 0 \qquad \text{Eq. (1.3.2)}$$

Based on these assumptions, and neglecting products of inertia per Eq. (1.1.2), the system Eq. (1.3.1') becomes

$$m\ddot{x}_c + \sum_i k_{x_i}\left(x_c - x_f\right) + \sum_i k_{x_i} a_{z_i}\left(\beta - \beta_f\right) - \sum_i k_{x_i} a_{y_i}\left(\gamma - \gamma_f\right) = F_x$$

18 • PASSIVE VIBRATION ISOLATION

$$I_x \ddot{\alpha} - \sum_i k_{y_i} a_{z_i} (y_c - y_f) + \sum_i k_{z_i} a_{y_i} (z_c - z_f) + \sum_i \left(k_{y_i} a_{z_i}^2 + k_{z_i} a_{y_i}^2 \right) (\alpha - \alpha_f)$$

$$- \sum_i k_{z_i} a_{y_i} a_{z_i} (\beta - \beta_f) - \sum_i k_{y_i} a_{x_i} a_{z_i} (\gamma - \gamma_f)$$

$$= M_x \ddot{y}_c + \sum_i k_{y_i} (y_c - y_f) - \sum_i k_{y_i} a_{z_i} (\alpha - \alpha_f) + \sum_i k_{y_i} a_{x_i} (\gamma - \gamma_f)$$

$$= F_y I_y \ddot{\beta} + \sum_i k_{z_i} a_{z_i} (x_c - x_f) - \sum_i k_{z_i} a_{x_i} (z_c - z_f) - \sum_i k_{z_i} a_{x_i} a_{y_i} (\alpha - \alpha_f)$$

$$+ \sum_i \left(k_{x_i} a_{z_i}^2 + k_{z_i} a_{x_i}^2 \right) (\beta - \beta_f) - \sum_i k_{x_i} a_{y_i} a_{z_i} (\gamma - \gamma_f)$$

$$= M_y \ddot{z}_c + \sum_i k_{z_i} (z_c - z_f) + \sum_i k_{z_i} a_{y_i} (\alpha - \alpha_f) - \sum_i k_{z_i} a_{y_i} a_{z_i} (\beta - \beta_f)$$

$$= F_z I_z \ddot{\gamma} - \sum_i k_{x_i} a_{y_i} (x_c - x_f) + \sum_i k_{y_i} a_{x_i} (y_c - y_f) - \sum_i k_{y_i} a_{x_i} a_{z_i} (\alpha - \alpha_f)$$

$$- \sum_i k_{x_i} a_{y_i} a_{z_i} (\beta - \beta_f) + \sum_i \left(k_{x_i} a_{y_i}^2 + k_{y_i} a_{x_i}^2 \right) (\gamma - \gamma_f) = M_z$$

Eq. (1.3.3)

Eq. (1.3.3) is still characterized by coupling between all six coordinates.

Specifics of vibration isolating installation of stationary machinery (as well as of many units of mobile machinery) allows one to assume that all isolators (mounts) are situated in a horizontal plane, or

$$a_{z_i} = const. \qquad \text{Eq. (1.3.4)}$$

If, in addition, the following conditions are satisfied:

$$\sum_i k_{z_i} a_{x_i} = 0, \sum_i k_{z_i} a_{y_i} = 0, \qquad \text{Eq. (1.3.5)}$$

then vertical (z) vibration of the system is *not coupled* with motions in other translational or rotational coordinates.

There are many ways to satisfy Eq. (1.3.5) by selecting stiffness constants of each isolator. The most natural way to comply with Eq. (1.3.5) is to find the object weight W distribution between the mounting points and to use isolators whose stiffness constants are proportional to the fraction W_i of the total weight acting on the corresponding i-th mounting point. Indeed, if each k_{zi} in Eq. (1.3.5) were replaced by W_i, then Eq. (1.3.5) becomes an identity since distances a_x, a_y are measured from the C.G.

Thus, if vertical stiffness k_{z_i} of i-th isolator/mount is proportional to vertical reaction $R_i = W_i$ of this mount to the weight load from the object,

$$k_{z_i} = AR_i, \ A = const, \qquad \text{Eq. (1.3.6)}$$

then Eq. (1.3.5) is satisfied since it reflects moment equilibrium of the object in two vertical planes X-Z and Y-Z.

If in addition to Eq. (1.3.5) the following conditions are satisfied:

$$\sum_i k_{x_i} a_{y_i} = 0, \sum_i k_{y_i} a_{x_i} = 0, \qquad \text{Eq. (1.3.7)}$$

then vibration along and about axis Z (coordinates z and γ) are uncoupled. These conditions are satisfied if, in addition to Eq. (1.3.5), ratios of k_x/k_z, k_y/k_z are constant for all isolators. If, in addition to Eq. (1.3.7), the following condition is satisfied:

$$\sum_i k_{z_i} a_{x_i} a_{y_i} = 0. \qquad \text{Eq. (1.3.8)}$$

then Eq. (1.3.3) is separating into two independent equations corresponding to coordinates z and γ, and two pairs of coupled equations x-β and y-α. Eq. (1.3.8) is satisfied if the object has at least one plane of symmetry. A further reduction of coupling is achieved if the plane of location of the isolators contains the C.G., or

$$a_{z_i} = 0 \qquad \text{Eq. (1.3.9)}$$

With Eqs. (1.3.7), (1.3.8), and (1.3.9) satisfied, vibratory motions along all coordinates are independent (uncoupled). Eq. (1.3.9) can be relatively easily realized if the object is isolated not directly from the floor, but is attached to a massive "inertia block," allowing a shift in the vertical C.G. position like in Figs. 4.10.1, 2. Then the inertia block can be placed on isolators at the new C.G. level. In some cases, the isolators can be supported by special brackets at the C.G. level even without the inertia block. Another way of achieving such an effect is by using inclined isolators in x-z and/or y-z planes (see Section 1.6.3). Coordinates x and/or y can also be uncoupled if angular stiffness values of isolators in

α and/or β directions, neglected in Eq. (1.3.3), are made very high (e.g., by using guideways in these directions or by using high angular stiffness isolators; see Section 4.7). In such cases angular vibrations are simply prevented from developing.

Compliance with Eqs. (1.3.7), (1.3.8), and (1.3.9) allows independent selection of all natural frequencies. Sometimes it is considered desirable to reduce the frequency range within which the six (or some lesser number of) natural frequencies are located, with about equal natural frequencies considered by some to be an ideal condition. It should be kept in mind, however, that closeness between the natural frequencies raises the probability of dynamic coupling. In this case the natural frequencies/modes of the vibration isolation system are, essentially, representing the partial subsystems of the overall isolation system. The example of a three-mass linear system in Section 1.5.1 below shows that the dynamic coupling σ between the subsystems is fast increasing when natural frequencies of the partial subsystems are approaching each other. The dynamic coupling is also developing due to nonlinearities of the isolators, which are always present in some degree. It is shown in Section 1.9 that even a slight nonlinearity induces coupling between the modes, which are uncoupled in the linear case, especially when the natural frequencies are close to one another. When the natural frequencies are equal, then coupling is unavoidable.

While Eq. (1.3.6) seems, at the first glance, a reasonably simple and straightforward one, this is not the case. Let's look deeper into this condition for two most frequently used vibration isolators: (a) constant stiffness isolators and (b) so-called constant natural frequency (CNF) isolators, whose axial (vertical) stiffness coefficients in the rated load range are proportional to axial (weight) load on the isolator (see Chapter 4).

(a.) If constant stiffness isolators are used, the total vertical (z-direction) stiffness $\sum k_{z_i}$ of all isolators is determined by the desired natural frequency in z-direction,

$$f_z \approx \frac{1}{2\pi}\sqrt{\frac{\sum k_{z_i}}{M}} \qquad \text{Eq. (1.3.10)}$$

where $\approx$ is used to reflect modal interactions in a strongly coupled system. For selecting individual isolators, the weight distribution between the mounting points should be known and then isolators that have stiffness

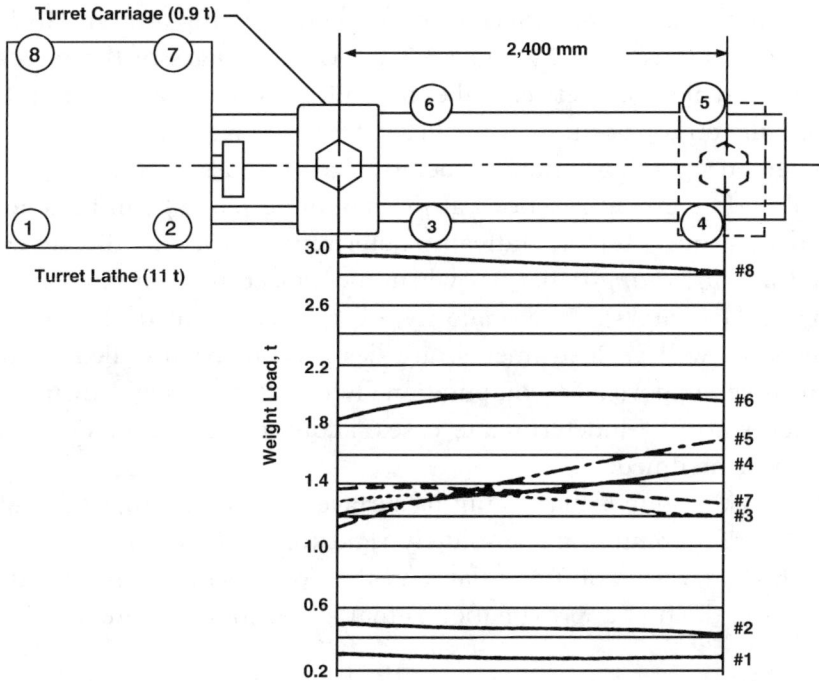

Fig. 1.3.2. Change of Weight Distribution Between Mounting Points of a Turret Lathe Due to Travel of Its Tool Carriage.

constants proportional to the weight loads at the mounting points have to be used.

The first step in determining the weight distribution is to determine the position of the center of gravity (C.G.) in the horizontal plane. With the exception of a relatively small number of cases where the object has two vertical planes of symmetry or a vertical axis of symmetry, the C.G. position can be determined either experimentally or by a calculation using detailed design drawings.

While experimental determination of the C.G. position for instruments and other small/light objects is feasible, e.g., see [1.2], neither of these approaches is usually applicable for the users of heavy machinery, and the manufacturers provide this information only in special cases, such as mass-produced engines for vehicles. Usually, the C.G. position is determined by an "expert judgment" by the user, with at least ±10–15% error. In many cases, especially for production machinery with heavy moving parts (tables, gantries, etc.) or capable of processing parts with widely differing masses, the position of the C.G. can shift significantly, adding to the above mentioned error for a given configuration. The latter statement is illustrated by Fig. 1.3.2, showing changes of the measured weight distribution between the mounting points for a turret lathe depending on the position of the heavy tool carriage.

After the C.G. position is determined (or assumed), the weight distribution (vertical reactions at the mounting points) can be uniquely calculated only for a statically determined case of the so-called *kinematic mounting*—that is, when the object is mounted on three supports (isolators). The majority of real-life industrial machinery objects, as well as instruments, are designed to be installed on more than three supports. To computationally determine weight distribution in such statically indeterminate cases, additional conditions should be known or assumed.

One assumption is that mounting surfaces both of the object and of the base/floor/foundation are absolutely rigid (or, at least, are much more rigid than the isolators) and flat, and the floor surface is horizontal. In such case, the first approximation is that all reactions are identical,

$$R_i \underset{i=1,\ldots,n}{} = \frac{W}{n} \qquad \text{Eq. (1.3.11)}$$

where n is number of mounts/isolators, and W is total weight of the object. While this assumption is acceptable if the mounting points are located at not very different distances from the C.G., the loading conditions can become very different if there is some unevenness (deviations from flatness) of the mounting surfaces, that is not negligible in comparison with the average mount deformation. This leads to uneven deformations of the mounts, thus to changing load distribution. The same effect would result from reduced stiffness of at least one mounting surface in at least one area.

If the mounting points are located at significantly different distances from the C.G., Eq. (1.3.11) is not acceptable. For example, the case of four mounting points in Fig. 1.3.3 can be analyzed assuming $R_1 = R_2$, $R_3 = R_4$; if b_1, b_2 are significantly different, then

$$R_1 = R_2 = W \frac{b_2}{2(b_1 + b_2)}; \; R_3 = R_4 = W \frac{b_1}{2(b_1 + b_2)} \qquad \text{Eq. (1.3.12)}$$

This distribution also can change if the mounting surfaces are not flat, if there are reduced stiffness areas on mounting surfaces, etc. [1.6]. Other

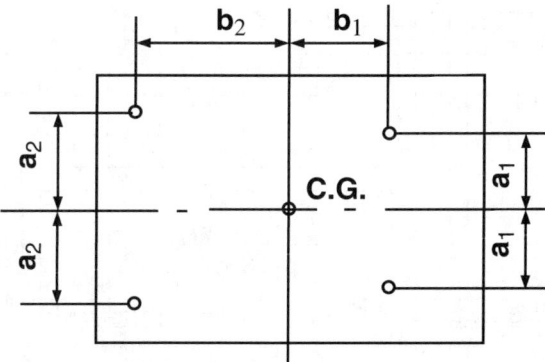

Fig. 1.3.3. A Typical Configuration of Mount Locations.

reasons for the indeterminable nature of the weight distribution are movement of heavy parts within the object (e.g., see Fig. 1.3.2) and wide variation of the workpiece weight in machine tools and CMMs. Conservatively, variation of weight distribution between the mounting points can be assumed to be at least ±35%. Fig. 1.3.4 [1.34] shows weight distribution between four mounting points of a 60-ton press before and after fine-tuning adjustments using special load-sensing isolators. One can see huge differences, significantly exceeding ±35%, between the weight fractions at each mounting points for these two conditions. Similar data is presented in Section 4.4.

After the weight distribution is established, the appropriate isolators compliant with Eqs. (1.3.6) and (1.3.10) have to be selected. Commercial lines of vibration isolators usually have ratios of the nominal (static) stiffness coefficients between the adjacent mounts in the line in the range of ~1.5–2.0. Machinery mounts for Navy ships have this ratio in the 1.2–2.0 range [1.7]. Assumption of an average ratio ~1.6 seems to be prudent. However, often "the adjacent mounts in the line" differ only by the rubber durometer, so their dynamic-to-static stiffness ratios K_{dyn} can be quite different. The difference between $K_{dyn} = 1.5$ and 4.0 are not unusual (e.g., see Sections 4.3.1 and 4.4.1). A conservative assumption of ±1.5 times variation can be adopted.

If elastomeric isolators are used, the generally accepted tolerance on hardness of rubber used in their flexible elements is ±5 durometer units, which is equivalent to ±17% variation in their stiffness (see Section 3.3.2).

Taking all these uncertainties into consideration, the scatter of stiffness of isolators selected in accordance with the calculated/assumed weight

24 • PASSIVE VIBRATION ISOLATION

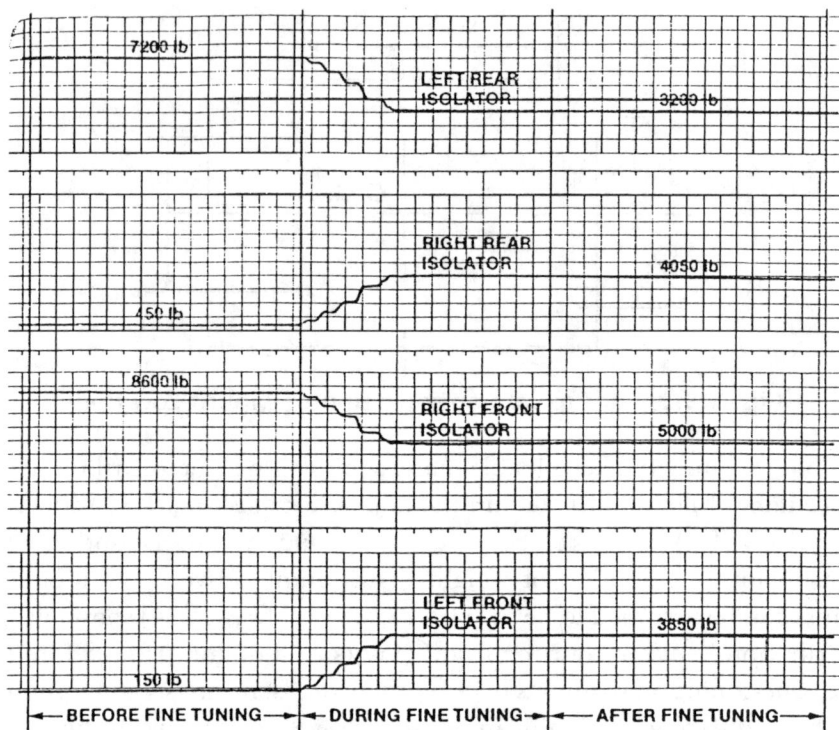

Fig. 1.3.4. Weight Supported by Each of Four Isolators before and after Fine-Tuning/Leveling Adjustments of 60-ton Press with Load-Sensing Isolators.

distribution is at least $\pm 1.15 \times 1.35 \times 1.6 \times 1.5 \times 1.17 = \pm 4.3$ times. It means, that Eq. (1.3.5) is not satisfied even in the first approximation if constant stiffness isolators are used (see Section 1.6.1).

(b.) CNF vibration isolators have a special nonlinear load-deflection characteristic so that their stiffness is proportional to the load on the isolator. For a real CNF isolator, its stiffness values in all directions are proportional to the load in the direction of its main axis (weight load), as expressed by Eq. (1.3.6). In the commercially realized CNF isolators, deviation from this proportionality does not exceed ± 10–15% (see Chapter 4). It is also shown in Chapter 4 that the tolerance on the rubber hardness does not noticeably influence the CNF characteristic. Thus, use of CNF isolators satisfies the de-coupling Eq. (1.3.5) with the scatter not exceeding ± 1.1–1.15 times. This provides an adequate decoupling, as demonstrated in Section 1.6.1.

Potentially, the nonlinear load-deflection characteristic of CNF isolators may create undesirable nonlinear vibratory effects such as excitation of sub- or super-harmonic vibration. However, due to relatively low vibration amplitudes in the majority of vibration isolation systems, such effects are not significant, especially when damping in the isolators is not very low (e.g., see Section 1.9).

If isolators, both constant stiffness and CNF, have flexible elements made from a material with a significant amplitude dependency on their dynamic stiffness (wire mesh and cable isolators, Sections 3.3.1, 4.6.3, and also some rubber blends, Section 3.3.2E), a degree of inter-coordinate coupling may develop due to different vibration amplitudes at different mounting points.

1.4 DYNAMICS OF SINGLE-DEGREE-OF-FREEDOM VIBRATION ISOLATION SYSTEM

The goals of isolation can be different. In some cases, the object is a source of steady vibrations or shocks transmitted to the base/foundation as dynamic or vibratory forces. Transmission of these forces to the base should be prevented or reduced (some examples: forging hammer; unbalanced rotor, see Section 2.3). In other cases, the object is vibration-sensitive and has to be protected from steady and/or transient vibrations transmitted from some vibration-producing objects via the base (an example: precision stepper, see Section 2.2). Some objects are vibration-sensitive during their working process while generating objectionable dynamic loads during their auxiliary motions (e.g., a precision surface grinder with a heavy reciprocating table or a CMM with start/stop motions of the heavy gantry). A similar but different case is represented by an internal combustion engine on a vehicle. The engine produces undesirable dynamic inputs to the occupants of the vehicle but, in its turn, has to be protected from the intense vibration caused by the road.

Accordingly, transmissibilities of vibratory forces and/or displacements (or accelerations) both from the object to the base and in the opposite direction may be of interest. These transmissibilities depend on correlations between the spectral components of the dynamic inputs to the system and natural frequencies and natural modes of the system shown in Fig.1.3.1. They also depend on stiffness constants of the isolators, on effective masses of the object and the base and, last but not least, on character and magnitudes of damping in the isolators. When the dynamic characteristics of the subsystems have to be considered in more detail— e.g., when wave resonances in the isolators as well as several natural modes of the object/base have to be considered—the subsystems cease to be "lumped parameters," as shown in Figs. 1.0.1 and 1.3.1 and can be characterized by their impedances, e.g. [1.8].

Considering all these factors in practical designs of isolation systems is extremely difficult even when powerful specialized software packages are used. While vehicle manufacturers use the most sophisticated (usually, proprietary) software packages and can afford huge manpower and computer power to design engine isolation systems (characteristics and positioning of the engine mounts), even more effort is usually made at the last stage of the vehicle development process wherein several dozens of engine mounts with slightly different parameters are fabricated and tried in the prototype vehicle in order to achieve an optimal performance.

It seems that a better (albeit, somewhat qualitative) understanding of the dynamic processes in a vibration isolation system can be achieved by analyzing basic, generic vibration isolation systems in order to better visualize influences of such critical factors as damping and intercoordinate coupling. The results of such analysis would allow better utilization of the available software and development of more robust isolation systems that are less sensitive to inevitable variations in the dynamic characteristics of the principal sub-systems as well as in spectral content and intensity of the excitations.

A single-degree-of-freedom, uni-axial vibration isolation system is a convenient model to clarify basic dynamic effects of vibration isolation installations. Typically, a vibration isolation system comprises three subsystems as presented in Fig. 1.0.1: object to be isolated (mass m); flexible connectors (vibration isolators); non-attached base/foundation, mass m_f. Usually, $m_f \gg m$, or at least $m_f > m$ (unless a heavy object is installed on a thin reinforced concrete floor plate). The vibration isolators in this system can deform only in one direction. Depending on character of energy dissipation in the flexible element and its design, the dynamic model of the uni-axial system can change. Typical for vibration isolation of stationary objects are four modifications of the dynamic model shown in Fig. 1.4.1. In Fig. 1.4.1(a) the isolator is represented by a parallel connection of an elastic element (*stiffness k*) and viscous damper whose

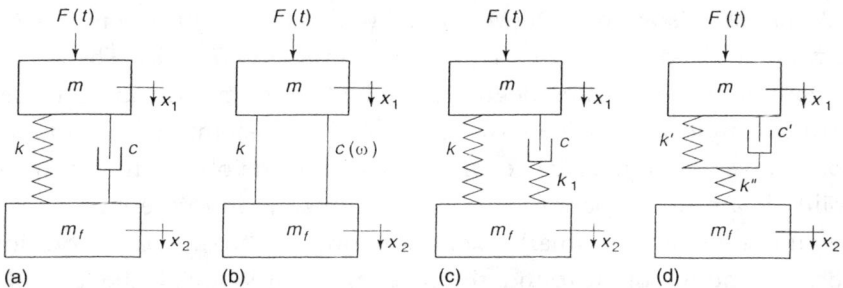

Fig. 1.4.1. Typical Models of General Uni-Axial Vibration Isolation System: (a) Isolator with Linear Stiffness k and Viscous Damper c; (b) Isolator Combining Stiffness k and Hysteretic Damping; (c) and (d) "Relaxation" Isolators.

resistance F_v to the relative motion x between its terminals is proportional to the relative velocity, $F_v = c\,(\dot{x}_1 - \dot{x}_2)$, where c is the *damping coefficient*. Pure viscous damper is a convenient idealization for analytical purposes but not a realistic element (with some exceptions, see Chapters 3, 4). The flexible element of the isolator in Fig. 1.4.1(b) combines elastic properties (stiffness k) and energy dissipation properties of a hysteretic type (material damping), e.g., see [1.6], [1.8], whereas the relative energy dissipation ψ (and logarithmic decrement δ) does not depend on vibration frequency ω. It is shown, e.g., in [1.6], that such hysteretic damping can be represented by a linear (viscous) damper with a variable damping coefficient

$$c_h = c(\omega) = \psi/2\pi\omega \qquad \text{Eq. (1.4.1)}$$

Hysteretic damping is characteristic for elasto-damping materials such as elastomers (rubbers), wire mesh, etc.

Figs. 1.4.1(c), (d) depict two embodiments of so-called relaxation suspension [1.9]. In Fig. 1.4.1(c), the isolator is represented as a parallel connection of elastic element k and viscous damper c in series with another elastic element $k_1 = Nk$; in Fig. 1.4.1(d), there is a series connection of a damped isolator $k' - c'$ and an undamped isolator $k'' = Nk'$. It is shown in [1.9] that these isolators are dynamically equivalent if $k = [N/(N+1)]\,k'$, and $c = [N/(N+1)]^2 c'$.

In some vibration control applications where the vibration amplitudes are very significant (e.g., vehicle suspensions), isolators with Coulomb friction dampers may be found. These do not satisfy requirements of typical isolation systems for precision machines and instruments due to their insensitivity to low-amplitude vibration.

28 • PASSIVE VIBRATION ISOLATION

A degree of isolation achieved in the uni-axial systems in Fig. 1.4.1 can be characterized by *absolute* and *relative transmissibilities*. The absolute transmissibility indicates degree of reduction of the dynamic force or vibratory motion provided by the isolation system. If the source of vibration is dynamic force generated inside of the object, then transmissibility is a ratio of the force amplitude transmitted to the foundation to the amplitude of the primary, excitation force. If the source of vibration is vibratory motion of the foundation, then transmissibility is the ratio of the vibratory amplitude of the object to the amplitude of vibratory motion of the foundation. The relative transmissibility is the ratio of the amplitude of the vibratory displacements between the object and the foundation (i.e., deformation amplitude of the flexible element) to the amplitude of the foundation motion.

1.4.1 Vibration Isolator with Viscous Damper, Fig. 1.4.1(a)

Equations of motion of this system are

$$m\ddot{x}_1 + c(\dot{x}_1 - \dot{x}_2) + k(x_1 - x_2) = F$$

$$m_f \ddot{x}_2 + c(\dot{x}_2 - \dot{x}_1) + k(x_2 - x_1) = 0 \qquad \text{Eq. (1.4.2)}$$

Multiplying the first equation by m_f, second equation by m, and subtracting the second one from the first one, results in

$$\frac{mm_f}{m + m_f}\ddot{\theta} + c\dot{\theta} + k\theta = \frac{m_f}{m + m_f}F, \qquad \text{Eq. (1.4.3)}$$

where $\theta = x_1 - x_2$, and $mm_f/(m + m_f) = m_{ef}$, "effective mass" of the system.

Energy dissipation (damping) characteristics of Eqs. (1.4.2) and (1.4.3) can be expressed in various formats:

$$\Delta = \frac{c}{2m_{ef}} = \frac{c(m + m_f)}{2mm_f}; \quad \zeta = \frac{c}{c_{cr}}; \quad c_{cr} = 2\omega_n \frac{mm_f}{m + m_f} = 2\sqrt{k\frac{mm_f}{m + m_f}};$$

$$\omega_n = \sqrt{\frac{k}{\frac{mm_f}{m+m_f}}}; \quad \delta = \frac{2\pi\frac{c}{c_{cr}}}{\sqrt{1 - \left(\frac{c}{c_{cr}}\right)^2}} \approx \frac{\pi c}{\omega_n \frac{mm_f}{m+m_f}}; \quad \mu_{res} = \frac{\pi}{\delta} = \frac{\omega_n mm_f}{c(m + m_f)};$$

$$\psi = \frac{\Delta \Pi}{\Pi} = \frac{2\pi c \omega_n}{k} \qquad \text{Eq. (1.4.4)}$$

Here c is damping coefficient; Δ is relative damping coefficient; c_{cr} is critical damping coefficient; ζ is relative damping; δ is logarithmic decrement; μ_{res} is resonance amplification factor; ω_n is natural angular frequency; Π – maximal potential energy of the vibratory system within one cycle; $\Delta\Pi$ – energy dissipated during the cycle (a function of vibration frequency ω for viscous damping systems); ψ – relative energy dissipation during one vibratory cycle. At resonance or for a free vibration process $\psi = 2\pi c\omega_n/k \cong 2\delta$.

To determine force (and identical displacement/acceleration) transmissibility at the harmonic (sinusoidal) exciting force $F = F_o \sin \omega t$, where F_o is the force amplitude, the excitation can be presented in complex variables as $F = F_o e^{j\omega t}$. Then deformation of the flexible connection is $\theta = \theta_o e^{j(\omega t + \phi)}$, where ϕ is phase shift between the force and the deformation. Substituting these into Eq. (1.4.3) and considering that the force acting on the foundation is $F_f = k\theta + c\dot{\theta}$, arrive to

$$F_f = \frac{\frac{m_f}{m+m_f} F_o(k + jc\omega)e^{j\omega t}}{\left(-\frac{mm_f}{m+m_f}\omega^2 + k\right) + jc\omega}, \qquad \text{Eq. (1.4.5)}$$

from which the absolute force transmissibility coefficient is

$$\mu_F = \frac{|F_f|}{|F_o|} = \frac{|x_1|}{|x_2|} = \frac{m_f}{m+m_f}\sqrt{\frac{k^2 + \omega^2 c^2}{\left(k - \frac{mm_f}{m+m_f}\omega^2\right)^2 + \omega^2 c^2}}$$

$$= \frac{m_f}{m+m_f}\sqrt{\frac{1+\left(2\zeta\frac{\omega}{\omega_n}\right)^2}{\left(1-\frac{\omega^2}{\omega_n^2}\right)^2 + \left(2\zeta\frac{\omega}{\omega_n}\right)^2}} = \frac{m_f}{m+m_f}\sqrt{\frac{1+\left(\frac{\delta}{\pi}\frac{\omega}{\omega_n}\right)^2}{\left(1-\frac{\omega^2}{\omega_n^2}\right)^2 + \left(\frac{\delta}{\pi}\frac{\omega}{\omega_n}\right)^2}}$$

$$\text{Eq. (1.4.6)}$$

To express absolute displacement transmissibility for "kinematic excitation" from the foundation, the first equation of Eq. (1.4.2) with $F = 0$ can be used. If vibration excitation from the foundation is $x_2 = x_{2_o} e^{j\omega t}$, then $x_1 = x_{1_o} e^{j(\omega t + \phi)}$, and

$$\frac{x_1}{x_2} = \frac{k + jc\omega}{(k - m\omega^2) + jc\omega} \qquad \text{Eq. (1.4.7)}$$

From Eq. (1.4.7), the absolute transmissibility is

$$\mu_x = \frac{|x_1|}{|x_2|} = \sqrt{\frac{1 + \left(2\zeta \frac{\omega}{\omega_n}\right)^2}{\left(1 - \frac{\omega^2}{\omega_n^2}\right)^2 + \left(2\zeta \frac{\omega}{\omega_n}\right)^2}} = \sqrt{\frac{1 + \left(\frac{\delta}{\pi} \frac{\omega}{\omega_n}\right)^2}{\left(1 - \frac{\omega^2}{\omega_n^2}\right)^2 + \left(\frac{\delta}{\pi} \frac{\omega}{\omega_n}\right)^2}} \quad \text{Eq. (1.4.8)}$$

It can be seen from Eqs. (1.4.6) and (1.4.8) that $\mu_F = \mu_x = \mu$ only for an infinitely large foundation ($m_f = \infty$).

To express the relative transmissibility μ_{rel}, the first equation of Eq. (1.4.2) with $F = 0$ can be written as

$$m(\ddot{\theta} + \ddot{x}_2) + c\dot{\theta} + k\theta = 0 \quad \text{Eq. (1.4.9)}$$

If $x_2 = x_{2_o} e^{j\omega t}$, $\theta = \theta_o e^{j(\omega t + \phi)}$, then

$$\frac{\theta}{x_2} = \frac{m\omega^2}{-m\omega^2 + jc\omega + k}, \quad \text{Eq. (1.4.10)}$$

and

$$\mu_{rel} = \frac{|\theta|}{|x_2|} = \sqrt{\frac{m^2 \omega^4}{\left(k - m\omega^2\right)^2 + \omega^2 c^2}} = \frac{\frac{\omega^2}{\omega_n^2}}{\sqrt{\left(1 - \frac{\omega^2}{\omega_n^2}\right)^2 + \left(2\zeta \frac{\omega}{\omega_n}\right)^2}}$$

$$= \frac{\frac{\omega^2}{\omega_n^2}}{\sqrt{\left(1 - \frac{\omega^2}{\omega_n^2}\right)^2 + \left(\frac{\delta}{\pi} \frac{\omega}{\omega_n}\right)^2}}. \quad \text{Eq. (1.4.11)}$$

Fig. 1.4.2 presents $\mu_F(\omega/\omega_n)_{m_f = \infty} = \mu_x$ at various damping levels. It can be seen that absolute transmissibility μ_F and μ_x are less than 1 only at the *frequency ratio* $\omega/\omega_n > \sqrt{2} \cong 1.4$, or *vibration isolation* (reduction of vibration transmission) begins only at relatively high frequencies. When damping increases, transmissibility at the isolation zone deteriorates. However, relative transmissibility μ_{rel} always decreases with increasing damping (see Fig. 1.4.3).

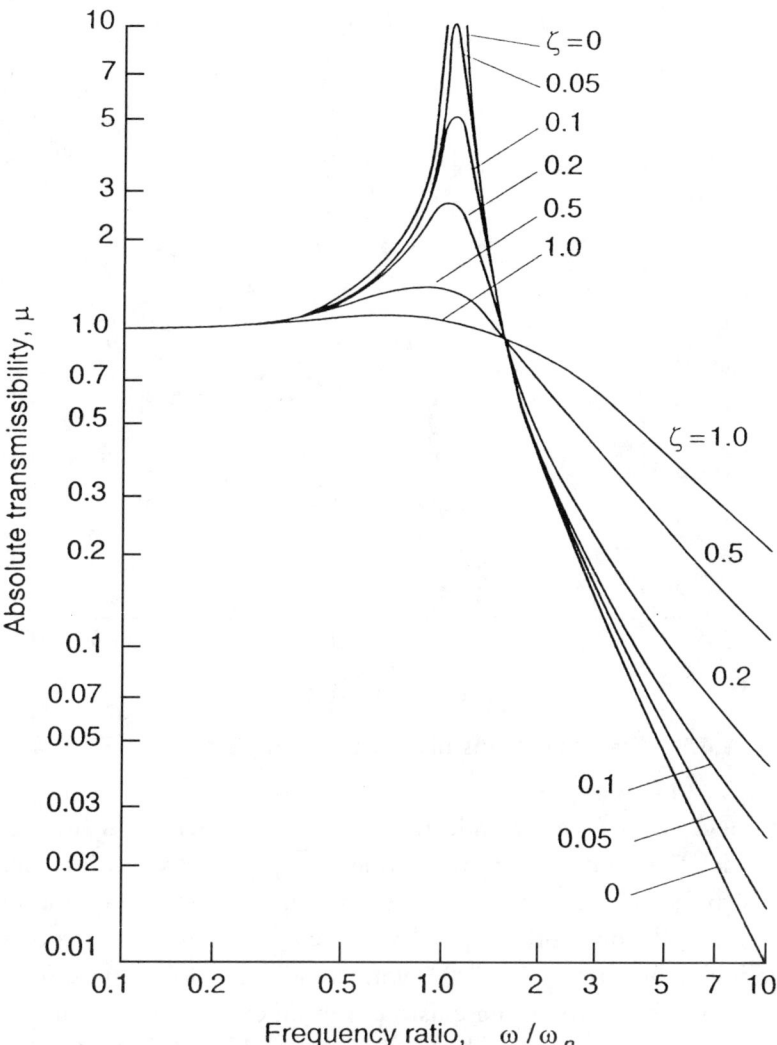

Fig. 1.4.2. Absolute Transmissibility of the System in Fig.1.4.1(a) ($m_f = \infty$).

1.4.2 Vibration Isolator with Internal (Hysteresis) Damping, Fig. 1.4.1(b)

The damping mechanism in elasto-damping materials (Chapter 3) is very different from the viscous damping mechanism. The log decrement in systems with such materials does not depend on natural frequency as in Eq. (1.4.4), but usually it has a more mild or a more complicated

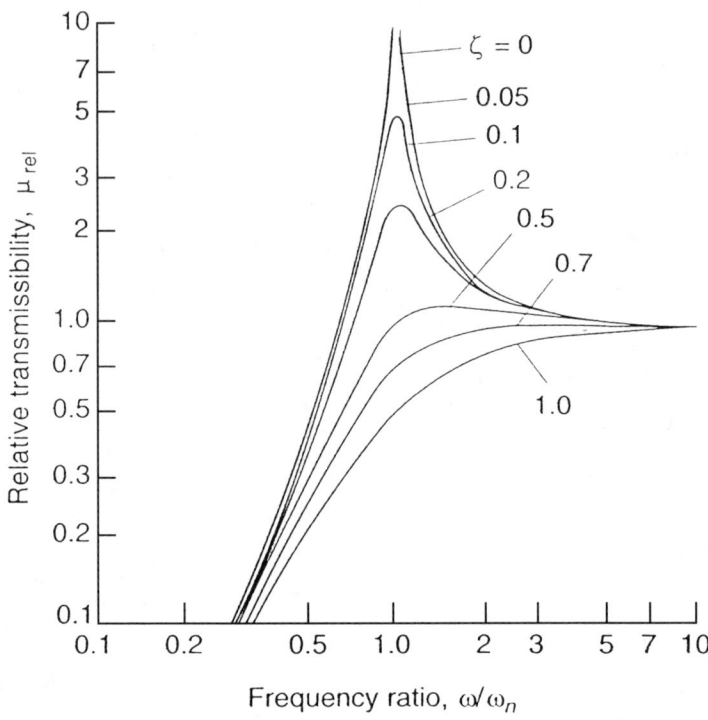

Fig. 1.4.3. Relative Transmissibility of the System in Fig.1.4.1(a).

dependence on frequency and, in many cases, depends on the vibration amplitude. For materials most frequently used for vibration isolation, such as rubber blends, in many cases and in the first approximation, the frequency and amplitude dependencies can be neglected for a narrow frequency range, up to ~50–100 Hz and small vibration amplitudes. There are many ways of expressing elasto-damping characteristics of materials with internal friction/energy dissipation. A widely used format of *"complex stiffness"* is based on the fact that deformation of such materials has a phase delay β relative to the stress causing this deformation. Thus the deformation is presented as comprising two components: *elastic* one (in phase with the stress) and *velocity-dependent* one, lagging the stress by 90° ($\pi/2$ rad), or directed along the imaginary axis. In such notation, the total modulus G is expressed as

$$G = G' + jG'' = G'[1 + j(G''/G')] = G'(1 + j\tan\beta), \qquad \text{Eq. (1.4.12)}$$

where G' is *elastic modulus*, G'' is *loss modulus*, and β is *loss angle*.

Accordingly, the complex stiffness of an element is

$$k = k' + jk'' = k'(1 + j\tan\beta) \qquad \text{Eq. (1.4.13)}$$

Equations (1.4.2) and (1.4.3) become, respectively,

$$m\ddot{x}_1 + k'(1 + j\tan\beta)(x_1 - x_2) = F$$

$$m_f \ddot{x}_2 + k'(1 + j\tan\beta)(x_2 - x_1) = 0 \qquad \text{Eq. (1.4.14)}$$

$$\frac{mm_f}{m + m_f}\ddot{\theta} + k'(1 + j\tan\beta)\theta = \frac{m_f}{m + m_f}F. \qquad \text{Eq. (1.4.15)}$$

If $F = F_o e^{j\omega t}$, then assuming that $\theta = \theta_o e^{j(\omega t + \phi)}$ and neglecting amplitude and frequency dependencies of G' and G'', arrive to

$$\left[-\frac{mm_f}{m + m_f}\omega^2 + k'(1 + j\tan\beta)\right]\theta_o e^{j(\omega t + \varphi)} = \frac{mm_f}{m + m_f}F_o e^{j\omega t} \qquad \text{Eq. (1.4.16)}$$

$$F_f = k'(1 + j\tan\beta)\theta = \frac{k'(1 + j\tan\beta)\frac{m_f}{m + m_f}}{-\frac{mm_f}{m + m_f}\omega^2 + k'(1 + j\tan\beta)}F_o e^{j\omega t} \qquad \text{Eq. (1.4.17)}$$

$$\mu_F = \frac{|F_f|}{|F|} = \frac{m_f}{m + m_f}\sqrt{\frac{(k')^2(1 + \tan^2\beta)}{\left(k' - \frac{mm_f}{m + m_f}\omega^2\right)^2 + \tan^2\beta}}$$

$$= \frac{m_f}{m + m_f}\sqrt{\frac{1 + \tan^2\beta}{\left(1 - \frac{\omega^2}{\omega_n^2}\right)^2 + \tan^2\beta}}$$

$$\text{Eq. (1.4.18)}$$

where $\omega_n = \sqrt{\frac{k'(m + m_f)}{mm_f}}$ is angular natural frequency. Comparing Eq. (1.4.17) with Eq. (1.4.5), one can conclude that $\tan\beta$ in Eq. (1.4.17) corresponds to $c\omega$ in Eq. (1.4.5). Thus, it can be concluded that Eq. (1.4.17) can be derived from an equation similar to Eq. (1.4.3) in which damping coefficient c is dependent on frequency, $c = (\tan\beta)/\omega$. Then

$$\psi = 2\pi\tan\beta; \quad c = \psi/2\pi\omega. \qquad \text{Eq. (1.4.19)}$$

34 • PASSIVE VIBRATION ISOLATION

With the above-made assumption that ψ does not depend on the vibration amplitude, decay of free vibration in the system with internal friction-based damping is characterized by a logarithmic decrement that is the constant for the given elasto-damping material and is independent of mass m of the object,

$$\delta = \psi/2 = \pi \tan \beta, \qquad \text{Eq. (1.4.20)}$$

or

$$\mu_F = \frac{m_f}{m + m_f} \sqrt{\frac{1 + \frac{\delta^2}{\pi^2}}{\left(1 - \frac{\omega^2}{\omega_n^2}\right)^2 + \frac{\delta^2}{\pi^2}}} \qquad \text{Eq. (1.4.18')}$$

The absolute transmissibility for kinematic excitation from the foundation is derived from the first part of Eq. (1.4.14) at $F = 0$, $x_2 = x_{2_o} e^{j\omega t}$, $x_1 = x_{1_o} e^{j(\omega t + \phi)}$. Thus,

$$\frac{x_1}{x_2} = \frac{k'(1 + j \tan \beta)}{-m\omega^2 + k'(1 + j \tan \beta)}, \qquad \text{Eq. (1.4.21)}$$

$$\mu_x = \frac{|x_1|}{|x_2|} = \sqrt{\frac{k'^2(1 + \tan^2 \beta)}{(k' - m\omega^2)^2 + k'^2 \tan^2 \beta}} = \sqrt{\frac{1 + \frac{\delta^2}{\pi^2}}{\left(1 - \frac{\omega^2}{\omega_n^2}\right)^2 + \frac{\delta^2}{\pi^2}}} \qquad \text{Eq. (1.4.22)}$$

The relative transmissibility is derived analogously to Eq. (1.4.10):

$$\frac{\theta}{x_2} = \frac{m\omega^2}{-m\omega^2 + k'(1 + j \tan \beta)}, \qquad \text{Eq. (1.4.23)}$$

$$\mu_{rel} = \frac{|\theta|}{|x_2|} = \sqrt{\frac{m^2 \omega^4}{(k' - m\omega^2)^2 + k'^2 \tan^2 \beta}} = \frac{\frac{\omega^2}{\omega_n^2}}{\sqrt{\left(1 - \frac{\omega^2}{\omega_n^2}\right)^2 + \frac{\delta^2}{\pi^2}}}.$$

$$\text{Eq. (1.4.24)}$$

By comparing Eqs. (1.4.6) and (1.4.8) with Eqs. (1.4.18) and (1.4.22), it can be seen that due to the absence of frequency-dependent terms in the numerators under the square-root sign in the latter two expressions, the high-frequency transmissibility for systems with damping is significantly

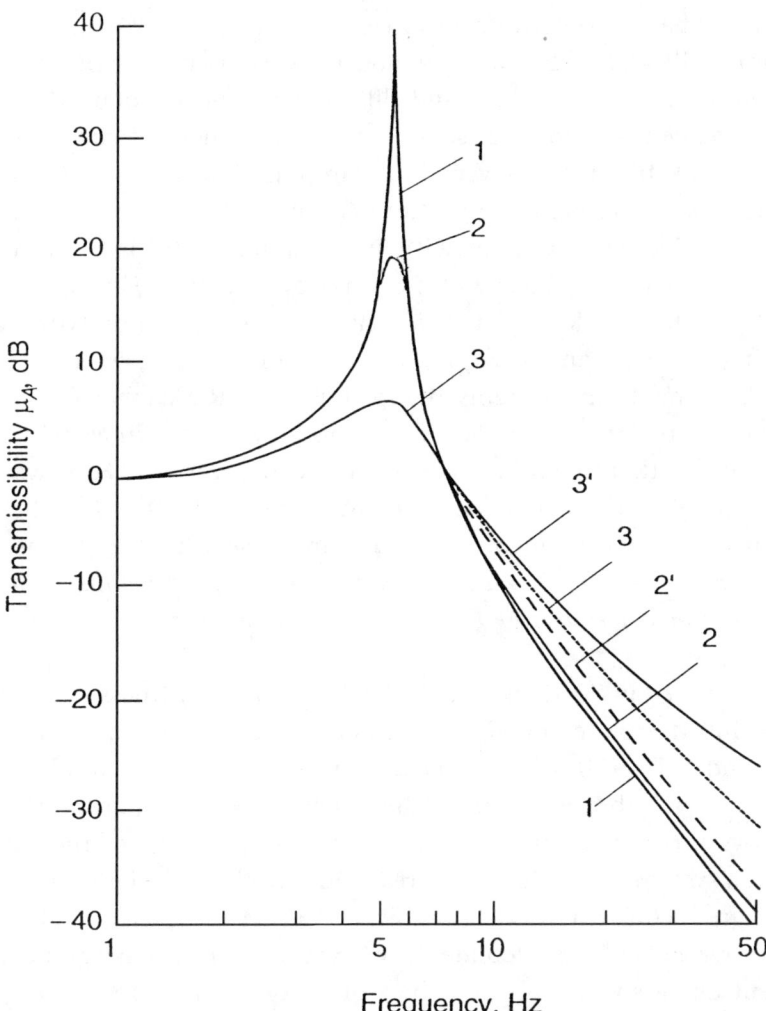

Fig. 1.4.4. Absolute Transmissibility of the System in Fig. 1.4.1(b) for Isolator Made of Various Rubber Blends Considering Frequency Dependencies of Their Stiffness and Damping. 1 – Unfilled NR; 2 – NR Filled with 50% ppw HAF Carbon Black; 3 – Thiocol RD; 2' – Isolator with Viscous Damping Having Same μ_{res} as 2; 3' – Same for 3; $m_f = \infty$.

reduced, thus improving vibration isolation without deteriorating the resonance behavior. In real-life systems the improvement may be less pronounced due to the above noted frequency dependencies of G' and G'' (see Chapter 3), but the improvement is still very significant. Fig. 1.4.4, which is based on information from [1.8] shows the absolute transmissibility plots for $m_f = \infty$ and for various rubber blends having different damping (δ) values and different G' and G'' dependencies on frequency. Plot 1 is for the isolator made from natural rubber without carbon black filler; it has very low damping ($\delta = 0.07$) and negligible dependence on frequency of G' and G'' (or δ) below 50 Hz. This case can be considered as isolator without damping. Plot 2 is for an isolator made from natural rubber with 50 parts by weight of HAF-type carbon black ($\delta = 0.31$) while plot 2' is for a "virtual" isolator with viscous damping and the same damping at the resonance (5 Hz). Plot 3 is for an isolator made from highly damped Thiocol RD rubber ($\delta = 1.55$ at resonance frequency 5 Hz and monotonously increasing with frequency, with three-fold increase between 1 and 50 Hz), while the plot 3' is for a "virtual" viscous damping isolator with the same damping at the resonance (5 Hz). It can be seen that in the isolation zone $\omega/\omega_n > 1.4$, *increase in the hysteresis damping results in a much less isolation deterioration* than a comparable increase in the viscous damping.

As discussed in Sections 3.3.1A and 3.3.2D, both stiffness and damping of real-life flexible elements with hysteretic damping (practically, wire mesh and rubber flexible elements) are dependent on amplitude and frequency of vibration. Since the most important parameters of a vibratory process are the natural frequency $\omega_n = 2\pi f_n$ of the vibration isolation system and height of the resonance peak, several iterations might be required when using Eqs. (1.4.18), (1.4.18'), (1.4.22), and (1.4.24). One of the approaches is to calculate the vibratory deformation θ of the flexible element using some realistic values of k' (or ω_n) and $\delta = \pi \tan \beta$. The obtained value of θ allows one to obtain new values of k' and δ, etc., until the desired convergence is achieved.

1.4.3 "Relaxation" Isolation System with Viscous Damper, Figs. 1.4.1(c), (d)

Equations of motion of the systems in Figs. 1.4.1(c), (d) for $m_f = \infty$ are, respectively [1.9]:

$$\left(\frac{mc}{Nk}\right)\dddot{x}_1 + m\ddot{x}_1 + c\left(\frac{N+1}{N}\right)\dot{\theta} + k\theta = F + \frac{c}{Nk}\dot{F} \qquad \text{Eq. (1.4.25)}$$

$$\left(\frac{m\frac{c'}{k'}}{N+1}\right)\dddot{x}_1 + m\ddot{x}_1 + c'\left(\frac{N}{N+1}\right)\dot{\theta} + k'\left(\frac{N}{N+1}\right)\theta = F + \left[\frac{c'}{(N+1)k'}\right]\dot{F}.$$
Eq. (1.4.26)

The formats of both equations are the same. Thus, they can be made dynamically equivalent by equating coefficients at like terms,

$$k = \left(\frac{N}{N+1}\right)k'; \quad c = \left(\frac{N}{N+1}\right)^2 c',$$
Eq. (1.4.27)

where $N = k_1/k$ for Fig. 1.4.1(c) and $N = k''/k'$ for Fig. 1.4.1(d). Because of Eq. (1.4.27), the transmissibility expressions given below for the system in Fig. 1.4.1(c) [1.10] are applicable also for Fig. 1.4.1(d) taking Eq. (1.4.27) into consideration. The absolute transmissibilities are

$$\mu_x = \mu_F = \sqrt{\frac{1 + 4\left(\frac{N+1}{N}\right)^2 \zeta^2 \frac{\omega^2}{\omega_n^2}}{\left(1 - \frac{\omega^2}{\omega_n^2}\right)^2 + \frac{4}{N^2}\zeta^2 \frac{\omega^2}{\omega_n^2}\left(N + 1 - \frac{\omega^2}{\omega_n^2}\right)^2}}$$
Eq. (1.4.28)

and the relative transmissibility is

$$\mu_{rel} = \sqrt{\frac{\frac{\omega^2}{\omega_n^2} + \frac{4}{N^2}\zeta^2 \frac{\omega^6}{\omega_n^6}}{\left(1 - \frac{\omega^2}{\omega_n^2}\right)^2 + \frac{4}{N^2}\zeta^2 \frac{\omega^2}{\omega_n^2}\left(N + 1 - \frac{\omega^2}{\omega_n^2}\right)^2}}$$
Eq. (1.4.29)

In Eqs. (1.4.28) and (1.4.29), $\omega_n = \sqrt{\frac{k}{m}}$ [which is the natural frequency of the system in Fig. 1.4.1(c) when $c = 0$]; $c_{cr} = 2\sqrt{km}$; $\zeta = c/c_{cr}$. Fig. 1.4.5 [1.11] compares the absolute transmissibility functions for a rigidly connected viscous damper [Fig. 1.4.1(a) or Fig. 1.4.1(c) with $N = \infty$] and an elastically connected viscous damper [Fig. 1.4.1(c)] for the relative damping value $\zeta = 0.2$. It can be seen that transmissibility of the latter at high frequencies (isolation zone) decreases much faster with increasing frequency ω than for the former (at 12 dB per octave vs. 6 dB per octave), but at the price of increased resonance amplification factor. For $c = 0$, the mass m is supported only by the main spring k, and the transmissibility function is the same as for the system in Fig. 1.4.1(a) [see Eqs. (1.4.6) and (1.4.8)]. When $c = \infty$, the transmissibility function is the same as for the

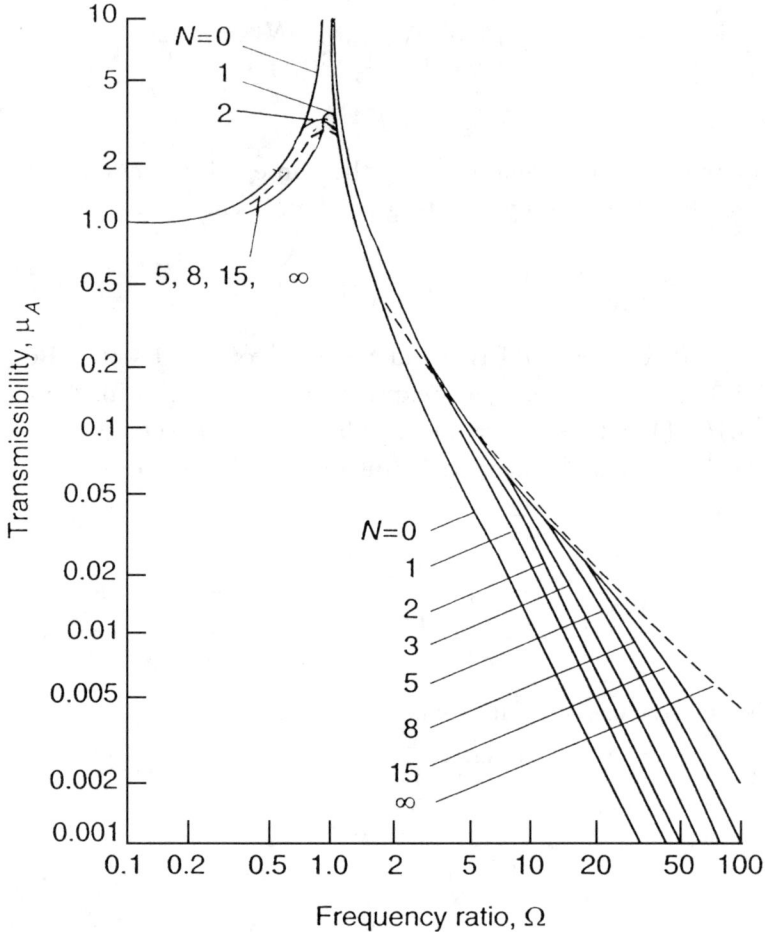

Fig. 1.4.5. Absolute Transmissibilities of Uni-Axial Systems in Fig. 1.4.1(c) with Rigidly ($N = \infty$) and Elastically Connected Viscous Damper; $\zeta = 0.2$.

case of $c = 0$ in Eqs. (1.4.6) and (1.4.8) but with the natural frequency $\omega'_n = \omega_\infty = \sqrt{\frac{k+k_1}{m}} = \omega_n \sqrt{N+1}$. For intermediate values of c, the transmissibility function is between these two extremes. The value of c, which produces the minimum resonance amplification, is called the *optimum damping*. At high excitation frequencies, transmissibility functions with any values of c are approaching the transmissibility function for $c = \infty$, and the absolute transmissibility at high frequencies is inversely proportional to the excitation frequency.

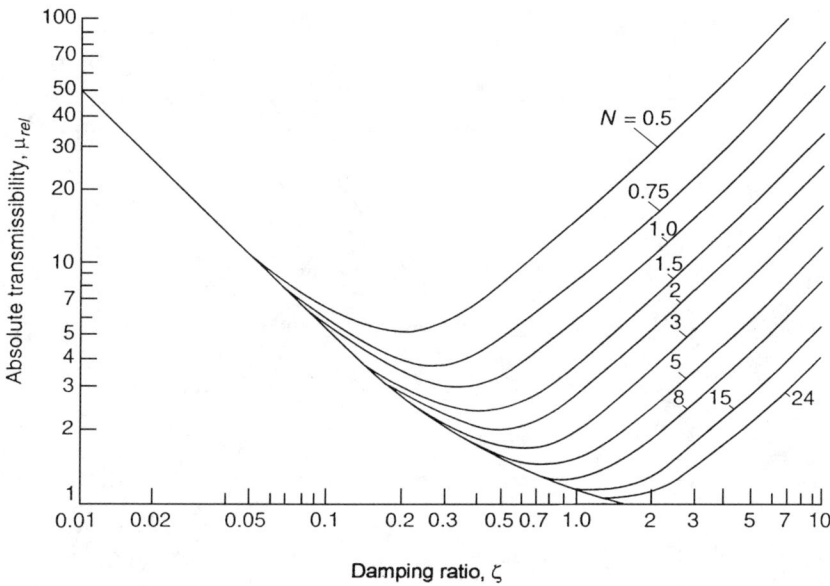

Fig. 1.4.6. Transmissibility at Resonance μ_{res} of the system in Fig. 1.4.1(c) vs. Stiffness Ratio N and Damping Ratio ζ.

The resonance amplification (transmissibility) μ_{res} as the function of the damping ratio ζ and the stiffness ratio N is plotted in Fig. 1.4.6 [1.11]. For $\zeta < \sim 0.1$, μ_{res} is weakly dependent on N, but at larger ζ, N determines the maximum μ_{res}. The minimum μ_{res} for a given N corresponds to the *optimum damping* situation. The resonance frequency ratios for the minimum μ_{res} are different for absolute (*abs*) and relative (*rel*) transmissibility; they are [1.11]

$$\left(\frac{\omega_{res}}{\omega'_n}\right)_{abs} = \sqrt{\frac{2(N+1)}{N+2}}; \quad \left(\frac{\omega_{res}}{\omega'_n}\right)_{rel} = \sqrt{\frac{N+2}{2}}. \quad \text{Eq. (1.4.30)}$$

The optimum (minimum) resonance amplification for both absolute and relative motion is

$$(\mu_{res})_{opt} = 1 + 2/N. \quad \text{Eq. (1.4.31)}$$

The damping ratios producing the minimum (optimum) values expressed by Eq. (1.4.31) are different for absolute and relative transmissibility and are

$$(\zeta_{opt})_{abs} = \frac{N}{4(N+1)}\sqrt{2(N+2)}; \quad (\zeta_{opt})_{rel} = \frac{N}{\sqrt{2(N+1)(N+2)}}$$

$$\text{Eq. (1.4.32)}$$

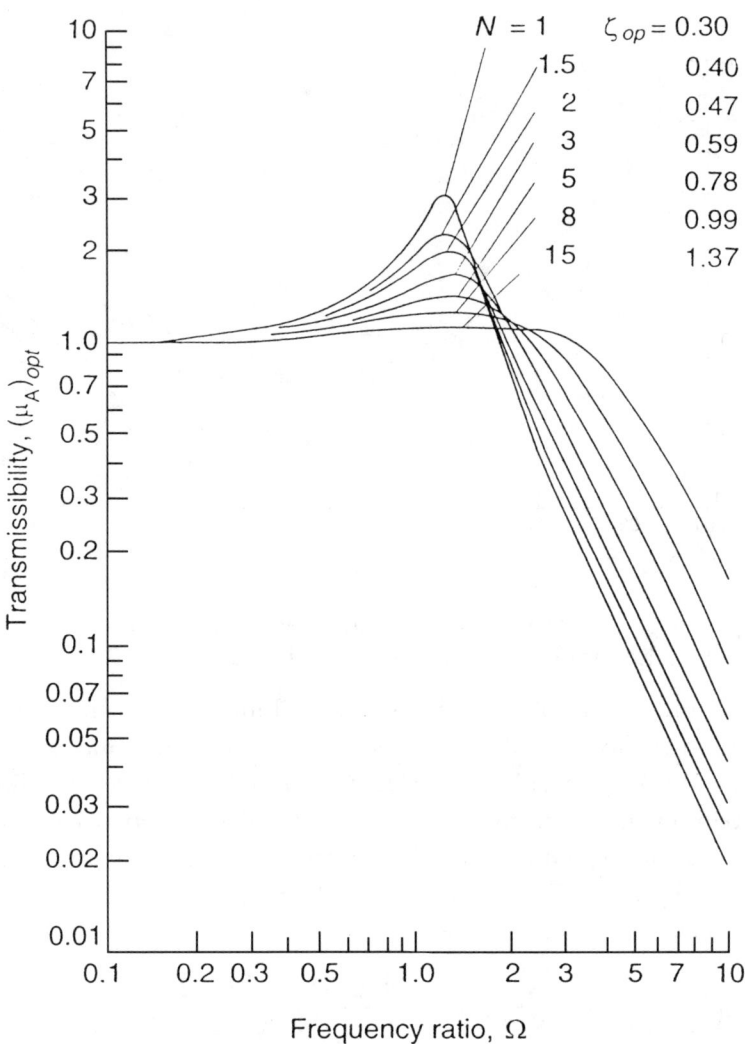

Fig. 1.4.7. Absolute Transmissibility of the System in Fig. 1.4.1(c) at Optimum Damping Ratios.

Plots of absolute and relative transmissibility at optimum damping values are shown in Figs. 1.4.7 and 1.4.8 [1.11], respectively. It can be seen that for low values of N (soft attachment of the damper) the resonance amplification is large, but isolation in the high-frequency range is very good (low absolute transmissibility). On the other hand, large N (stiff attachment of the damper) reduces μ_{res}, but leads to deterioration of isolation efficiency at the high frequencies. For the relative transmissi-

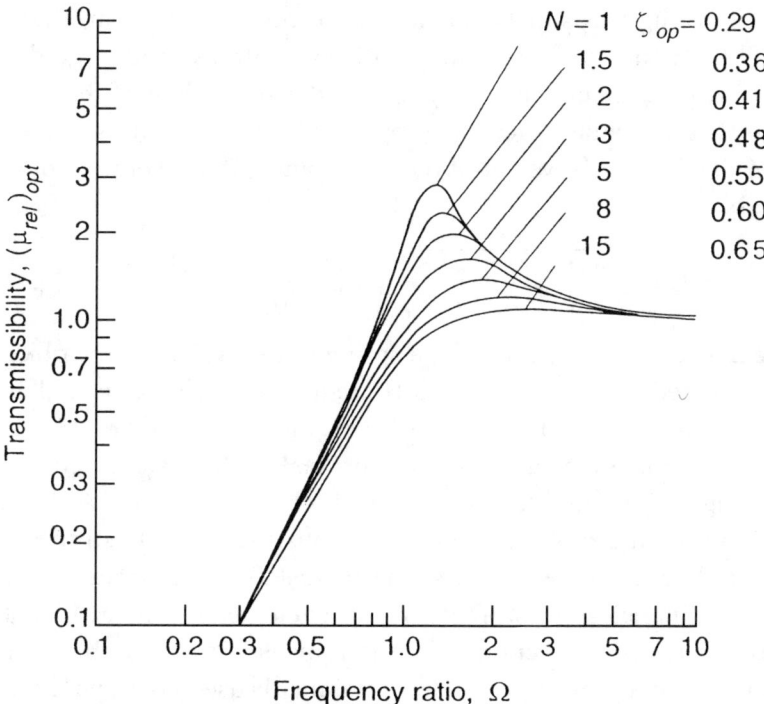

Fig. 1.4.8. Relative Transmissibility of the System in Fig. 1.4.1(c) at Optimum Damping Ratios.

bility, the effects of N variation are significantly pronounced only at the resonance.

The transmissibility functions for damping values other than optimal are given in [1.11].

1.4.4 Single-Degree-of-Freedom System with Motion Transformation

The vibration isolation systems described in Section 1.4.3 and in Section 1.5.2 improve performance of the basic single-degree-of-freedom systems (Section 1.4.1 and Section 1.4.2) at high frequencies by having steeper declines of the transmissibility plot at excitation frequencies that are significantly higher than the natural frequency of the system. A seemingly more straightforward way of reducing high-frequency transmissibility is by reducing the natural frequency of the basic system. Until recently, this approach was considered unrealistic for isolation of

42 • PASSIVE VIBRATION ISOLATION

vibration along z-axis or having a component in z-direction, since a low natural frequency f_z is associated with low stiffness and large deformations of the elastic elements (springs) under the weight of the supported objects. With f_z being determined by Eq. (1.3.10), and mass of the object $M = W/g$, where W is weight of the object and g is acceleration of gravity, it can be written that

$$f_z = \sqrt{k_z/M} = \sqrt{k_z g/W} = \sqrt{g/\Delta} \approx \frac{5}{\sqrt{\Delta}}, \qquad \text{Eq. (1.4.33)}$$

where Δ is deflection in cm of the spring under weight of the object, and f_z is expressed in Hz. Thus, $f_z = 5$ Hz requires weight-induced deflection 1 cm = 10 mm (0.4 in.), and $f_z = 1$ Hz requires deflection of 25 cm = 250 mm (~10 in.). Such large deflections make the system very "shaky," exhibiting large amplitude motions, both translational and rocking, especially when the object is light and thus the absolute stiffness of the spring is low. In many cases, such an "instability" associated with low natural frequencies/low stiffness of spring elements is objectionable also for horizontal x-, y-directions, although this is less of a problem than for the vertical direction. Here, the term stability is used not in its "academic" meaning, but to indicate relatively small excursions of the body. When such low frequency must be realized in a passive system, the mass of the object should be artificially enlarged by installing it on an expensive and bulky "inertia block" (usually made of reinforced concrete), so that the combined stiffness of the springs could be proportionally increased and stability of the system assured.

To realize low vertical (or horizontal) natural frequencies, it was proposed that relatively high stiffness springs even for light objects can be used by introducing "virtual inertia" elements [1.12]. This effect is achieved by using motion transformation systems or "motion transformers." A typical system of this kind is shown in Fig. 1.4.9. The system in Fig. 1.4.9 comprises object of isolation 1 having mass m, isolator (spring) 2, damper 3, and motion transformer composed of flywheel/nut 4, which is restrained in the axial direction (it has same axial displacements as object 1) and supported for rotation in relation to object 1 by thrust bearings 5. Flywheel 4 is engaged with threaded bar 6 attached to vibrating foundation 7. The presence of the motion transformer results in generation of additional dynamic forces caused by inertia of the flywheel/nut and by friction in the threaded connection 4–6.

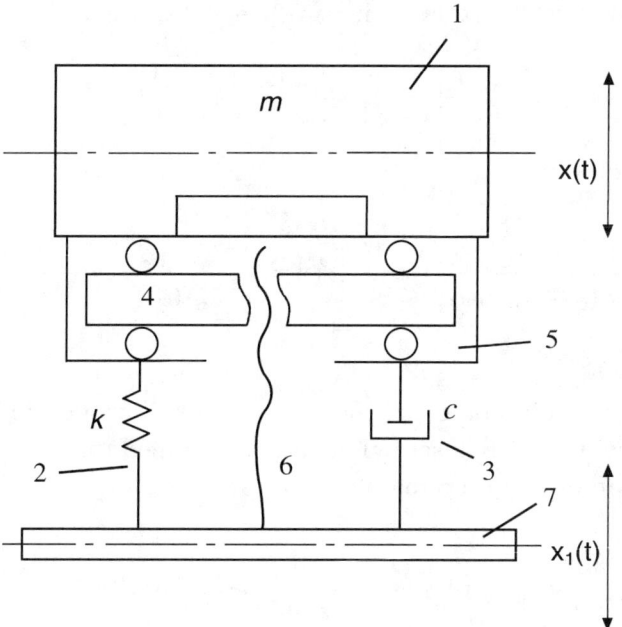

Fig. 1.4.9. Vibration Isolation System with Motion Transformation.

When object 1 moves in the axial direction, flywheel 4 is forced to rotate. Axial force from the object to the flywheel should overcome inertia torque from the flywheel, $T = I\ddot{\varphi}$, where φ is rotational angle of the flywheel and $\ddot{\varphi}$ is its angular acceleration. The resulting axial force is

$$P = T/r\tan(\alpha \pm \gamma) = \frac{I\ddot{\varphi}}{r\tan(\alpha \pm \gamma)}, \qquad \text{Eq. (1.4.34)}$$

where I is moment of inertia of the flywheel, α is helix angle of the thread, r is median radius of the thread, γ is friction angle (friction coefficient in the thread is $f = \tan\gamma$) and its sign depends on direction of sliding in the thread ("minus"—towards equilibrium position, "plus"—from the equilibrium position). Vibration of the system can thus be described by two equations of motion for these two phases of motion—to and from the equilibrium position, respectively,

$$m\ddot{x} + kx + c\dot{x} + \frac{I\ddot{\varphi}}{r\tan(\alpha - \gamma)} = kx_1 + c\dot{x}_1$$

$$m\ddot{x} + kx + c\dot{x} + \frac{I\ddot{\varphi}}{r\tan(\alpha + \gamma)} = kx_1 + c\dot{x}_1. \qquad \text{Eq. (1.4.35)}$$

From kinematics of a screw pair $\phi = \frac{x-x_1}{r\tan\alpha}$, thus finally,

$$\left[m + \frac{I}{r^2 \tan(\alpha - \gamma)\tan\alpha}\right]\ddot{x} + c\dot{x} + kx = \frac{I}{r^2 \tan(\alpha - \gamma)\tan\alpha}\ddot{x}_1 + c\dot{x}_1 + kx_1$$

$$\left[m + \frac{I}{r^2 \tan(\alpha + \gamma)\tan\alpha}\right]\ddot{x} + c\dot{x} + kx = \frac{I}{r^2 \tan(\alpha + \gamma)\tan\alpha}\ddot{x}_1 + c\dot{x}_1 + kx_1$$

Eq. (1.4.36)

Since the system has practical significance only if friction in the thread is low (e.g., by using a ball screw), it can be assumed that $\gamma = 0$. Then only one equation of motion remains,

$$\left[m + \frac{I}{r^2 \tan^2\alpha}\right]\ddot{x} + c\dot{x} + kx = \frac{I}{r^2 \tan^2\alpha}\ddot{x}_1 + c\dot{x}_1 + kx_1. \qquad \text{Eq. (1.4.37)}$$

The expression in the bracket is the effective mass of the object, with the second term $m' = (I/r^2 \tan^2\alpha)$ being a "contribution" from the flywheel. This term is responsible for the specifics of the system with the motion transformer. For small r and α (commercially available ball screws), it can easily be one or even two orders of magnitude larger than m even for a very small flywheel. The natural frequency of this system is

$$f_n = \frac{1}{2\pi}\sqrt{\frac{k}{m + \frac{I}{r^2 \tan^2\alpha}}}, \qquad \text{Eq. (1.4.38)}$$

and can be very low while k is quite large.

Transmissibility of the system per Eq. (1.4.37) is

$$\mu = \frac{|x|}{|x_1|} = \sqrt{\frac{(1 - \eta\Omega^2)^2 + 4n^2\Omega^2}{[1 - (1+\eta)\Omega^2]^2 + 4n^2\Omega^2}}, \qquad \text{Eq. (1.4.39)}$$

where $\omega_o^2 = 2\pi f_o^2 = k/m$ (ω_o is the natural frequency of the system without motion transformer); $\Omega = \omega/\omega_o$; $c/m = 2n$; $\eta = m'/m$. If $\eta = 0$, Eq. (1.4.39)

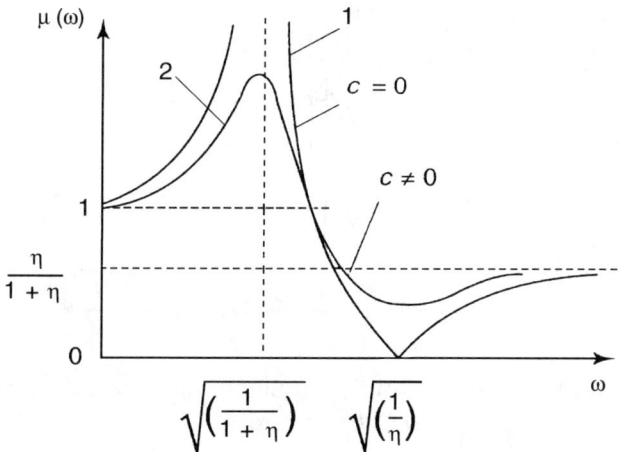

Fig. 1.4.10. Transmissibility of System in Fig. 1.4.9 for $m' = m/3$ and Various Damping. 1 – $c = 0$; 2 – $c > 0$.

becomes identical to Eq. (1.4.8). If $c = n = 0$, magnitude of μ in Eq. (1.4.39) becomes

$$|\mu| = \frac{\left|1 - \eta\Omega^2\right|}{\left|1 - (1+\eta)\Omega^2\right|}. \qquad \text{Eq. (1.4.40)}$$

Fig. 1.4.10 shows plots of $\mu(\Omega)$ for $\eta = 0.33$. When $c = n = 0$ (line 1), the system has sharp resonance at $\Omega_1^2 = 1/(1 + \eta)$, and transmissibility becomes zero at $\Omega_2^2 = 1/\eta$. Thus, the natural frequency is lower than for the system without the motion transformation per Eq. (1.4.38), and also there is an effect of "dynamic absorption" of vibration at the frequency ratio Ω_2. The latter effect is well-known for two-degrees-of-freedom systems but is unusual for single-degree-of-freedom systems. With increasing frequency ratio from Ω_2 to $\Omega = \infty$, transmissibility approaches to an asymptote at the level

$$\mu_\infty = \eta/(1+\eta), \qquad \text{Eq. (1.4.41)}$$

as can be easily found from Eq. (1.4.40). When damping is present in the system ($c > 0$), the resonance peak is reduced and the full vibration absorption does not develop (although the asymptote does not shift), as shown by line 2 in Fig. 1.4.10.

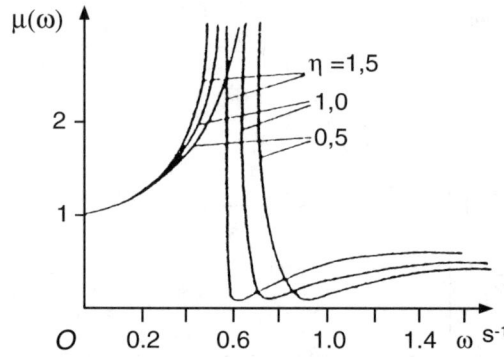

Fig. 1.4.11. Transmissibility of System in Fig. 1.4.9 for Various $\eta = m'/m$ and $c = 0$.

The above analysis shows that the vibration isolation system with a motion transformer exhibits a lower natural frequency without reducing spring stiffness, thus allowing realization of stable low-frequency systems. However, the obtained effect is not equivalent to a conventional system, such as one in Fig. 1.4.1(a) having low natural frequency, since transmissibility of the system with the motion transformer does not monotonously decrease with the increasing frequency ratio but is limited by the asymptote at $\eta/(1 + \eta)$. Use of the "dynamic absorption" effect is limited since it occurs very close to the resonance (especially at large η) and is not very pronounced due to always-present damping in the system. Still, the system in Fig. 1.4.9 can be beneficially utilized. Fig. 1.4.11 compares transmissibility at $c = 0$ for three values of η. While the natural frequency decreases with the increasing η, the height of the asymptote increases with the increasing η in accordance with Eq. (1.4.41). For very large values of η, the natural frequency can be made very low even with very stiff isolators, but the height of the asymptote approaches $\mu = 1$, thus making the system useless for isolation purposes. However, with relatively small values of η, a noticeable reduction in the natural frequency and a significant reduction in transmissibility μ across the frequency range, starting from low frequencies, can be achieved. It can be found from Eq. (1.4.40) that the isolation starts ($\mu < 1$) when the frequency ratio Ω exceeds

$$\Omega_3 = \sqrt{\frac{2}{1 + 2\eta}} \qquad \text{Eq. (1.4.42)}$$

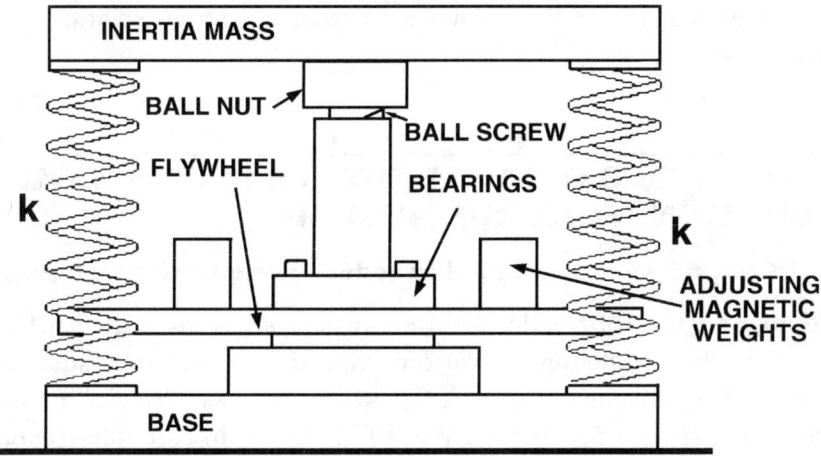

Fig. 1.4.12. Vibration Isolation System with Motion Transformer (f_n = 11 Hz with Disengaged Flywheel, 0.55 Hz with Engaged Motion Transformer).

The transmissibility curve crosses the asymptote at Ω_4,

$$\Omega_4 = \sqrt{\frac{1+2\eta}{2(\eta+\eta^2)}} \,.\qquad \text{Eq. (1.4.43)}$$

For example, for $\eta = 1$ (the effective mass $m + m' = 2m$), the natural frequency is reduced by 1.41 times, while the transmissibility μ cannot exceed $\eta/(1+\eta) = 1/(1+1) = 0.5$. For a conventional vibration isolation system having natural frequency $f_n = 5$ Hz, isolation starts from $f = 5 \times 1.41 = \sim 2.0$ Hz and transmissibility is reduced to 0.5 after $f = 8.66$ Hz. For this system combined with the motion transformer at $\eta = 1$, isolation starts from $f = 4.08$ Hz and transmissibility curve crosses $\mu = 0.5$ level at $f = 4.33$ Hz. Thus, a relatively small contribution from the flywheel resulted in a dramatic reduction of the frequency from which a significant isolation effect begins. In many cases, halving vibration transmission at low frequencies may fully satisfy the vibration requirements. Also, a motion transformation system with a large "virtual" mass may replace a costly massive foundation for some vibration-producing objects, e.g. see Section 2.3.1A. In each practical case, an optimization can be performed to decide what reduction of

f_n is beneficial. Fig. 1.4.12 shows a realized vibration isolation system with $\eta = 400$.

1.5 DYNAMICS OF TWO-DEGREES-OF-FREEDOM VIBRATION ISOLATION SYSTEM

1.5.1 Generic Two-Degrees-of-Freedom System with Damping

The two-degrees-of-freedom three-mass system in Fig. 1.5.1(a) is important for modeling of various vibration isolation systems. For analysis, it is convenient to break this system into two *"partial"* two-mass subsystems, Fig. 1.5.1(b). Analysis of dynamic interactions (dynamic coupling) between these partial subsystems clarifies physical meaning of the dynamic processes in the system in Fig. 1.5.1a. Equations of motion of the system in Fig. 1.5.1 are as follows:

$$m_1\ddot{x}_1 + c_{12}(\dot{x}_1 - \dot{x}_2) + k_{12}(x_1 - x_2) = F_1(t)$$

$$m_2\ddot{x}_2 + c_{12}(\dot{x}_2 - \dot{x}_1) + k_{12}(x_2 - x_1) + c_{23}(\dot{x}_2 - \dot{x}_3) + k_{23}(x_2 - x_3) = F_2(t)$$

$$m_3\ddot{x}_3 + c_{23}(\dot{x}_3 - \dot{x}_2) + k_{23}(x_3 - x_2) = F_3(t) \qquad \text{Eq. (1.5.1)}$$

This is a nominally three-degrees-of-freedom system with one practically unimportant "zero" mode (mode with zero natural frequency). This mode can be eliminated by transforming coordinates in these equations into relative coordinates $\theta_1 = x_1 - x_2$ and $\theta_2 = x_2 - x_3$. It can be accomplished by multiplying the first and the third equations by m_2 and subtracting from them the second equation multiplied by m_1 and m_3, respectively. Thus,

$$\frac{m_1 m_2}{m_1 + m_2}\ddot{\theta}_1 + c_{12}\dot{\theta}_1 + k_{12}\theta_1 - \frac{m_1}{m_1 + m_2}c_{23}\dot{\theta}_2 - \frac{m_1}{m_1 + m_2}k_{23}\theta_2$$

$$= \frac{m_2}{m_1 + m_2}F_1 - \frac{m_1}{m_1 + m_2}F_2$$

$$\frac{m_2 m_3}{m_2 + m_3}\ddot{\theta}_2 + c_{23}\dot{\theta}_2 + k_{23}\theta_2 - \frac{m_3}{m_2 + m_3}c_{12}\dot{\theta}_1 - \frac{m_3}{m_2 + m_3}k_{12}\theta_1$$

$$= -\frac{m_3}{m_2 + m_3}F_2 + \frac{m_2}{m_2 + m_3}F_3, \qquad \text{Eq. (1.5.2)}$$

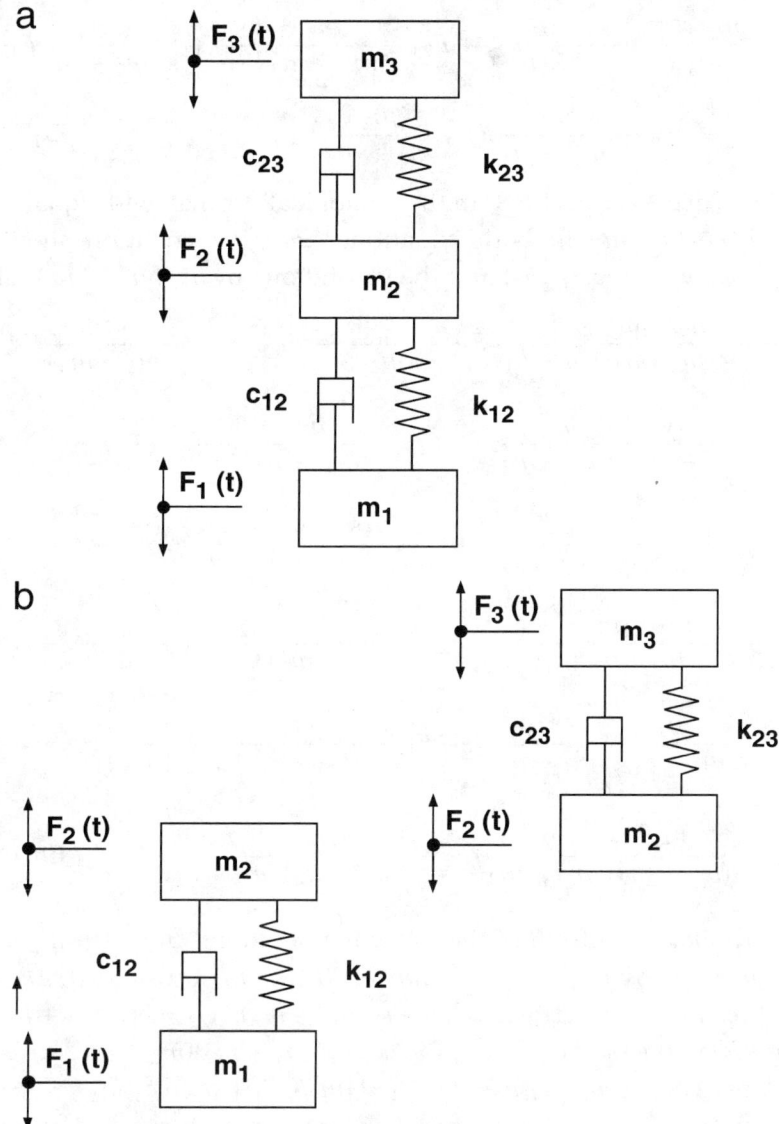

Fig. 1.5.1. (a) Generic Two-Degrees-of-Freedom System; and (b) and Its Partial Subsystems.

or, since the forces in the elastic connections are $F_{12} = k_{12}\theta_1$, $F_{23} = k_{23}\theta_2$,

$$\frac{m_1 m_2}{m_1 + m_2}\frac{1}{k_{12}}\ddot{F}_{12} + \frac{c_{12}}{k_{12}}\dot{F}_{12} + F_{12} - \frac{m_1}{m_1+m_2}\frac{c_{23}}{k_{23}}\dot{F}_{23} - \frac{m_1}{m_1+m_2}F_{23}$$

$$= \frac{m_2}{m_1+m_2}F_1 - \frac{m_1}{m_1+m_2}F_2$$

50 • PASSIVE VIBRATION ISOLATION

$$\frac{m_2 m_3}{m_2 + m_3} \frac{1}{k_{23}} \ddot{F}_{23} + \frac{c_{23}}{k_{23}} \dot{F}_{23} + F_{23} - \frac{m_3}{m_2 + m_3} \frac{c_{12}}{k_{12}} \dot{F}_{12} - \frac{m_3}{m_2 + m_3} F_{12}$$

$$= -\frac{m_3}{m_2 + m_3} F_2 + \frac{m_2}{m_2 + m_3} F_3 \,. \qquad \text{Eq. (1.5.3)}$$

In order to express Eq. (1.5.3) in the "canonical" format, with equal *elastic coupling coefficients* in both equations, the first equation should be multiplied by $m_3/(m_2 + m_3)$ and the second one by $m_1/(m_1 + m_2)$. Then

$$\frac{m_1 m_2 m_3}{(m_1 + m_2)(m_2 + m_3)} \frac{1}{k_{12}} \ddot{F}_{12} + \frac{m_3}{(m_2 + m_3)} \frac{c_{12}}{k_{12}} \dot{F}_{12} + \frac{m_3}{(m_2 + m_3)} F_{12}$$

$$- \frac{m_1}{(m_1 + m_2)} \frac{m_3}{(m_2 + m_3)} \frac{c_{23}}{k_{23}} \dot{F}_{23} - \frac{m_1}{(m_1 + m_2)} \frac{m_3}{(m_2 + m_3)} F_{23}$$

$$= \frac{m_2}{(m_1 + m_2)} \frac{m_3}{(m_2 + m_3)} F_1 - \frac{m_1}{(m_1 + m_2)} \frac{m_3}{(m_2 + m_3)} F_2$$

$$\frac{m_1 m_2 m_3}{(m_1 + m_2)(m_2 + m_3)} \frac{1}{k_{23}} \ddot{F}_{23} + \frac{m_1}{(m_1 + m_2)} \frac{c_{23}}{k_{23}} \dot{F}_{23} + \frac{m_1}{(m_1 + m_2)} F_{23}$$

$$- \frac{m_1}{(m_1 + m_2)} \frac{m_3}{(m_2 + m_3)} \frac{c_{12}}{k_{12}} \dot{F}_{12} - \frac{m_1}{(m_1 + m_2)} \frac{m_3}{(m_2 + m_3)} F_{12}$$

$$= -\frac{m_1}{(m_1 + m_2)} \frac{m_3}{(m_2 + m_3)} F_2 + \frac{m_1}{(m_1 + m_2)} \frac{m_2}{(m_2 + m_3)} F_3 \qquad \text{Eq. (1.5.4)}$$

It is convenient to introduce the following notations: $\beta_{11} = [m_1 m_2 m_3/(m_1 + m_2)(m_2 + m_3)] 1/k_{12}$, $\beta_{22} = [m_1 m_2 m_3/(m_1 + m_2)(m_2 + m_3)] 1/k_{23}$ are *generalized masses* (inertias); $\alpha_{11} = m_2/(m_2 + m_3)$, $\alpha_{22} = m_1/(m_1 + m_2)$ are *generalized stiffnesses*; $\alpha_{12} = \alpha_{23} = -m_1 m_3/(m_1 + m_2)(m_2 + m_3)$ are *elastic (coordinate) coupling coefficients*; $\xi_{11} = [m_3/(m_2 + m_3)] c_{12}/k_{12}$, $\xi_{22} = [m_1/(m_1 + m_2)] c_{23}/k_{23}$ are *generalized damping coefficients*; $\xi_{12} = [m_1/(m_1 + m_2)][m_3/(m_2 + m_3)] c_{12}/k_{12}$, $\xi_{23} = -[m_1/(m_1 + m_2)][m_3/(m_2 + m_3)] c_{23}/k_{23}$ are damping (first derivative) coupling coefficients; $F'_1 = [m_2/(m_1 + m_2)][m_3/(m_2 + m_3)] F_1 - [m_1/(m_1 + m_2)][m_3/(m_2 + m_3)] F_2$, $F'_2 = -[m_1/(m_1 + m_2)][m_3/(m_2 + m_3)] F_2 + [m_1/(m_1 + m_2)][m_2/(m_2 + m_3)] F_3$ are *generalized excitation forces*; and $\gamma = \sqrt{\alpha_{12}^2/\alpha_{11}\alpha_{22}} = \sqrt{[m_1 m_3/(m_1 + m_2)(m_2 + m_3)]}$ is generalized *dimensionless coefficient of elastic coupling*. Damping ratios for the partial subsystems are $c_{12}/(c_{cr})_{1p} = [c_{12}/2n_1(m_1 m_2/m_1 + m_2)] =$

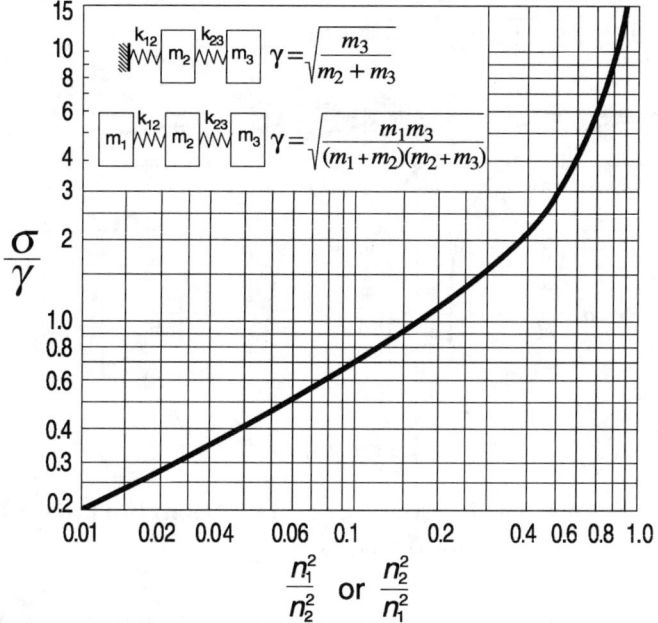

Fig. 1.5.2. Coupling Coefficient σ vs. Ratio of Partial Natural Frequencies n_1/n_2 for System in Fig. 1.5.1.

$\xi_{11}/2n_1\beta_{11}$; $c_{23}/(c_{cr})_{2p} = [c_{23}/2n_2(m_2m_3/m_2 + m_3)] = \xi_{22}/2n_2\beta_{22}$, where $n_1 = \sqrt{(m_1 + m_2/m_1m_2)k_{12}} = \sqrt{\alpha_{11}/\beta_{11}}$, $n_2 = \sqrt{(m_2 + m_3/m_2m_3)k_{23}} = \sqrt{\alpha_{22}/\beta_{22}}$ are *partial angular natural frequencies* of the respective (partial) subsystems, Fig. 1.5.1(b).

Dynamic interaction between the partial systems can be conveniently characterized by the *coupling coefficient* [1.13]

$$\sigma = 2\gamma \frac{n_1 n_2}{|n_1^2 - n_2^2|} = 2\gamma \frac{\frac{n_2}{n_1}}{\left|1 - \frac{n_2^2}{n_1^2}\right|},\qquad \text{Eq. (1.5.5)}$$

Fig. 1.5.2 shows σ/γ plotted vs. n_1/n_2; it should be noted that $\sigma(n_2/n_1) = \sigma(n_1/n_2)$.

The greater σ is, the greater the interaction between the partial subsystems. At small σ the dynamic processes in the partial subsystems in Fig. 1.5.1(b) are, practically, independent and these subsystems can be analyzed separately. In this case, two natural frequencies of the complete system in Fig. 1.5.1(a) are very close to the respective partial natural frequencies. Fig. 1.5.3 has plots for finding natural frequencies ω_{n_1} and ω_{n_2}

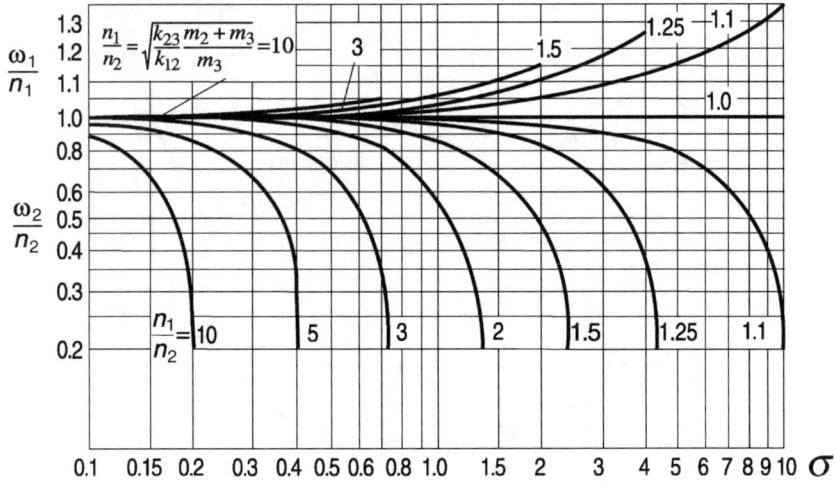

Fig. 1.5.3. Natural Frequencies ω_n of System in Fig. 1.5.1(a) vs. Coupling Coefficient σ and n_1/n_2.

of the three-mass system in Fig. 1.5.1(a) if partial natural frequencies n_1 and n_2 and coefficient σ are known.

Assuming solution of Eq. (1.5.4) with zero excitation to be $F_{12} = F_{O1}e^{\lambda t}$, $F_{23} = F_{O2}e^{\lambda t}$, its characteristic equation becomes

$$(\beta_{11}\lambda^2 + \xi_{11}\lambda + \alpha_{11})(\beta_{22}\lambda^2 + \xi_{22}\lambda + \alpha_{22}) - (\xi_{12}\lambda + \alpha_{12})(\xi_{23}\lambda + \alpha_{23}) = 0.$$
Eq. (1.5.6)

If damping in the partial subsystems is small, or [see Eq. (1.4.4)]

$$\frac{c_{12}}{(c_{cr})_{1p}} = \frac{\xi_{11}}{2\beta_{11}}\frac{1}{n_1} = \frac{\Delta_{1p}}{n_1} \ll 1, \quad \frac{c_{23}}{(c_{cr})_{2p}} = \frac{\xi_{22}}{2\beta_{22}}\frac{1}{n_2} = \frac{\Delta_{2p}}{n_2} \ll 1,$$

where $\Delta_{1p, 2p}$ are *relative damping coefficients* for the partial subsystems, then the damping coupling is insignificant and Eq. (1.5.6) can be written as

$$(\beta_{11}\lambda^2 + \xi_{11}\lambda + \alpha_{11})(\beta_{22}\lambda^2 + \xi_{22}\lambda + \alpha_{22}) - \alpha_{12}^2 = 0,$$
Eq. (1.5.7)

since $\alpha_{12} = \alpha_{23}$, or

$$(\lambda^2 + 2\Delta_{1p}\lambda + n_1^2)(\lambda^2 + 2\Delta_{2p}\lambda + n_2^2) - \gamma^2 n_1^2 n_2^2 = 0.$$
Eq. (1.5.8)

With the assumed smallness of damping in the system (and in its partial subsystems),

$$\lambda_{1,2} \cong j\omega_{1,2} - \Delta_{1,2},$$
Eq. (1.5.9)

where $\omega_{1,2}$ are natural frequencies of the full system in Fig. 1.5.1(a) without damping and $\Delta_{1,2}$ are relative damping coefficients for vibratory modes of the full system. Substituting Eq. (1.5.9) into Eq. (1.5.8), neglecting small quantities $\Delta_{1,2}^2$, $\Delta_{1,2}\Delta_{1p}$, $\Delta_{1,2}\Delta_{2p}$, and considering that ω are solutions of the frequency equation without damping,

$$(n_1^2 - \omega^2)(n_2^2 - \omega^2) - \gamma^2 n_1^2 n_2^2 = 0, \quad \text{Eq. (1.5.10)}$$

arrive to equation for $\Delta_{1,2}$

$$(n_1^2 - \omega^2)\Delta_{2p} + (n_2^2 - \omega^2)\Delta_{1p} - (n_1^2 - n_2^2 - 2\omega^2)\Delta_{1,2} = 0. \quad \text{Eq. (1.5.11)}$$

Substituting $\omega_{1,2}$ from Eq. (1.5.10) into Eq. (1.5.11) and considering Eq. (1.5.5), obtain

$$\Delta_1 = \frac{1}{2}\frac{\left(1+\sqrt{1+\sigma^2}\right)\Delta_{1p} - \left(1-\sqrt{1+\sigma^2}\right)\Delta_{2p}}{\sqrt{1+\sigma^2}};$$

$$\Delta_2 = \frac{1}{2}\frac{-\left(1-\sqrt{1+\sigma^2}\right)\Delta_{1p} + (1+\sqrt{1+\sigma^2})\Delta_{2p}}{\sqrt{1+\sigma^2}} \quad \text{Eq. (1.5.12)}$$

Values of Δ_1 and Δ_2 determine values of log decrement $\delta_{1,2} = 2\pi(\Delta_{1,2}/\omega_{1,2})$ and resonance amplification factors $\mu_{1,2} = \omega_{1,2}/2\Delta_{1,2}$ for both vibratory modes. Correlations between $\Delta_{1,2}$, $\Delta_{1p,2p}$, and σ are plotted in Fig. 1.5.4.

An important characteristic of the three-mass system is distribution of elastic forces between elastic connections k_{12} and k_{23} at natural frequencies ω_1 and ω_2 (modal coefficients for elastic forces) $K_1 = (F_{23}/F_{12})_{\omega 1_n}$, $K_2 = (F_{23}/F_{12})_{\omega 2_n}$. These coefficients are also determined mainly by σ. From Eq. (1.5.6) it can be derived that

$$K_{1,2} = \sqrt{\frac{\beta_{11}}{\beta_{22}}}\frac{1 \mp \sqrt{1+\sigma^2}}{2\sigma} = \sqrt{\frac{k_{23}}{k_{12}}}\frac{1 \mp \sqrt{1+\sigma^2}}{2\sigma} \quad \text{Eq. (1.5.13)}$$

These expressions are plotted in Fig. 1.5.5. It can be seen that at small σ, vibration at resonances develops only in one partial subsystem since $K_1 >> 1$, ($K_2(\| << 1$).

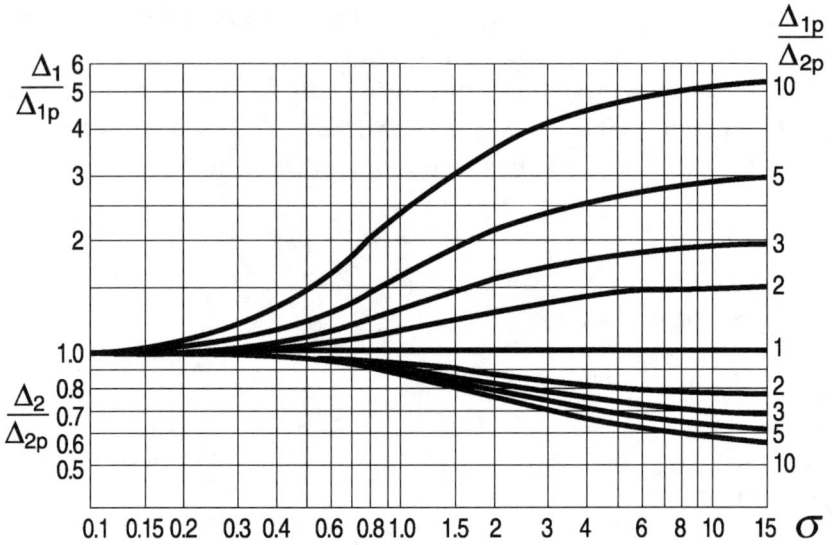

Fig. 1.5.4. Relative Damping Coefficients Δ_1, Δ_2 for Two Natural Modes of System in Fig. 1.5.1(a) vs. Damping in Partial Subsystems Δ_{1p}, Δ_{2p} and Coupling Coefficient σ.

In some cases, e.g., for determining reactions to pulse excitations, it is useful to transform Eq. (1.5.4) into normal coordinates $\tilde{F}_1$ and $\tilde{F}_2$. Lagrange function for the system in Fig. 1.5.1(a) with elastic forces as coordinates is

$$L = T - \Pi + \sum_{i=1,2} F'_i(t) F_{i,i+1} = \frac{1}{2}\beta_{11}\dot{F}_{12}^2 + \frac{1}{2}\beta_{22}\dot{F}_{23}^2 - \frac{1}{2}\alpha_{11}F_{12}^2 - \frac{1}{2}\alpha_{22}F_{23}^2$$

$$- \frac{1}{2}\alpha_{12}F_{12}F_{23} + F'_1(t)F_{12} + F'_2(t)F_{23} \quad \text{Eq. (1.5.14)}$$

Here T is kinetic energy, Π is potential energy. Considering that

$$F_{12} = \tilde{F}_1 + \tilde{F}_2, \quad F_{23} = K_1\tilde{F}_1 + K_2\tilde{F}_2 \quad \text{Eq. (1.5.15)}$$

and also considering that in normal coordinates both kinetic energy T and potential energy Π are expressed as sums of squares, it can be written

$$L = \frac{1}{2}\beta_1\dot{\tilde{F}}_1^2 + \frac{1}{2}\beta_2\dot{\tilde{F}}_2^2 - \frac{1}{2}\alpha_1\tilde{F}_1^2 - \frac{1}{2}\alpha_2\tilde{F}_2^2 + F'_1(t)(\tilde{F}_1 + \tilde{F}_2)$$

$$+ F'_2(t)(K_1\tilde{F}_1 + K_2\tilde{F}_2), \quad \text{Eq. (1.5.16)}$$

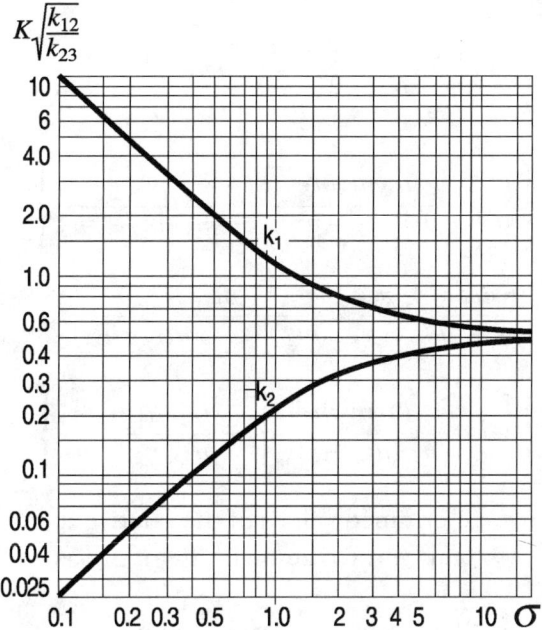

Fig. 1.5.5. Modal Coefficients for System in Fig. 1.5.1(a) vs. σ.

where α, β are generalized stiffnesses and inertias of the system in normal coordinates. Lagrange equations of motion for this system are

$$\frac{d}{dt}\frac{\partial L}{\partial \dot{\tilde{F}}_i} = \frac{\partial L}{\partial \tilde{F}_i} \quad (I = 1, 2) \qquad \text{Eq. (1.5.17)}$$

from where

$$\beta_1 \ddot{\tilde{F}}_1 + \alpha_1 \tilde{F}_1 = F'_1(t) + K_1 F'_2(t)$$

$$\beta_2 \ddot{\tilde{F}}_2 + \alpha_2 \tilde{F}_2 = F'_1(t) + K_2 F'_2(t) \qquad \text{Eq. (1.5.18)}$$

Since $\alpha_1/\beta_1 = \omega_{n1}^2$, $\alpha_2/\beta_2 = \omega_{n2}^2$, Eq. (1.5.18) can be rewritten as

$$\frac{1}{\omega_{n1}^2}\ddot{\tilde{F}}_1 + \tilde{F}_1 = \frac{1}{\alpha_1}[F'_1(t) + K_1 F'_2(t)]$$

$$\frac{1}{\omega_{n2}^2}\ddot{\tilde{F}}_2 + \tilde{F}_2 = \frac{1}{\alpha_2}[F'_1(t) + K_2 F'_2(t)] \qquad \text{Eq. (1.5.19)}$$

where

$$\alpha_1 = \frac{m_1}{m_1+m_2}\frac{m_3}{m_2+m_3}K_1^2 - 2\frac{m_1 m_3}{(m_1+m_2)(m_2+m_3)}K_1$$

$$= \frac{m_1 m_2}{m_1+m_2}\left\{\left[\omega_1^2 - \frac{m_1 m_2 m_3}{m_1(m_2+m_3)}\right]^2 + \frac{m_3}{m_1}\frac{m_1 m_2 m_3}{m_2+m_3}\right\}$$

$$\alpha_2 = \frac{m_1}{m_1+m_2}\frac{m_3}{m_2+m_3}K_2^2 - 2\frac{m_1 m_3}{(m_1+m_2)(m_2+m_3)}K_2$$

$$= \frac{m_1 m_2}{m_1+m_2}\left\{\left[\omega_2^2 - \frac{m_1 m_2 m_3}{m_1(m_2+m_3)}\right]^2 + \frac{m_3}{m_1}\frac{m_1 m_2 m_3}{m_2+m_3}\right\} \quad \text{Eq. (1.5.20)}$$

Using the above-determined values of the relative damping coefficients Δ_1, Δ_2, Eq. (1.5.19) can be rewritten with damping as

$$\frac{1}{\omega_{n1}^2}\ddot{\tilde{F}}_1 + 2\frac{\Delta_1}{\omega_{n1}^2}\dot{\tilde{F}}_1 + \tilde{F}_1 = \frac{1}{\alpha_1}[F_1'(t) + K_1 F_2'(t)]$$

$$\frac{1}{\omega_{n2}^2}\ddot{\tilde{F}}_2 + 2\frac{\Delta_1}{\omega_{n2}^2}\dot{\tilde{F}}_2 + \tilde{F}_2 = \frac{1}{\alpha_2}[F_1'(t) + K_2 F_2'(t)] \quad \text{Eq. (1.5.21)}$$

Eq. (1.5.21) describes dynamic motion of two independent (uncoupled) systems. Damping coupling coefficients omitted in transition from Eq. (1.5.4) to Eq. (1.5.21) would be significant only if damping in one of the partial subsystems were a large fraction of the critical damping. After $\tilde{F}_1$ and $\tilde{F}_2$ are determined (e.g., for a pulse excitation), actual forces in the elastic connections can be analytically or graphically determined using Eq. (1.5.15).

1.5.2 Vibration Isolator with Intermediate Mass (Two-Stage Isolator)

If mass m_3 is very large ($m_3 \approx \infty$), and mass m_1 represents the object being isolated, the system $k_{12} - m_2 - k_{23}$ in Fig. 1.5.1(a) represents a vibration isolator with an intermediate mass. Such isolators provide an improved transmissibility function in the high-frequency range due to influence of the inertia force interacting with vibratory motions of elastic elements k_{12},

k_{23}. Two natural frequencies of the isolation system without damping can be calculated using Eq. (1.5.10). The natural frequencies are

$$\omega_{1,2} = n_1 \sqrt{\frac{1}{2}\left(\frac{n_2^2}{n_1^2}+1+\frac{k_{23}}{k_{12}}\right) \pm \sqrt{\frac{1}{4}\left(\frac{n_2^2}{n_1^2}+1+\frac{k_{23}}{k_{12}}\right)^2 - \frac{n_2^2}{n_1^2}}} \quad \text{Eq. (1.5.22)}$$

where $n_1 = \sqrt{(k_{12}+k_{23})/m_2}$, $n_2 = \sqrt{[k_{12}k_{23}/(k_{12}+k_{33})](1/m_1)}$ are partial natural frequencies representing the sub-systems in which m_1 is fixed and m_2 is removed, respectively. In Eq. (1.5.22) the natural frequency ω_1 (for the "minus" sign) is lower than both n_1 and n_2, while the natural frequency ω_2 (for the "plus" sign) is greater than both n_1 and n_2.

Both force and absolute displacement transmissibility (not considering damping) is

$$\mu = \frac{1}{\frac{n_1^2}{n_2^2}\left(\frac{\omega}{n_1}\right)^4 - \left[1 + \left(1+\frac{k_{23}}{k_{12}}\right)\frac{n_1^2}{n_2^2}\right]\left(\frac{\omega}{n_1}\right)^2 + 1} \quad \text{Eq. (1.5.23)}$$

For excitation frequency $\omega \gg n_1, n_2$, $\mu \to (n_1 n_2/\omega^2)^2$. Thus, at high excitation frequencies transmissibility decreases as the fourth power of the excitation frequency (see Fig. 1.5.6), while for the simple isolators (without an intermediate mass), it decreases only as the second power of the excitation frequency, e.g., see Eqs. (1.4.6), (1.4.8), and (1.4.11). On the other hand, high transmissibility is observed at two natural frequencies ω_1 and ω_2, rather than at one resonance frequency for the system with a simple isolator. The frequency ratio ω_2/ω_1 will advantageously be a minimum (thus ω_2 will appear to be the closest to ω_1) when the stiffness ratio $\alpha = k_2/k_1 = 1 + \beta$, where $\beta = m_2/m_1$. For this stiffness ratio [1.8],

$$\frac{\omega_2}{\omega_1} = \frac{1+\sqrt{1+\beta}}{\sqrt{\beta}}, \quad \text{Eq. (1.5.24)}$$

and the force transmissibility is

$$\mu_F = \frac{1+\delta^2/\pi^2}{\sqrt{\left[\frac{\beta\lambda\Omega^4}{2+\beta} - 2\lambda\Omega^2 + 1 - \delta^2/\pi^2\right]^2 + (2\delta/\pi)^2(1-\lambda\Omega^2)^2}}, \quad \text{Eq. (1.5.25)}$$

where $\lambda = (1+\beta)/(2+\beta)$, δ is the log decrement (assumed to be of the same

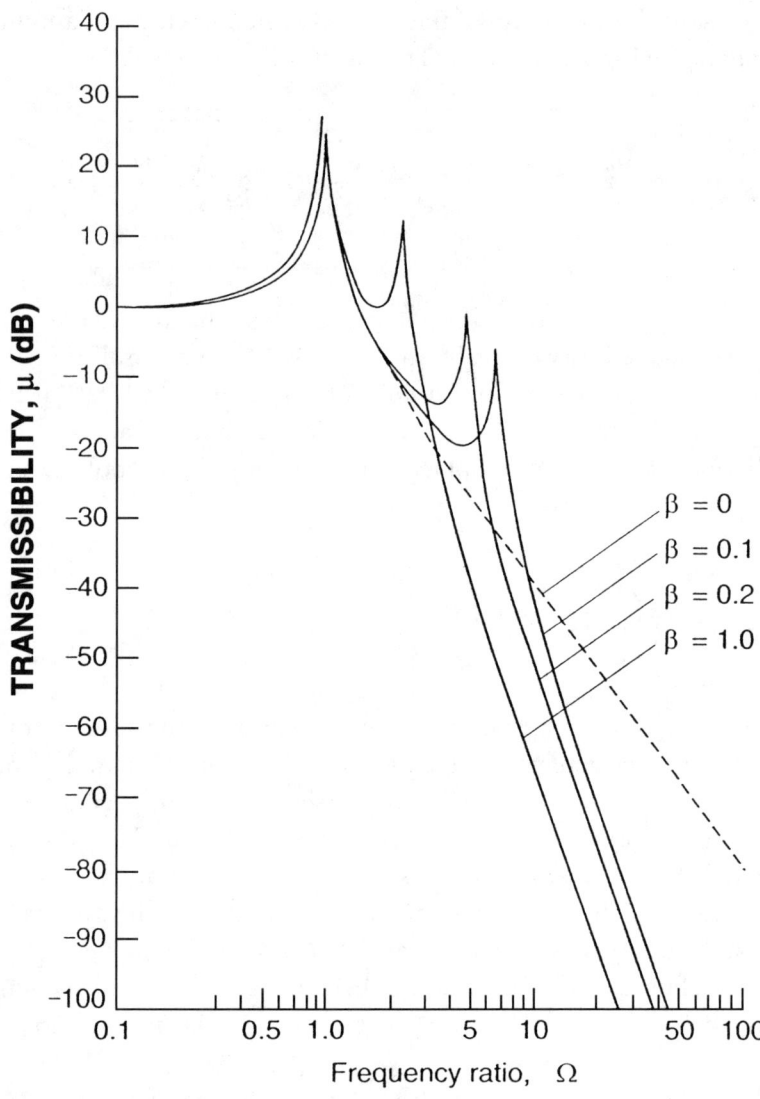

Fig. 1.5.6. Transmissibility of Uni-Axial Isolation System with Intermediate Mass m_2 as a Function of $\beta = m_2/m_1$.

magnitude for both k_{12} and k_{23}), and $\Omega = \omega_1/n_1$. At high frequencies and $\beta \leq 0.5$

$$\mu_F = \frac{\left(1 + \delta^2/\pi^2\right)(2 + \beta)^2}{\beta(1 + \beta)\Omega^4} \approx \frac{4\left(1 + \delta^2/\pi^2\right)}{\beta\Omega^4}, \qquad \text{Eq. (1.5.26)}$$

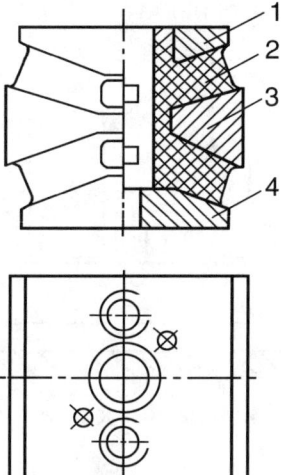

Fig. 1.5.7. Intermediate Mass Integrated into Vibration Isolator; 1, 4—Upper, Lower Base; 2—Rubber Flexible Element; 3—Intermediate Mass.

or the transmissibility is inversely proportional to the size of the intermediate mass m_2.

Transmissibility plots of a low-damped system with various intermediate mass values ($\beta = 0.1; 0.2; 1.0$) are shown in Fig 1.5.6 in comparison with a conventional system, $\beta = 0$ [1.8]. While the advantage of introducting the intermediate mass for high-frequency transmissibility is clearly visible, this advantage is pronounced only at quite high frequencies even for not-too-small masses. Thus, for $\beta = 0.1$, the transmissibility advantage starts at a frequency that is about ten times higher than the natural frequency of the base system (without the intermediate mass). Such systems can be useful, for example, if it is necessary to mount the object, such as a machinery unit, on a non-rigid foundation having many high-frequency resonances and low damping. A small intermediate mass may then be integrated into a vibration isolator design, e.g., see Fig. 1.5.7. Another application of this type of system is to mount the object on an inertia block or a massive mounting structure, possibly together with other objects, using vibration isolators between the objects and the mounting structure and use the inertia block as an effective intermediate mass, like in Fig. 1.5.8, which presents installation of three diesel generators on a massive intermediate structure [1.8].

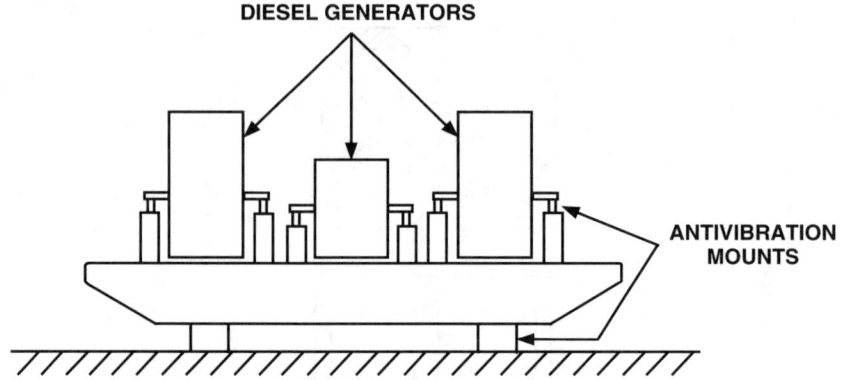

Fig. 1.5.8. Use of Inertia Block as an Intermediate Mass.

In such cases the transmissibility improvement starts at relatively low frequencies.

Since the characteristic feature of the isolation system with an intermediate mass is a radical reduction of transmissibility in the high-frequency range, it is very important that the intermediate mass be rigid and not have internal resonances within the frequency range of interest. Even an inadequate stiffness of mounting feet on the intermediate mass structure can result in parasitic resonances distorting the $1/\omega^4$ trend of the transmissibility plot in the high-frequency range. Fig. 1.5.9 [1.8] compares transmissibility plots of a basic "mass-spring" isolation system (broken line), system with intermediate mass ($\beta = 0.2$) that can be considered as an absolutely rigid body (solid line), and the same system where the mass m_2 has two mounting feet, each with mass $m_F = 0.0125 m_2$ and stiffness $k_F = 5k_2$ (chain line).

If n intermediate masses connected by spring elements are used instead of one ("multi-stage isolation"), than the high-frequency transmissibility decline at the rate of $1/\Omega^{2(n+1)}$ will be realized, but at a price of having $n + 1$ resonances. Also, the starting frequency of the fast decline in transmissibility is shifting toward higher frequencies with each added mass. A similar multi-stage system designed and constructed for isolation of the "gravity wave antenna" at its fundamental frequency 1,100 Hz is described in [1.14]. At this frequency, isolation of 272 dB (reduction of vibration at 1,100 Hz by $\sim 10^{14}$) has been achieved.

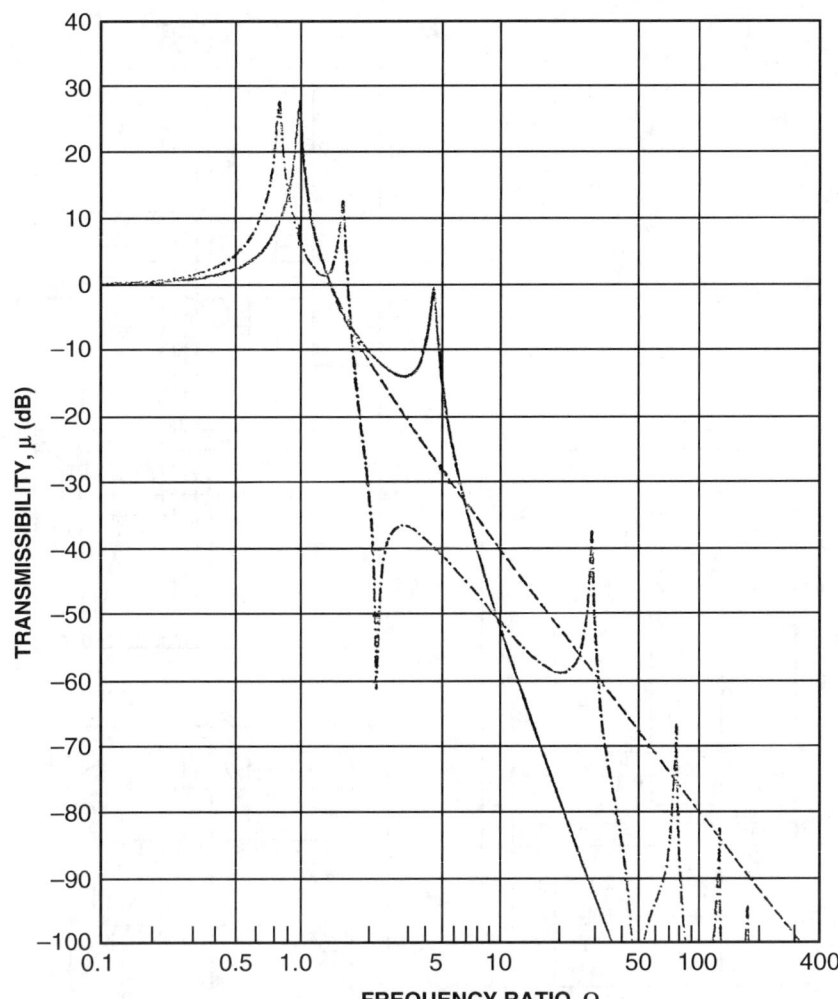

Fig. 1.5.9. Influence of Non-Rigidity in the System on Behavior of Vibration Isolation System with Intermediate Masses.

1.6 THREE-DEGREES-OF-FREEDOM (PLANAR) VIBRATION ISOLATION SYSTEM

It was shown in Section 1.3 that a vibration isolation system having one plane of symmetry is represented in the general case by two systems of three second-order differential equations. The most interesting are vibratory motions within the plane of symmetry, comprising coupled vibratory motions in two linear and one angular coordinates.

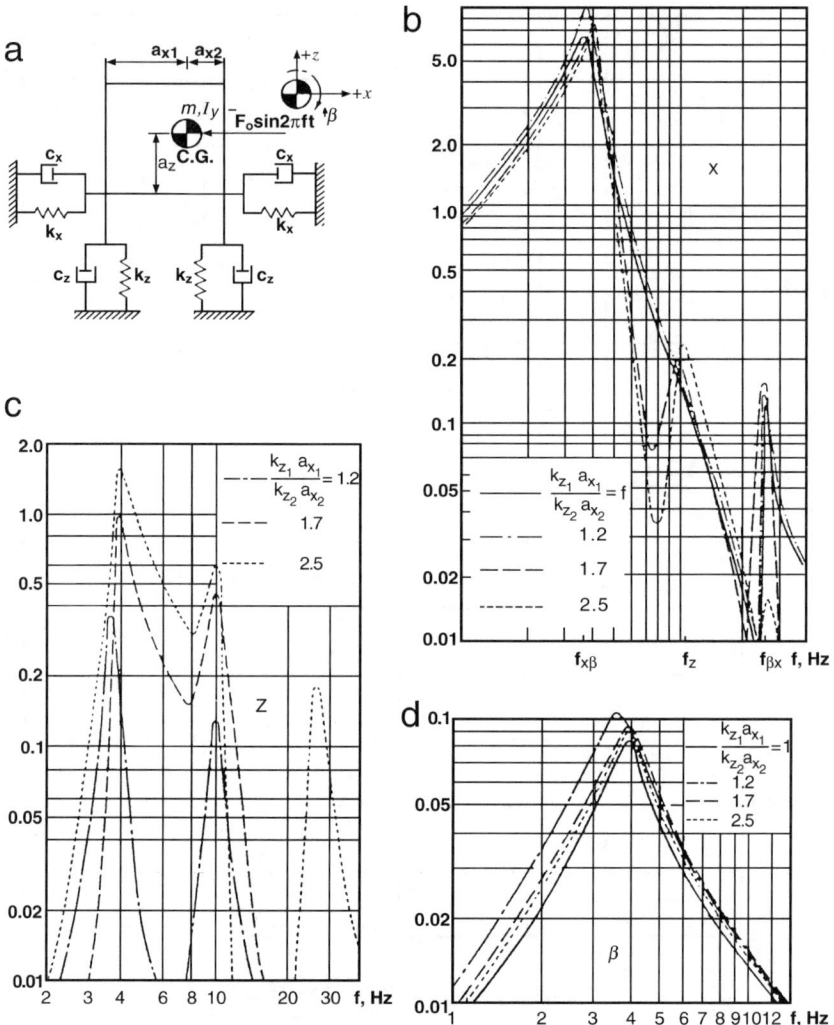

Fig. 1.6.1. (a) Planar Three-Degree-of-Freedom System; (b) and Amplitude-Frequency Characteristics of Coordinate Vibratory Displacements z; (c) x, (d) β for Horizontal Force Excitation Through C.G.

1.6.1 Dynamics of Planar Isolation System

A typical planar vibration isolation system is presented in Fig. 1.6.1a. Although the stiffness constants of both elastic mounts in this model are identical, the model represents a general case since different stiffnesses of the isolators can be simulated by shifting the C.G. position. If the mount stiffnesses are proportional to the weight loads on these mounts, it is equivalent, for the purpose of the dynamic analysis, to the central position of C.G. in Fig. 1.6.1(a). The equations of motion for this system are

$$m\ddot{x} + 2k_x(x - x_f) + 2c_x(\dot{x} - \dot{x}_f) - 2k_x a_z(\beta - \beta_f) - 2c_x a_z(\dot{\beta} - \dot{\beta}_f) = \sum F_x$$

$$I_y \ddot{\beta} - 2k_x a_z(x - x_f) - 2c_x a_z(\dot{x} - \dot{x}_f) - k_z(a_{x2} - a_{x1})(z - z_f) + [2k_x a_z^2$$

$$+ k_z(a_{x1}^2 + a_{x2}^2)](\beta - \beta_f) + [2c_x a_z^2 + c_z(a_{x1}^2 + a_{x2}^2)](\dot{\beta} - \dot{\beta}_f) = \sum M_F + \sum M$$

$$m\ddot{z} + 2k_z(z - z_f) + 2c_z(\dot{z} - \dot{z}_f) - k_z(a_{x2} - a_{x1})(\beta - \beta_f)$$

$$- c_z(a_{x2} - a_{x1})(\dot{\beta} - \dot{\beta}_f) = \sum F_z, \qquad \text{Eq. (1.6.1)}$$

where ΣF_x, ΣF_z are sums of force projections on the respective coordinate directions; ΣM_F is the sum of moments relative to C.G. from all forces acting on the body (object); ΣM is the sum of external moments acting on the object.

Eq. (1.6.1) was simulated in [1.15] on an analog computer for the object whose geometric and inertia parameters correspond to a medium-size lathe (mass $m = 2,300$ kg; $I_y = 270$ kg-m^2; $\rho = 0.343$ m; $a_{x1} + a_{x2} = 0.7$ m; $a_z = 0.75$ m; ratio between vertical and horizontal stiffness of isolators varying as $\eta_x = k_z/k_x = 0.5$–10). The object was excited by a horizontal force acting at the C.G. level [see Fig. 1.6.1(a)]. Figs. 1.6.1(b), (c), (d) show amplitude-frequency characteristics of coordinates x, z, and α, respectively of this system at various $a_{x1}/a_{x2} = 1.0$–2.5 and at $\eta_x = 1$. For other values of η_x the obtained correlations change insignificantly. The results of the simulation led to the following conclusions:

1. Natural frequencies of a planar vibration isolation system for a given $a_{x1} + a_{x2}$ are only slightly dependent on a_{x1}/a_{x2}, even for a significant asymmetry (see Fig. 1.6.2). They are largely determined by the value of η_x. Thus, *the natural frequencies of an asymmetrical planar vibration isolation system are close to the natural frequencies of the symmetrical system*, also see Eq. (1.6.8).
2. When the symmetry condition $a_{x1} = a_{x2}$ for $k_{z1} = k_{z2}$ (or $k_{z1}a_{x1} = k_{z2}a_{x2}$ for $k_{z1} \neq k_{z2}$) is satisfied, the excitation force acting in the x-direction does not excite vibration in the z-direction. The rocking

64 • PASSIVE VIBRATION ISOLATION

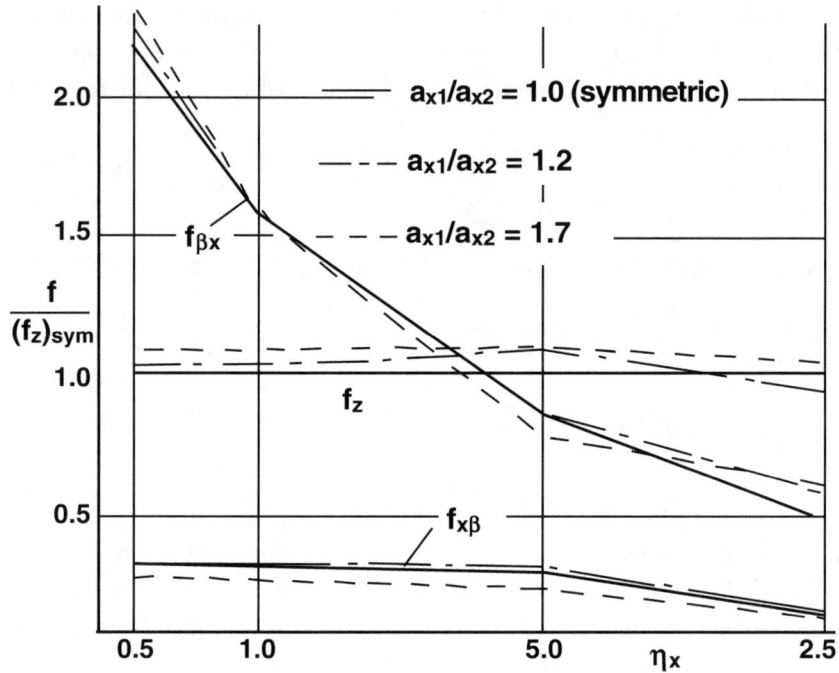

Fig. 1.6.2. Natural Frequencies vs. Degree of Asymmetry.

(horizontal–angular) vibration has two resonances at frequencies $f_{x\beta}$ and $f_{\beta x}$. When the symmetry condition is not satisfied, the third resonance of the rocking vibration appears at the frequency f_z (vertical natural frequency) and, what is even more undesirable, intensive *vertical vibrations* of the mass m (caused by the *horizontal excitation*) develop. Amplitude of this vibration at $k_{z1}a_{x1}/k_{z2}a_{x2} = 1.2$ is $0.35/6.7 \approx 0.05$ of the maximum resonance amplitude of the rocking vibration; at $k_{z1}a_{x1}/k_{z2}a_{x2} = 1.7$ it is $0.95/8 \approx 0.12$ of the maximum rocking amplitude; and at $k_{z1}a_{x1}/k_{z2}a_{x2} = 2.5$ it is $1.6/6.5 \approx 0.25$; it is assumed that $k_{x1}/k_{x2} = k_{z1}/k_{z2} = k_1/k_2$. The results are analogous in cases where the excitation force is vertical, or where there is a kinematic excitation from the foundation. This type of coupling, where the excitation in one direction causes a significant response in the perpendicular direction, is undesirable in many cases of vibration isolation. One such example—isolation of precision vibration-sensitive equipment (see Section 2.2). Frequently, precision equipment units are two-to-three times more sensitive to horizontal than to vertical floor vibrations, whereas the vertical floor vibration amplitudes are 1.3–1.5 times greater than horizontal. Since vibration isolation of precision equipment frequently involves resonance regimes, it can be conservatively assumed that the resonance amplitude of horizontal vibrations of the unit due to vertical excitation should not exceed $1/(3.0)(1.5) \approx 0.22$ of the resonant amplitude of vertical vibrations. Interpolation between the above-given values, suggests a maximum permissible asymmetry factor $k_{z1}a_{x1}/k_{z2}a_{x2} \leq 2.3$. An allowable value of the asymmetry factor for other conditions can be derived analogously.

3. Response of the system at the higher rocking mode ($f_{\beta x}$) to the horizontal force excitation is one-to-two orders of magnitude less than for the lower rocking mode ($f_{x\beta}$). The opposite is true for torque excitation of the angular coordinate β. This conclusion agrees with the results in [1.1], based on the study of a system similar to one in Fig. 1.6.1(a), with the assumption that the relative damping values in the isolators in x- and z-directions are the same, or $(c/c_{cr})_x = (c/c_{cr})_z = c/c_{cr}$, where $(c_{cr})_x = 2\sqrt{k_x m}$; $(c_{cr})_z = 2\sqrt{k_z m}$. When the system is excited by horizontal vibratory displacement of the base (foundation), it was found that at $c/c_{cr} = 0.01$–0.1 transmissibility at the second rocking natural frequency was about $\sim$10 times less than at the first (lower) rocking natural frequency. At higher damping values, $c/c_{cr} > 0.15$, resonance at the higher rocking natural frequency $f_{\beta x}$ is not noticeable at all. For excitation of the object m by a rotating force vector (unbalanced rotor) it was found that in the case of zero damping, width of the resonance peak on the amplitude-frequency characteristic at the natural frequency $f_{\beta x}$ is 10–12 times less than at the natural frequency $f_{x\beta}$. Even without damping the "thin" high-frequency resonance has a much lower probability of occurrence since it is very sensitive to small deviations of the excitation frequency, so introduction of even small damping would eliminate this resonance peak altogether.

1.6.2 Natural Frequencies of Planar Vibration Isolation System

Natural frequencies are very important characteristics of a vibration isolation system. In many cases when major spectral components of the excitations are known, values of natural frequencies represent the necessary and adequate information about the system. It was demonstrated above that magnitudes of natural frequencies for an asymmetric planar vibration isolation system are very close to the magnitudes of natural frequencies for the symmetrical system. Thus, it is important to be able to easily calculate natural frequencies for the symmetrical system.

The planar symmetrical vibration isolation system as shown in Fig. 1.6.1(a) at $a_{x1} = a_{x2} = a_x$ has three natural frequencies—vertical f_z and two natural frequencies of coupled rocking (horizontal-angular) vibration $f_{x\beta}$ and $f_{\beta x}$. The vertical natural frequency is easily determined from Eq. (1.4.33).

The natural frequencies of the coupled rocking vibrations in relation to the vertical natural frequency are

$$\frac{f_{x\beta}}{f_z}, \frac{f_{\beta x}}{f_z} = \frac{1}{\sqrt{2}} \sqrt{ \frac{\Sigma k_x}{\Sigma k_z}\left(1 + \frac{a_z^2}{\rho_y^2}\right) + \frac{a_x^2}{\rho_y^2} \mp \sqrt{\left[\frac{\Sigma k_x}{\Sigma k_z}\left(1 + \frac{a_z^2}{\rho_y^2}\right) + \frac{a_x^2}{\rho_y^2}\right]^2 - 4\frac{\Sigma k_x}{\Sigma k_z}\frac{a_x^2}{\rho_y^2}}},$$

Eq. (1.6.2)

where $\rho_y = \sqrt{\frac{I_y}{m}}$ is radius of inertia of the object about the y-axis.

The coupled rocking vibrations (e.g., $x - \beta$) can be presented as a purely translational horizontal vibratory motion of the C.G. combined with purely rotational vibratory motion about the Y-axis passing through the C.G. For the symmetrical system shown in Fig. 1.6.3 acted upon by force F_z, amplitudes of these motions are [1.16]

$$x_o = \frac{F_x}{4k_z} \frac{\left(\frac{a_z^2}{\eta_x \rho_y^2} + \frac{a_x^2}{\rho_y^2} - \frac{ea_z}{\eta_x \rho_y^2}\right) - \frac{f^2}{f_z^2}}{\frac{f^4}{f_z^4} - \left(\frac{1}{\eta_x} + \frac{a_z^2}{\eta_x \rho_y^2} + \frac{a_x^2}{\rho_y^2}\right)\frac{f^2}{f_z^2} + \frac{a_x^2}{\eta_x \rho_y^2}},$$

Eq. (1.6.3)

$$\beta_o = \frac{F_x}{4\rho_y k_z} \frac{\frac{e}{\rho_y}\frac{f^2}{f_z^2} + \left(\frac{a_z}{\eta_x \rho_y} - \frac{e}{\eta_x \rho_y}\right)}{\frac{f^4}{f_z^4} - \left(\frac{1}{\eta_x} + \frac{a_z^2}{\eta_x \rho_y^2} + \frac{a_x^2}{\rho_y^2}\right)\frac{f^2}{f_z^2} + \frac{a_x^2}{\eta_x \rho_y^2}}.$$

Eq. (1.6.4)

However, these coupled rocking vibrations also can be presented as purely angular vibratory motion about the axes not passing through the C.G. The

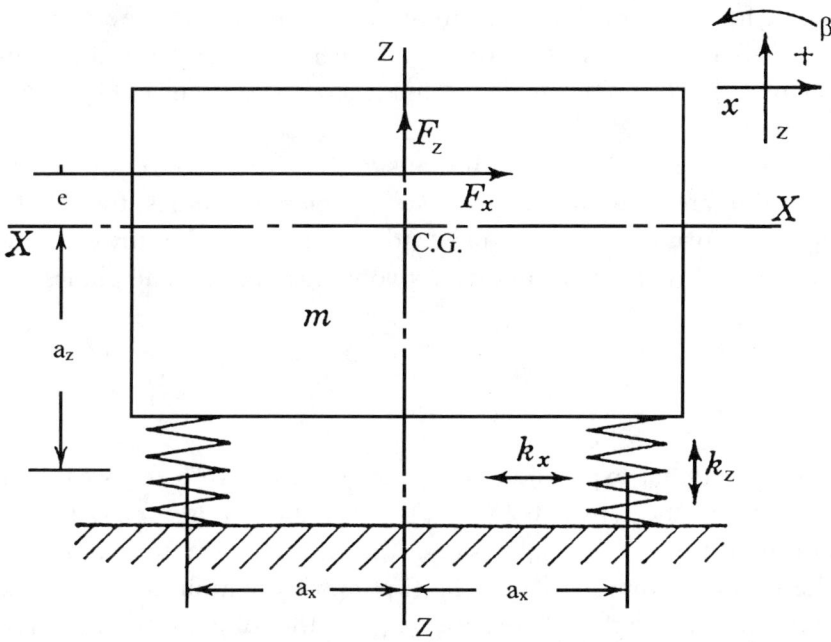

Fig. 1.6.3. Presentation of Coupled Vibrations of Isolated Object as Purely Angular Vibrations.

distance from the C.G. to the center of rocking O for a horizontal force excitation (Fig. 1.6.3) is [1.16]

$$l_o = -\frac{a_x^2 + \frac{a_z(a_z-e)}{\eta_x} - \rho_y^2 \frac{f^2}{f_z^2}}{\frac{a_z-e}{\eta_x} + e\frac{f^2}{f_z^2}}, \quad \text{Eq. (1.6.5)}$$

where e is distance from the line of action of the horizontal force to the C.G., f is frequency of the excitation. Thus, positioning of the center of rotation depends on inertia and geometry of the object as well as on position and frequency of the excitation force. The distance l_o is negative when the axis of rotation is above the C.G. and positive when it is below the C.G.

In many cases, only the lowest natural frequency $f_{x\beta}$ is of importance. Vibrations on the higher natural frequency $f_{\beta x}$ are usually insignificant (see Section 1.6.1). Also, since usually this natural frequency is rather high, $f_{\beta x} \approx (2.5-5)f_{x\beta}$, the *"center of rocking"* at this frequency is close to the C.G. as can be found from Eq. (1.6.5) at $e = 0, f = f_{\beta x}$. Thus, vibrations at this frequency have a quasi-angular character and are hardly excited by dynamic forces or translational vibrations of the base. However, they can be excited by torques. The rocking angles for $e = 0$ at the natural frequencies $f_{x\beta}, f_{\beta x}$ are normal coordinates for the systems in Figs. 1.6.1(a) and 1.6.3.

It was suggested [1.3] that considering general trends of mass distribution and magnitudes of $a_{x,y}/a_z$ in typical machines, the following simplified formula can be used for estimating the lower natural frequencies of rocking vibrations in two vertical coordinate planes

$$f_{x\beta}, f_{y\alpha} \cong f_z \sqrt{\frac{0.9}{\eta_{x,y} + \left(\frac{a_{x,y}}{a_z}\right)^2}}. \qquad \text{Eq. (1.6.6)}$$

The error of this expression does not exceed 5–10% for variation of parameters in the ranges $0.25 \leq \eta_{x,y} \leq 4$ and $0 \leq a_{x,y}/a_z \leq 1$. The most important feature of this expression is the fact that the natural frequencies of the lower rocking modes in the first approximation do not explicitly depend on the radii of inertia $\rho_{x,y}$ of the object. Some implicit dependencies are expressed by the ratio of the mount coordinates. This expression is applicable when an equal number of mounts are placed on each side of C.G., and for all mounts at the left from C.G. $a_{x1} = const$, and for mounts right of C.G. $a_{x2} = const$. Eq. (1.6.6) enables quick calculation of the required stiffness ratio of vibration isolators for a given installation if the required natural frequencies are specified (see Chapter 2). The stiffness ratio is

$$\eta_{x,y} = \frac{0.9}{\left(f_{x\beta}, f_{y\alpha}/f_z\right)^2} - \left(\frac{a_{x,y}}{a_z}\right)^2 \qquad \text{Eq. (1.6.7)}$$

Influence of the asymmetry on the magnitude of the low rocking natural frequency can be evaluated using the following expression derived by Banakh [1.17], using a general method also developed by her [1.18]:

$$\left(\frac{f_{x\beta}}{f_z}\right)_{asym} = \left(\frac{f_{x\beta}}{f_z}\right)_{sym} + \varepsilon; \quad \varepsilon = \left(\frac{a_{x_2} - a_{x_1}}{2a_z}\right)^2 \frac{\left[1 - \left(\frac{f_{x\beta}}{f_z}\right)_{sym} \eta_x\right]^2}{1 + \left[1 - \left(\frac{f_{xx}}{f_z}\right)_{sym} \eta_x\right]^2 \frac{\rho_y^2}{a_z^2}}.$$

$$\text{Eq. (1.6.8)}$$

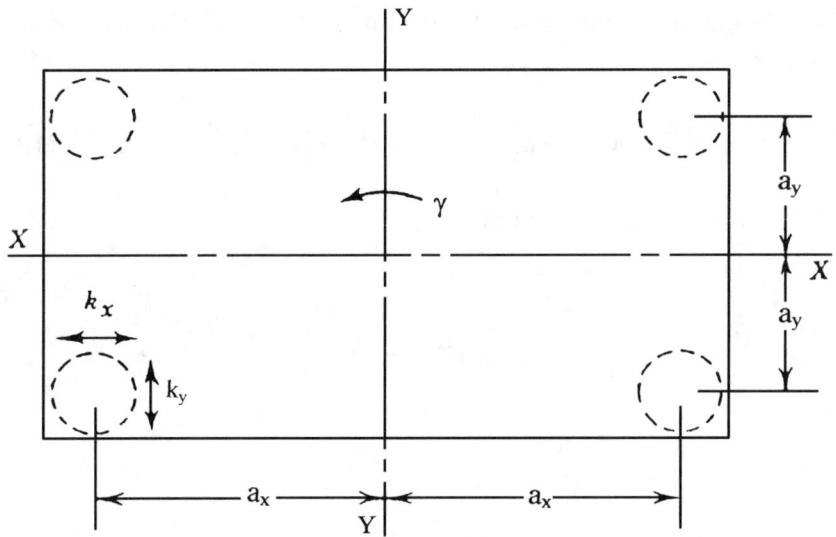

Fig. 1.6.4. Plane View on Vibration Isolation System.

A similar expression can be written for $f_{y\alpha}$. The subscript "*sym*" relates to the symmetrical system, "*asym*"—to the asymmetrical system. Analysis using Eq. (1.6.8) of the above case illustrated by Fig. 1.6.1(b) at $a_{z2}/a_{z1} = 1.7$, $\eta_z = 1$, resulted in $\varepsilon \cong 0.006$ at $(f_{x\beta}/f_z)_{sym} = 0.35$, thus supporting the conclusions in Section 1.6.1.

As shown in Section 1.3, if an isolation system has a vertical plane of symmetry, it breaks down into a planar isolation system in the plane of symmetry and a system of coupled-between-themselves rocking (horizontal-angular) vibrations in the perpendicular plane, as well as torsional vibrations about the vertical axis.

The natural frequency of the torsional vibration about the vertical axis can be expressed by analyzing a view from above onto the isolation system (see Fig. 1.6.4). The torsional stiffness of the system k_γ is

$$k_\gamma = \frac{M_z}{\gamma} = \frac{4P\sqrt{a_x^2 + a_y^2}}{\dfrac{\Delta}{\sqrt{a_x^2+a_y^2}}} = 4\frac{P}{\Delta}(a_x^2 + a_y^2) = 4k(a_x^2 + a_y^2), \qquad \text{Eq. (1.6.9)}$$

where $k = k_x = k_y$ is stiffness of vibration isolators in the horizontal direction. It is assumed that the isolators are symmetrical relative to axis z. Moment of inertia of the installed object relative to axis z, per Eq. (1.1.3), is

$$I_z = M\rho_z^2 \approx M\frac{1}{20}(4a_x^2 + 4a_y^2) = M\frac{1}{5}(a_x^2 + a_y^2), \qquad \text{Eq. (1.6.10)}$$

from where the natural frequency is

$$f_\gamma = \frac{1}{2\pi}\sqrt{\frac{k_\gamma}{I_z}} \approx \frac{1}{2\pi}\sqrt{\frac{k_r(a_x^2 + a_y^2)}{M\frac{1}{5}(a_x^2 + a_y^2)}} = \frac{1}{2\pi}\sqrt{\frac{4k_z}{\eta}\frac{5}{M}} = \frac{1}{2\pi}\sqrt{\frac{4k_z}{M}}\sqrt{\frac{5}{\eta}} = f_z\sqrt{\frac{5}{\eta}},$$

$$\text{Eq. (1.6.11)}$$

where $\eta = k_z/k_{x,y}$

1.6.3 Planar Vibration Isolation System with Inclined Mounts

In many cases of vibration isolation systems, dynamic coupling between horizontal and angular (about perpendicular horizontal axis) vibrations is undesirable. The decoupling can also be necessary to create conditions for independent variation of the corresponding natural frequencies in order to make them closer to one another, etc. Section 1.3 shows that this type of a decoupling in a system with one plane of symmetry can be achieved by placing the vibration isolators in the horizontal plane passing through the C.G. However, it is not easy to design this type of system. For example, such installation of a production or coordinate-measuring machine can be achieved only by using a heavy and expensive foundation (inertia) block. Another approach to achieve this decoupling is by using vibration isolators with inclined principal stiffness axes ("inclined mounts"), Fig. 1.6.5 [1.16]. It was mentioned above that in some cases the inter-coordinate coupling is desirable. For example, pneumatic isolators have high damping in z-direction but low damping in x-, y-directions, e.g. see Section 3.5. In such cases a significant coupling between vertical and rocking modes "transfers" some damping into the rocking modes.

In Fig. 1.6.5, the origin of the coordinate frame coincides with the *"stiffness center"* (also sometimes called the *"center of percussion"* or the *"elastic"* center). The stiffness center is such a point, that a force passing through it causes a purely translational displacement of the

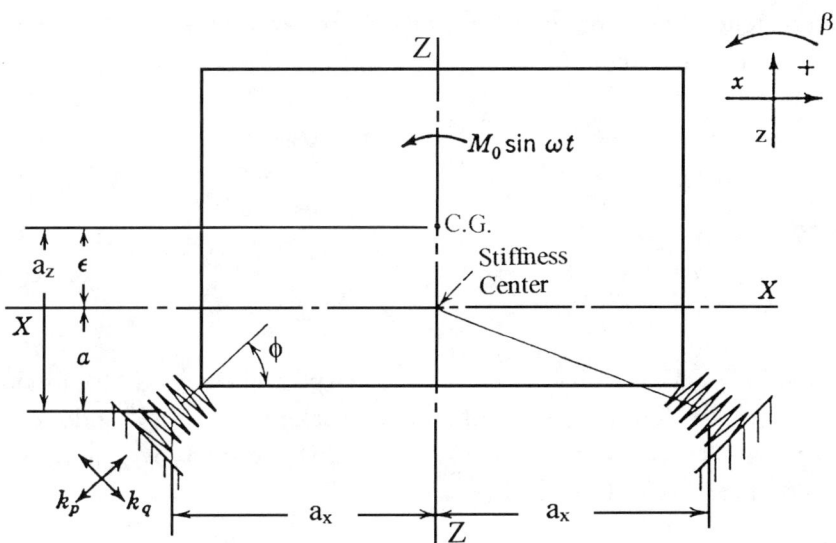

Fig. 1.6.5. Vibration Isolation System with Inclined Mounts.

installed object, without any angular displacements. A moment about the stiffness center causes a purely rotational motion, without any translational displacement. The principal elastic axes of the isolators p, q are inclined by angle ϕ relative to the coordinate axes; stiffness coefficients of the isolators in the directions p, q are k_p, k_q, respectively.

If an excitation moment $M = M_o \sin \omega t$ acts on the object in Fig. 1.6.5, then displacement of its C.G. in the direction of the x-axis is $x_{C.G.} = x - \varepsilon \beta$ and the corresponding acceleration is $\ddot{x}_{C.G.} = \ddot{x} - \varepsilon \ddot{\beta}$. Translational displacement x of the stiffness (elastic) center generates only translational elastic force, $k_x x$; angular displacement β generates only elastic moment, $k_\beta \beta$. Thus, equations of motion are

$$m(\ddot{x} - \varepsilon \ddot{\beta}) = -k_x x$$

$$m\rho_{ec}^2 \ddot{\beta} - m\varepsilon \ddot{x} = -k_\beta \beta + M_o \sin \omega t \qquad \text{Eq. (1.6.12)}$$

Here $\rho_{ec} = \sqrt{\rho_y^2 + \varepsilon^2}$ is the radius of inertia of mass m relative to the elastic center, k_x, k_β are translational, angular stiffness of the isolator system; ε is distance between the C.G. and the elastic center.

At a steady harmonic vibration regime, vibration amplitudes obtained from Eq. (1.6.12) are

$$x_o = \frac{-M_o \varepsilon \omega^2}{m[\rho_{ec}^2(\omega^2 - \omega_\beta^2)(\omega^2 - \omega_x^2) - \varepsilon^2 \omega^2]}$$

$$\beta_o = \frac{-M_o(\omega^2 - \omega_\beta^2)}{m[\rho_{ec}^2(\omega^2 - \omega_\beta^2)(\omega^2 - \omega_x^2) - \varepsilon^2 \omega^2]} \qquad \text{Eq. (1.6.13)}$$

where $\omega_x = \sqrt{k_x/m}$, $\omega_\beta = \sqrt{k_\beta/m\rho_{ec}^2}$ are "partial" natural frequencies. Actual natural frequencies $\omega_{x\beta}$ of coupled rocking vibratory modes can be determined by solving for ω the equation obtained by equating the denominator in Eq. (1.6.13) to zero,

$$\frac{\omega_{x\beta}}{\omega_x} = \sqrt{\frac{1 + \lambda_1^2 \pm \sqrt{(1 + \lambda_1^2)^2 - 4\lambda_1^2[1 - (\frac{\varepsilon}{\rho_{ec}})^2]}}{2[1 - (\frac{\varepsilon}{\rho_{ec}})^2]}}, \qquad \text{Eq. (1.6.14)}$$

where λ_1 is a dimensionless coefficient

$$\lambda_1 = \frac{\frac{a_x}{\rho_{ec}}\sqrt{\frac{k_q}{k_p}}}{\cos^2\varphi + \frac{k_q}{k_p}\sin^2\varphi}, \qquad \text{Eq. (1.6.15)}$$

$$\omega_x = \sqrt{\frac{k_x}{m}} = \sqrt{\frac{4k_p}{m}\left(\cos^2\varphi + \frac{k_q}{k_p}\sin^2\varphi\right)} \qquad \text{Eq. (1.6.16)}$$

The stiffness center coincides with the C.G. ($\varepsilon = 0$) at the value of ϕ, which can be obtained by considering a small horizontal displacement Δx of the object body in the x-direction and equating to zero elastic moment relative to the y-axis, developed by this action (which is perpendicular to the plane of Fig. 1.6.5),

$$-4a_x k_p(\Delta x)\sin\phi\cos\phi + 4a_x k_q(\Delta x)\sin\phi\cos\phi + 4ak_p(\Delta x)\cos^2\phi$$

$$+ 4ak_q(\Delta x)\sin^2\phi = 0. \qquad \text{Eq. (1.6.17)}$$

This condition corresponds to decoupling of the coupled rocking (horizontal-angular) vibratory modes. In this case the resultant of the

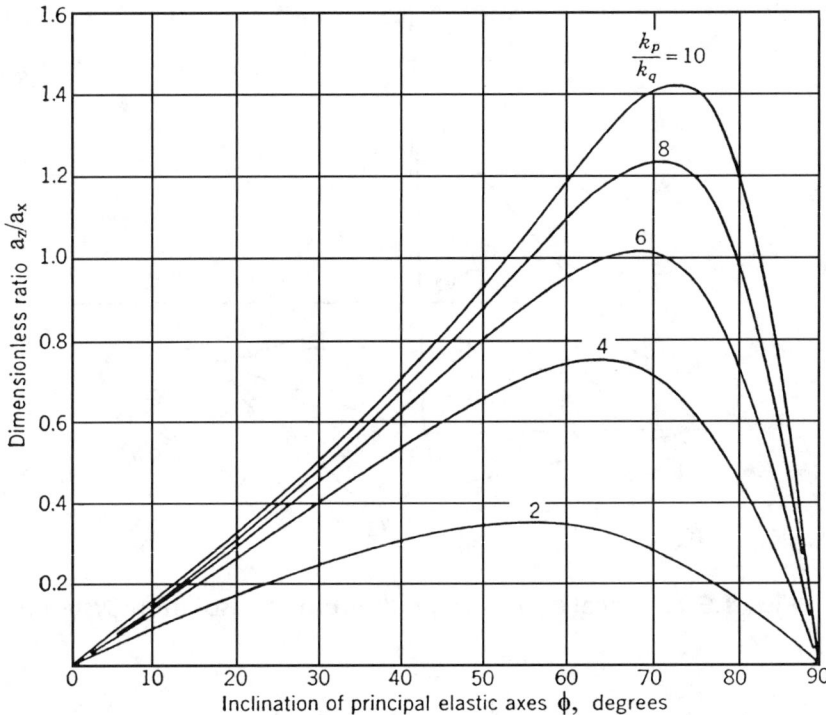

Fig. 1.6.6. Dynamic Decoupling Using Inclined Isolators.

elastic reactions, acting on the object from the isolators when the object is displaced in the x-direction, passes through the C.G. Also, the resultant of the elastic reactions to a rotation of the object is a moment about the C.G. Thus, the decoupling condition is, from Eq. (1.6.17),

$$\frac{k_q}{k_p} = \frac{\frac{a_x}{a_z} + \cot\varphi}{\frac{a_x}{a_z} - \tan\varphi} \qquad \text{Eq. (1.6.18)}$$

This decoupling condition is illustrated in Fig. 1.6.6 [1.16]. When this is satisfied, the natural frequency of translational vibration is

$$\omega_{x\beta} = \omega_x, \qquad \text{Eq. (1.6.19)}$$

and the natural frequency of angular vibration is

$$\omega_{\beta x} = \omega_x \lambda_1 = \omega_\beta = \frac{a_x}{\rho_y}\sqrt{\frac{4k_p}{m}\frac{1}{\cos^2\varphi + \frac{k_q}{k_p}\sin^2\varphi}}. \qquad \text{Eq. (1.6.20)}$$

74 • PASSIVE VIBRATION ISOLATION

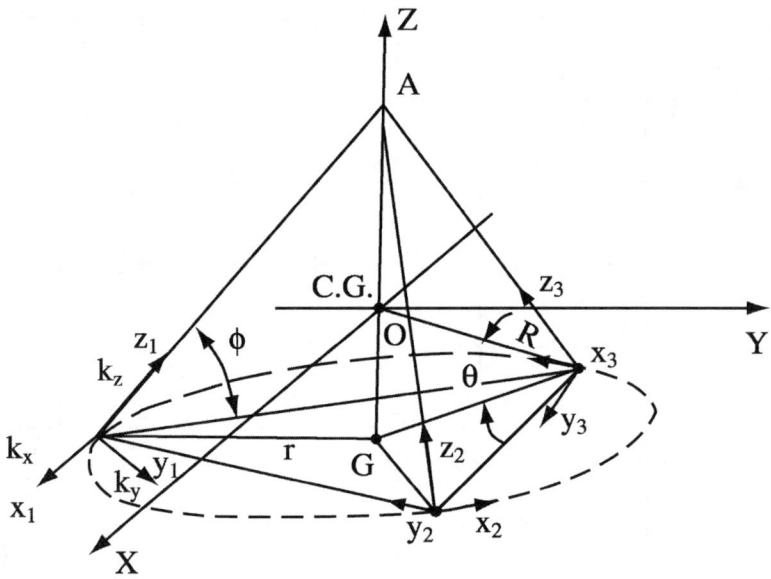

Fig. 1.6.7. Focal "Ring-Shaped" Vibration Isolation System.

A disadvantage of decoupling by using inclined vibration isolators is the need for individual selection of the required inclination angle ϕ for each object, unless this is done by the manufacturer for a mass-produced object, e.g., for a car engine. Also, such "geometric" decoupling does not provide for the complete decoupling if the mass distribution within the object may vary, e.g., due to the presence of moving components or changing weight of the accessories for the car engine. If the vibration is excited by stopping/reversal of a moving heavy mass, the complete decoupling would be realized only if the resultant of the dynamic forces passes through C.G. Otherwise, the dynamic exertion can be presented as a combination of a force passing through C.G. and of a moment, thus causing simultaneous translational and angular vibrations. Such an effect can be called *"coupling by the excitation forces."*

1.6.3A Focal Equi-Frequency System

An interesting modification of the planar system with inclined isolators is the "Focal Equi-Frequency System" [1.19]. The system shown in Fig. 1.6.7 has all its six modes uncoupled, with the same natural frequencies, which is claimed to be beneficial for automotive engine suspensions. The focal ring-shaped vibration isolation system comprises a set of three or more identical isolators/mounts placed at the apexes of a regular polygon inscribed into a circle of radius r parallel to XOY plane (broken line), and having its center in G on the Z-axis. Principal stiffness axes $z_1, z_2,\ldots$ of all mounts are intersecting in point A on the Z-axis and are inclined by angle ϕ to the X–Y plane. Axes $x_1, x_2,\ldots$ of all mounts are tangential to the circle. Principal stiffness coefficients of the mounts are k_x, k_y, k_z along their respective principal axes x, y, z. The coordinate frame origin O coincides with the C.G. of the isolated object. Angle θ is between the line "C.G.—center of a mount," and the plane of the mounts' location. The planar (three-degrees-of-freedom) system is decoupled if the following equation is satisfied

$$\left(k_z \cos^2 \phi + k_y \sin^2 \phi\right) \tan \theta = (k_x - k_y) \sin \phi \cos \phi. \qquad \text{Eq. (1.6.21)}$$

Decoupling of the six-degrees-of-freedom system in Fig. 1.6.7 requires

$$(k_z \cos^2 \phi + k_y \sin^2 \phi + \kappa_x) \tan \theta = (k_x - k_y) \sin \phi \cos \phi. \qquad \text{Eq. (1.6.22)}$$

If

$$\delta_o = (k_x - k_y)/2k_z; \quad \delta_{xy} = k_x/k_y,$$

then Eq. (1.6.22) can be written as

$$[\cos^2 \phi + 2\delta_o(1 + \delta_{xy})/(1 + \delta_{xy} - 2\delta_o)] \tan \theta = \sin \phi \cos \phi. \qquad \text{Eq. (1.6.23)}$$

Eq. (1.6.23) can be satisfied in many ways, but to achieve equal translational stifnesses in all directions and three equal angular stiffnesses, the respective equations are

$$\cos^2 \phi = 2(1 - \delta_o)(1 + \delta_{xy})/3(1 + \delta_{xy} + 2\delta_o), \qquad \text{Eq. (1.6.24)}$$

$$(1 + \delta_{xy} - 4\delta_o\delta_{xy})/(1 + \delta_{xy} - 2\delta_o) - \cos^2 \phi = \tan \theta \sin \phi \cos \phi. \qquad \text{Eq. (1.6.25)}$$

The equation providing for equality of natural frequencies of translational and angular motions is

$$\frac{\rho^2}{R^2} = \frac{3\delta_{xy} \cos^2 \theta}{1 + \delta_{xy}} \frac{2\delta_o}{1 + 2\delta_o} \qquad \text{Eq. (1.6.26)}$$

76 • PASSIVE VIBRATION ISOLATION

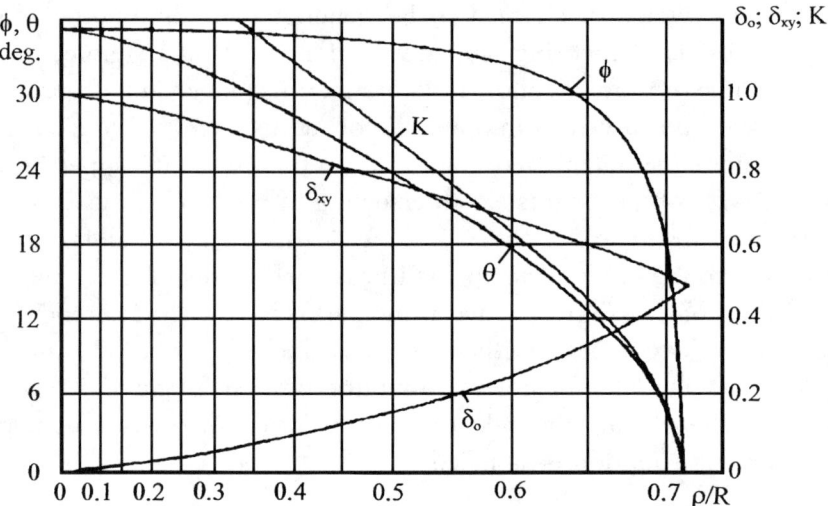

Fig. 1.6.8. Parameters of the Focal Suspensions Providing for Six Identical Natural Frequencies for the Objects Having Equal Principal Moments of Inertia.

where ρ is radius of inertia of the object (engine) and R is distance between a mount and the C.G. Solving of Eqs. (1.6.23) to (1.6.26) provides for parameters of mounts ensuring six identical natural frequencies for an object having equal moments of inertia about all principal axes of inertia. If the object has different moments of inertia about different axes, the natural frequencies will be different, but in a relatively narrow range. If the angular stiffnesses of the system in different directions are identical, the range can be estimated as

$$\frac{f_{max}}{f_{min}} = \sqrt{\frac{I_{max}}{I_{min}}}, \qquad \text{Eq. (1.6.27)}$$

where f_{max}, f_{min} are the highest and the lowest natural frequencies, and I_{max} and I_{min} are the greatest and the smallest principal moments of inertia.

Since this system employs inclined mounts, vertical forces on the object generate transverse reactions on the supporting structure. Coefficient K in Fig. 1.6.8 is the ratio of the transverse reaction to the applied vertical force. The transverse reactions have to be accommodated by the increasing stiffness of the supporting structure (e.g., by using a cradle on a car). The

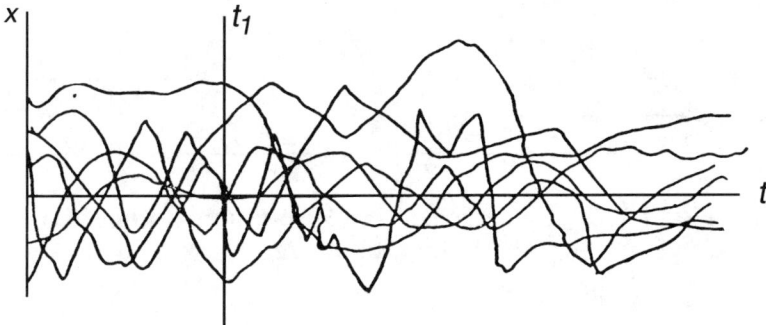

Fig. 1.7.1. Seven Realizations of a Stochastic Function.

axial stiffness k_z of each mount is selected in accordance with the required natural frequency (for all six modes) f_n as

$$k_z = \frac{12\pi^2 f_n^2 M}{N(1+2\delta_o)} , \qquad \text{Eq. (1.6.28)}$$

where M is mass of the object, N is number of mounts.

1.7 VIBRATION ISOLATION SYSTEM UNDER RANDOM EXCITATION

1.7.1 Definitions

For random vibratory excitation, at each moment of time the magnitudes of its parameters—such as displacement, velocity, and acceleration—are unpredictable. Such excitations are generated, for example, when a vehicle is moving along the road due to the roughness of the road, due to variations of gas dynamic processes in jet or rocket engines, due to exertions from various processing equipment units on the factory floor (see Section 2.1), etc. The time history of a single occurrence (*realization*) of a *random (stochastic)* vibratory process is unpredictable. However, if the process is repeated under the same conditions, it is possible to forecast a probability that the value of a certain parameter of the process would be within a certain range. Fig. 1.7.1 illustrates several realizations of a random vibratory process. The magnitude of each realization at time t_1 in Fig. 1.7.1 is a random value, different for each

78 • PASSIVE VIBRATION ISOLATION

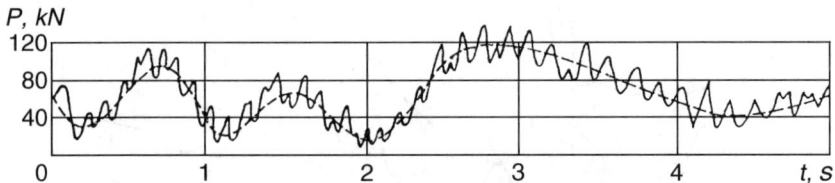

Fig. 1.7.2. Dynamic Load in Suspension System of a Working Bulldozer.

realization. The time histories, like those shown in Fig. 1.7.1, represent the origin (source) of random vibration (such as road profile or engine thrust variation). Vibration isolation systems usually connect mechanical structures. The mechanical structures of both the supporting surface (foundation) and the object can be described as multi-degrees-of-freedom dynamic systems acting as filters, thus enhancing some and supressing other components of the random excitation. As a result, the random vibratory processes usually have some pronounced spectral components whose amplitudes and phases may vary as random variables. An example of such filtered random vibration process is illustrated in Fig. 1.7.2, showing dynamic loads in the suspension system of a bulldozer; a dominant frequency ~ 7 Hz is clearly visible. Such a process can be described as narrow-band random vibration mixed with less intensive broad-band random vibration. The vibration isolation system also serves as a powerful filter (e.g., as shown in Section 2.2).

Characteristics of random vibration at each frequency $\omega = \omega_n$ can be described by *spectral power density* (SPD) $S(\omega_n)$

$$S(\omega) = \lim_{\Delta\omega \to 0} \frac{\overline{\Delta x^2}}{\Delta \omega} = \frac{\overline{dx^2}}{d\omega} . \qquad \text{Eq. (1.7.1)}$$

where *mean square value* of the random variable x is

$$\overline{x^2} = \frac{1}{2T} \int_{-T}^{T} x^2(t) dt = C_o^2 C_o^2 + \sum_{n=1}^{\infty} |C_n|^2 . \qquad \text{Eq. (1.7.2)}$$

Here $2T$ is length (duration) of the realization and C_n^* is amplitude of the nth spectral component in the Fourier series decomposition of the process. The mean square value of a random function can also be expressed via $S(\omega)$ as

$$\overline{x^2} = \int_0^\infty S(\omega)d\omega \ .\qquad \text{Eq. (1.7.3)}$$

The magnitude of $S(\omega)$ is commonly determined for acceleration in units of g per increments of frequency, or g^2/Hz. Envelopes of $S(\omega)$ are specified for testing critical equipment units subjected to random vibration. When $S(\omega) = const$ in a broad frequency range, the process is called *"white noise"* in this range. Some of the test specifications are given in [1.20]. For example, computers and sensitive instruments for military applications should be tested for $S(\omega) = 0.1$ g^2/Hz in the frequency range 10–2,000 Hz with the *root mean square* value $\sqrt{\overline{x^2}} = 14gRMS$.

For vibration isolation of precision and vibration-sensitive equipment, instead of specifying the envelope of $S(\omega)$, it is convenient to specify envelope of maximum floor displacement amplitudes at the potential installation sites for the equipment, or to specify envelopes of the floor vibratory velocities for various classes of buildings designed to house precision equipment (see Section 2.2).

1.7.2 Travel of Random Vibration Through Dynamic System

Effectiveness of vibration isolation from random vibration inputs is judged by the mean-square acceleration spectral power density on the object being protected while excited by the random vibratory input within the required envelope of $S(\omega)$. The practical capabilities of the vibration isolation system are limited since use of soft (more effective) isolators results in greater deflections of the isolators and, thus, in larger excursions of the object being protected. The allowable excursions are usually limited by packaging conditions for the object.

Output $y(t)$ of a linear dynamic system at any input excitation $x(t)$, acting from time $t = -\infty$, either deterministic or stochastic, can be expressed by the Duhamel integral as [1.21]

$$y(t) = \int_{-\infty}^{t} x(\tau)h(t-\tau)d\tau = \int_0^\infty x(t-\xi)h(\xi)d\xi \ ,\qquad \text{Eq. (1.7.4)}$$

where $h(t)$ is response of the system to the *delta function pulse* $\delta(t)$, or the *impulse response function* of the system. At a harmonic input excitation $x = x_0 e^{j\omega t}$

$$y(t) = \int_0^\infty x_0 e^{j\omega(t-\tau)} h(\tau) d\tau = x_0 e^{j\omega t} \int_0^\infty e^{j\omega\tau} h(\tau) d\tau = H(\omega) x_0 e^{j\omega t} \quad \text{Eq. (1.7.5)}$$

Here $H(\omega)$ is the *transfer function* of the system,

$$H(\omega) = \int_0^\infty e^{-j\omega\tau} h(\tau) d\tau . \quad \text{Eq. (1.7.6)}$$

Thus, the system response to a harmonic excitation is determined by multiplying the latter by the transfer function representing the spectral characteristics of the system. Since $h(\tau) = 0$ at $\tau < 0$, the lower integration limit in Eq. (1.7.6) can be replaced by $-\infty$ without changing the result. This shows that $H(\omega)$ is the Fourier transform of the impulse response function $h(\tau)$.

If the input excitation has a rich spectrum, e.g., if it's a random process, then it can be resolved into Fourier series. In the general case, it can be presented by Fourier integral as

$$S(\omega) = \int_{-\infty}^\infty x(t) e^{-j\omega t} dt \quad \text{Eq. (1.7.7)}$$

and function $x(t)$ can be expressed by an inverse Fourier transform,

$$x(t) = \frac{1}{2\pi} \int_{-\infty}^\infty S(\omega) e^{j\omega t} d\omega . \quad \text{Eq. (1.7.8)}$$

and characterized by spectrum $S(\omega)_{in}$. SPD function of the output signal for the absolute acceleration of the object is

$$S(\omega)_{out} = S(\omega)_{in} |H(\omega)|^2, \quad \text{Eq. (1.7.9)}$$

and the mean square of the output signal can be expressed via the mean square value of the input signal as

$$\overline{y^2} = \int_0^\infty S(\omega)_{out} d\omega = \int_0^\infty S(\omega)_{in} |H(\omega)|^2 d\omega . \quad \text{Eq. (1.7.10)}$$

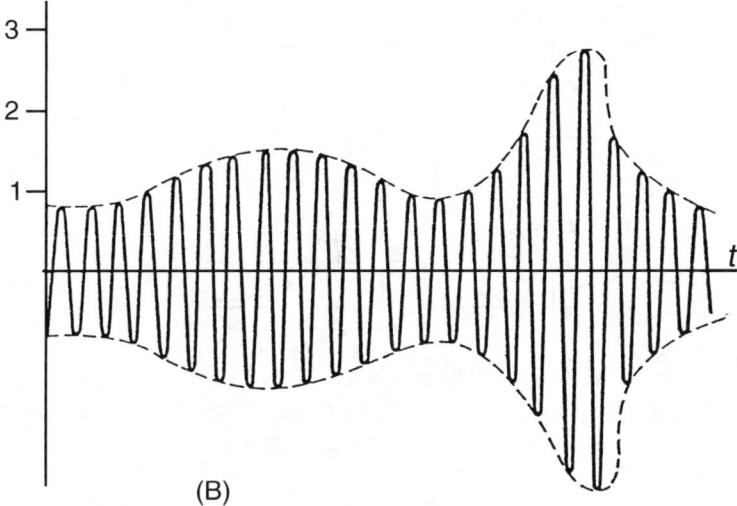

Fig. 1.7.3. Output Signal of SDOF System Acted Upon by a Random Vibratory Input.

Thus, the mean square value, which is a statistical characteristic of the output signal, is calculated by integrating by frequency the SPD function of the excitation signal modified by the transfer function of the system.

The SPD for the relative displacement between the object and the base for the acceleration excitation from the base is [1.22]

$$S(\omega)_{rel} = \frac{1}{\omega^4} |\mu_{rel}(j\omega)|^2 S(\omega)_{in} .$$ Eq. (1.7.11)

In the practice of vibration isolation, precision/vibration-sensitive equipment frequently has to be protected from random or quasi-random external excitations. This relates to isolation of measuring and computing apparatuses on-board rocket-propelled spacecraft and missiles, isolation of precision production and measuring equipment in factories (see Section 2.2), etc. The isolation system includes both the isolation system proper (vibration isolators and, possibly, inertia block) and the dynamic system of the protected object. In some cases, the isolation system proper can include other elements, such as intermediate mass (see Section 1.5.2), etc. It is important to understand how the external random excitation is filtered by the various subsystems of the isolation system. If the input signal is random and unpredictable, after passing a reasonably low-damping single-degree-of-freedom isolation system, the response will

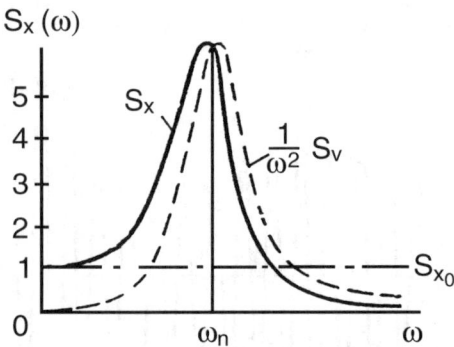

Fig. 1.7.4. Filtering Action of a SDOF System on White Noise Displacement Excitation (1); 2 – PSD of Output Vibratory Displacement; 3 – PSD of Output Vibratory Velocity.

have a narrow frequency band pattern, like that shown in Fig. 1.7.3 or in Fig. 1.7.2. Fig. 1.7.4 compares PSD functions of the random input signal [vibratory displacement, $S(\omega) = S_{xo}$] and of the output signal (vibratory displacement S_x or vibratory velocity S_v). The latter is "filtered" through the low-damping single-degree-freedom system so that the significant vibratory energy passes through the system only in the frequency band $\omega_n \pm \zeta\omega_n$, in the resonance zone of the system with the natural frequency ω_n and the damping ratio ζ. The output signal resembles a sinusoid having frequency ω_n, whose amplitude randomly changes in time. Dynamic systems with high damping pass vibratory energy in a wider frequency band, but their resonance amplitudes are much lower. For a multi-degrees-of-freedom vibration isolation system, the object mounted on a vibrating base would experience several such signals at natural frequencies of the system. This understanding is very important for evaluating propagation of random excitation across the vibration isolation system when the latter comprises several weakly coupled subsystems.

As was stated above, information on the range of displacements caused by the narrow band relative vibrations in isolators, is important for designing/packaging of vibration isolation systems. For low-damping systems ($\zeta \ll 1$), the mean-square of relative displacements is [1.21]

$$\overline{x^2_{rel}} = S(\omega)_{in} \int_0^\infty |\mu_{rel}|^2 d\omega = \frac{\pi}{4\zeta} \frac{S(\omega)_{in}}{\omega_n^3} \qquad \text{Eq. (1.7.12)}$$

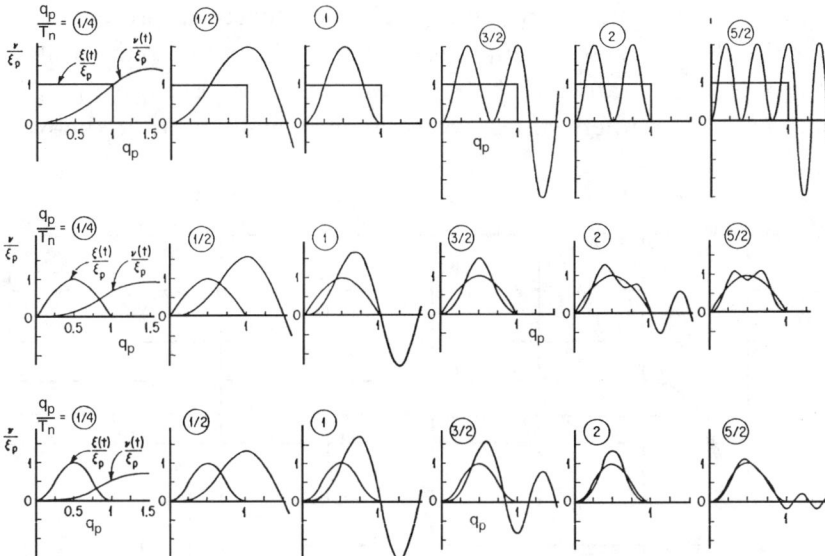

Fig. 1.8.1. Time Histories of Response of SDOF System to Various Pulse Excitations: (a) Rectangular Pulse; (b) Half-Sine Pulse; (c) Versed-Sine Pulse.

1.8 VIBRATION ISOLATION SYSTEM UNDER PULSE EXCITATION

Evaluation of the response of vibration isolation systems to pulse excitations generated either inside the isolated object (e.g., vibration-producing machine) or transmitted from the base/foundation is very important for designing an effective isolation system. In many cases, pulse excitations are more dangerous than steady vibration excitations. Frequently, it is more difficult to evaluate a system's response to pulse excitations because interaction of both forced motions and free vibration of the system excited by the pulse should be considered. While the response can be computed for each specific case, it is desirable to develop some general criteria for designing isolation systems for pulse excitations, as well as quick "first approximation" techniques. The *"shock spectrum,"* or *"response spectrum,"* approach is useful for accomplishing this task [1.23], [1.24].

The response spectrum is a graphical presentation of a selected quantity in the response process with reference to a certain selected

84 ● PASSIVE VIBRATION ISOLATION

quantity in the excitation process. The shock spectrum presents ratio of the peak value of the selected parameter in the response to the peak value of the selected parameter of the excitation pulse. Both are plotted as functions of a dimensionless parameter, usually a ratio between a

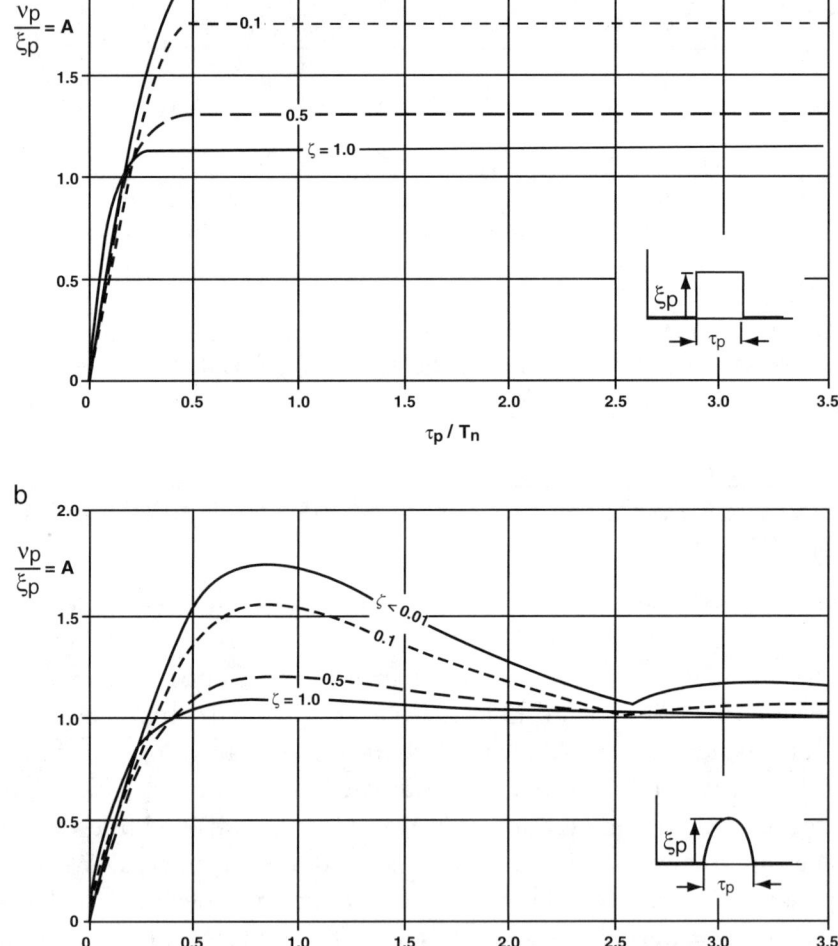

Fig. 1.8.2. Absolute Response "Shock" Spectra of a SDOF System to Various Pulse Excitations: (a) Rectangular pulse; (b) Half-Sine Pulse; (c) Versed-Sine Pulse; (d) Ramp Pulse; (e) Decaying Sinusoid.

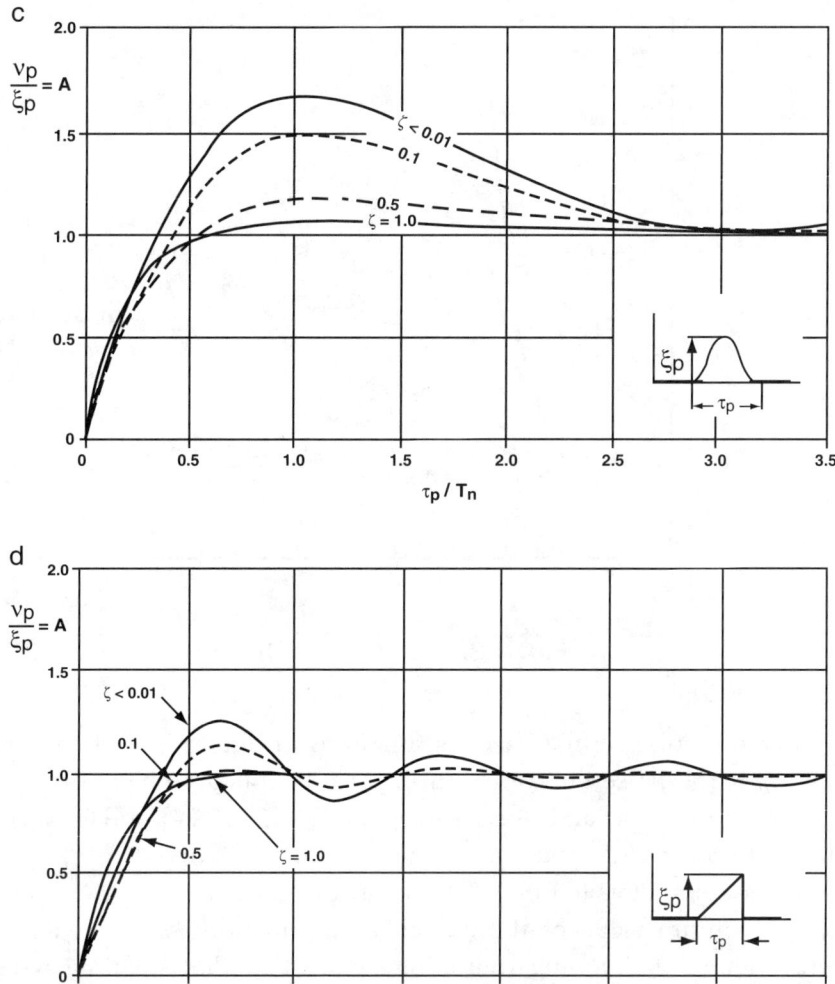

Fig. 1.8.2. (*continued*)

significant time measure in the excitation pulse and the *natural period* of the responding system. Clearly, these definitions show that the "response" and "shock" spectra are not the frequency spectra as defined in vibration textbooks. However, these terms are historically widely used in literature. Fig. 1.8.1 gives time histories of response of a single-degree-of-freedom system without damping [relative values $\nu(t)/\xi_p$, where $\nu(t)$ is instantaneous value of the response of the isolated object and ξ_p is maximum

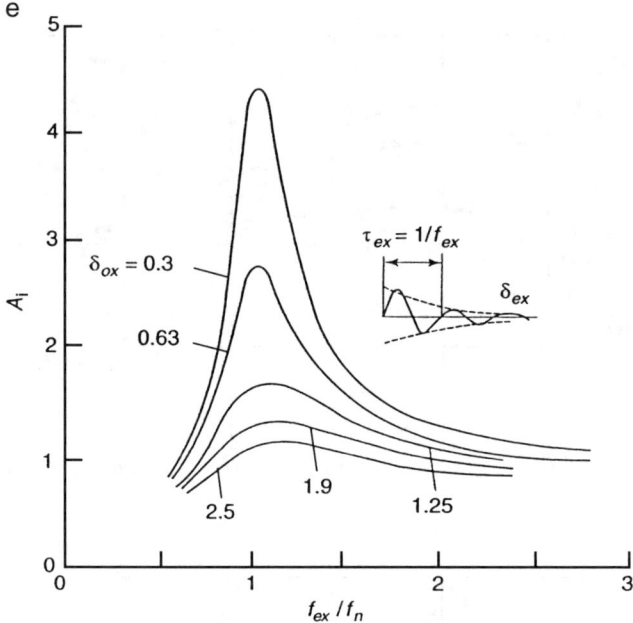

Fig. 1.8.2. (*continued*)

height of the pulse on the base, as functions of dimensionless time t/τ_p, where τ_p is the pulse duration]. In Fig. 1.8.1 plots are given for typical rectangular, half-sine, and versed-sine pulses [1.23], [1.24], and for various ratios between *pulse duration* τ_p and natural period $T_n = 1/f_n$ of the isolation system. Plots in Fig. 1.8.2 show shock spectra $A = \nu_p/\xi_p$, where ν_p is the maximum value of the response, as functions of τ_p/T_n, and for several values of damping (damping ratio ζ) in the isolation system. Figs. 1.8.2(a), (b), (c), (d) give "shock spectra" for rectangular, half-sine, versed-sine, and "ramp" pulses, respectively [1.25]. Fig. 1.8.2(e) [1.26] shows the shock spectrum for "decaying sine wave" pulse for the single-degree-of-freedom system. The shock spectrum in Fig. 1.8.3 [1.27] is for relative vibration between the base and the isolated object due to a verse-sine pulse excitation,

$$A_{rel} = |\nu(t) - \xi(t)|_{peak}]/\xi_p. \qquad \text{Eq. (1.8.1)}$$

Here both parameters ξ and ν should represent the same parameters of the dynamic system [e.g., if ξ can be a displacement pulse coming from the foundation m_f in Fig.1.4.1(a), then ν is forced displacement of

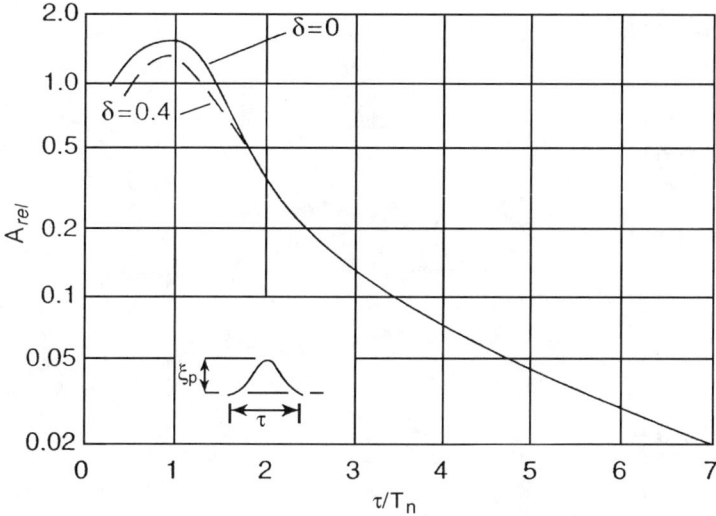

Fig. 1.8.3. Relative Vibration "Shock Response Spectrum" for a Versed-Sine Pulse.

mass m; or if ξ is the force pulse acting on mass m, then ν is force transmitted to the foundation mass m_f]. Plots in Figs. 1.8.1 and 1.8.2 are computed for the case of $m_f = \infty$; if m_f is finite, then the response ν and shock spectrum A should be modified. The modified value of the shock spectrum A_f is

$$A_f = \frac{m_f}{m + m_f} A .$$
Eq. (1.8.2)

Fig. 1.8.1 shows that for $\tau_p/T_n \leq 0.5$, the maximum response amplitude occurs after the pulse ends, while for longer pulses it occurs within the pulse duration time. For the design of vibration isolation systems, it is important that *in the first approximation* the first (and the most intense) part of the time history of the response is verse-sine-like pulse for many shapes of the excitation pulse, especially if $\tau_p \ll T_n$. The duration τ_r of this *response* pulse can be determined from Fig. 1.8.4 [1.28]. This approximation is useful for evaluating propagation of pulses across the isolation system since usually only the peak value of the response and shape and duration of its first pulse are of interest (e.g., see Section 2.2.4). More accurate analysis can be performed by calculating the spectral content of the pulse excitation functions and then applying Eq. (1.7.4).

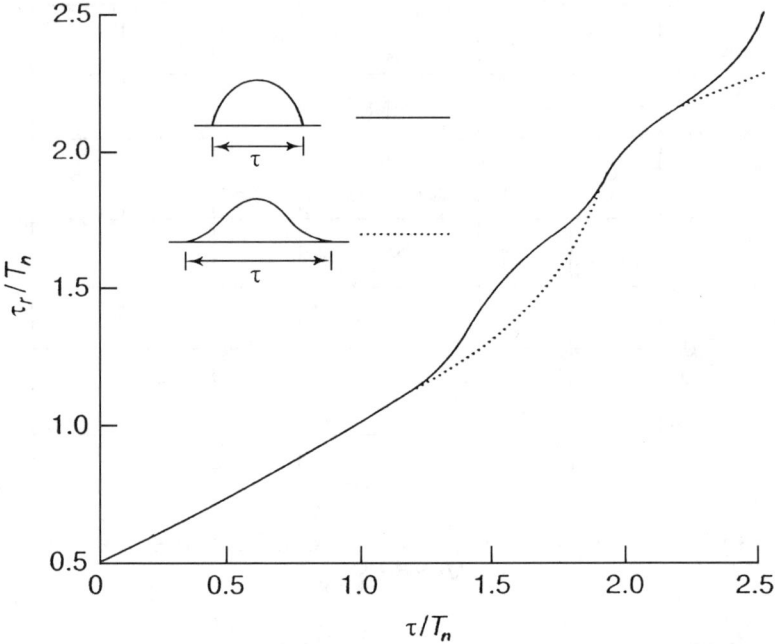

Fig. 1.8.4. Duration τ_r of the (Approximately) Versed-Sine Response for Half-Sine and Versed-Sine Excitation Pulses (Duration τ_p).

For undamped systems, shock spectra for half-sine (A_1) and versed-sine (A_2) pulses can be expressed analytically as [1.23]

$$A_1 = \frac{T_n/\tau_p \cos \pi\tau_p/T_n}{(T_n/2\tau_p)^2 - 1}; \quad A_2 = \frac{\sin \pi\tau_p/T_n}{1 - (\tau_p/T_n)^2}. \qquad \text{Eq. (1.8.3)}$$

For $\tau_p \ll T_n$ (e.g., the typical case for vibration isolation of forging hammers),

$$A_1 \cong \frac{4\tau_p}{T_n} \cos \frac{\pi\tau_p}{T_n}; \quad A_2 \cong \pi\tau_p/T_n \qquad \text{Eq. (1.8.4)}$$

Presence of damping in the dynamic system under pulse excitation results, first, in fast decaying of the response at $t > \tau_p$, and in some reduction of the peak value of the response function (Fig. 1.8.2). Within

~5%, influence of damping on the shock spectrum values for half-sine and verse-sine pulses can be expressed as

$$A_d \cong A_o(1 - 0.14\delta), \qquad \text{Eq. (1.8.5)}$$

where A_d is shock spectrum value considering damping; and A_o—without considering damping.

1.9 NONLINEARITY IN VIBRATION ISOLATION SYSTEMS

Nonlinearity in vibration isolation systems, as in other dynamic systems, is usually pronounced at large vibration amplitudes. However, steady vibration amplitudes in vibration isolation systems for stationary objects (production and measuring machines, internal combustion engines), as well as for aggregates on surface and air vehicles, are usually much smaller than their static deformations. Accordingly, in most cases consideration of specifically nonlinear effects does not add much to the results of dynamic analysis performed in a linear approximation. Still, in some cases nonlinear effects have to be considered: e.g., vibration isolation systems that can be subjected to high-level incidental intense pulse excitations may require snubbers that introduce highly nonlinear processes; use of nonlinear isolators with low damping may result in such nonlinear effects as subharmonic resonances; an additional inter-coordinate coupling may develop in the isolation system due to nonlinear effects.

The term "nonlinearity" in the above paragraph is used to indicate a nonlinear character of the load-deflection characteristic of isolators ("static nonlinearity"). Other types of nonlinearity are dependencies of effective stiffness and damping of isolators on amplitude and frequency of vibration. The amplitude dependency can be a property of the material of the flexible element (see Chapter 3), or it can result from use of Coulomb friction connections in isolator designs. The amplitude and frequency dependencies of both stiffness and damping are typical properties of elastomeric and plastic materials frequently used in vibration isolator designs. All types of such "dynamic nonlinearities" (both amplitude and frequency dependencies) can be pronounced and important in any amplitude range.

A comprehensive methodology of designing vibration isolation systems with snubbers, which, in addition to attenuation of deterministic vibratory forces, are required to withstand severe incidental shocks, is given in [1.29]. The case in [1.29] relates to installation of a compressor producing near-harmonic excitation at 60 Hz. This excitation should be attenuated 15-fold to protect adjacent precision instruments, thus requiring a relatively low natural frequency ($f_n \cong 15$ Hz) and a low-damping suspension. In addition, the system should tolerate severe environmental shocks (a saw-tooth acceleration pulse with magnitude 50 g and duration 18 ms) and white-noise vibration input with 16.6 g root mean square (RMS) level of acceleration in the 0–2,000 Hz range (the required test parameters per [1.20]). In order for a conventional linear vibration isolation system to survive such test conditions, it has to allow for unacceptably large deflections of the elastic element (spring). The proposed approach was to allow only such dynamic deflections of the isolation springs, that are adequate for performing the main task of the vibration isolation system—protection of precision instruments. To accommodate the severe test conditions, the suspended object (compressor) comes in contact with relatively stiff and highly damped snubbers. Consequently, at high-intensity excitations, the system is much stiffer and becomes a highly nonlinear vibro-impact system. It is shown in [1.29] that the best dynamic quality (the lowest overloads) in this type of system is achieved by minimizing the criterion

$$\Pi = \dot{X}\ddot{X} = V^2 \Psi(\varsigma), \qquad \text{Eq. (1.9.1)}$$

where $\dot{X}, \ddot{X}$ are, respectively, maximum velocity, acceleration of the object after impacting (and rebounding from) the snubber with velocity V, and $\Psi(\varsigma)$ is function of damping in the snubber, which reaches its minimum when its damping ratio is $\varsigma \cong 0.8$ (loss factor $\tan \beta \cong 0.4$, log decrement $\delta \cong 1.29$).

The most widely used vibration isolators whose flexible elements have static nonlinearity are "constant natural frequency" (CNF) isolators (see Section 1.3 and also Sections 3.3.2 and 4.5). Nonlinearity of the load-deflection characteristic of CNF isolators is a positive factor in several respects. Besides providing decoupling between various modes in the vibration isolation system (as shown in Section 1.3), these isolators have low sensitivity to process irregularities in their manufacturing (as shown in Section 4.5), and thus can be more accurately specified. Since stiffness of such isolators varies in a very broad range with changing loads, they are "self-snubbing" and provide for performance that is same as or better than that of the snubbers described in the above paragraph. Another interesting feature of nonlinear isolators is the fact that at large vibration amplitudes across this type of isolator, the amplitude-frequency characteristic of the isolation system "bends" (e.g., Fig. 1.9.1). This effect leads to some reduction of resonance amplitudes without adding damping to the system.

On the other hand, at high-excitation amplitudes development of subharmonic vibrations in systems with nonlinear, e.g., CNF, elastic connectors is possible. Subharmonic vibrations are those with natural frequencies $\omega_n = 2\pi f_n$ of the system, but these vibrations are excited by spectral components of external excitation having much higher frequencies: $2\omega_n$, $3\omega_n$, etc, e.g. see [1.30]. Development of subharmonic regimes may result in deterioration of isolation effectiveness, especially for high-frequency excitations. Subharmonic vibrations were observed in vibration isolation systems with low damping CNF isolators (e.g., based on conical springs).

In order to analyze nonlinear dynamic processes in vibration isolation systems with CNF isolators, such as generation of subharmonic vibration regimes, the equation of motion of this type of system has to be derived. The load (P) – deflection (z) characteristic of a nonlinear vibration isolator represented in Fig. 1.9.2(a) by a conical coil spring is shown in Fig. 1.9.2(b). Point A in Fig. 1.9.2(b) corresponds to loading of the flexible element in Fig. 1.9.2(a) by static (weight) load P_A. Stiffness k_A, and natural frequency ω_A at point A are (since mass m_A generating weight load P_A is $m_A = P_A/g$)

$$k_A = \left(\frac{dP}{dz}\right)_A; \quad \omega_A = \sqrt{\frac{g(dP/dz)_A}{P_A}}. \qquad \text{Eq. (1.9.2)}$$

Thus, if the CNF characteristic develops at all points of the P-z plot, the ratio $(dP/dz)/P = const$.

To obtain a specified natural frequency ω_o, the plot $P = P(z)$ must satisfy the differential equation

$$(dP/dz)/P = \frac{\omega_o^2}{g}, \qquad \text{Eq. (1.9.3)}$$

whose solution is

$$\frac{g}{\omega_o^2} \ln P = z + C \quad \text{or} \quad P = e^{\frac{\omega_o^2(z+C)}{g}}. \qquad \text{Eq. (1.9.4)}$$

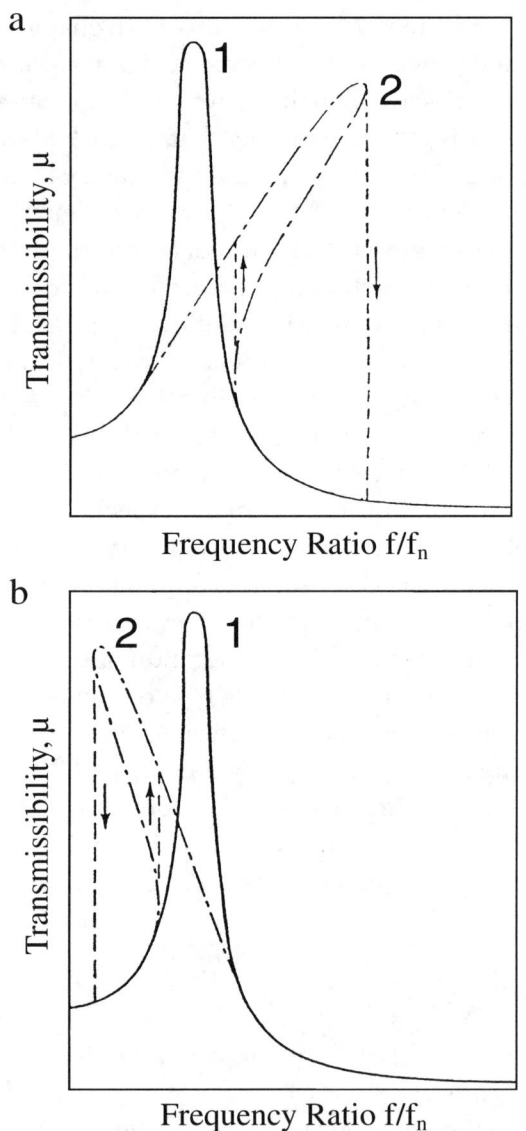

Fig. 1.9.1. Resonance Characteristics of Linear (lines 1) and Nonlinear (lines 2) Systems with the Same Damping: (a) Nonlinear Isolators with Hardening Load-Deflection Characteristic; (b) Nonlinear Isolators with Softening Load-Deflection Characteristic.

At any C, functions $P(z)$ from Eq. (1.9.4) do not pass through the point $P = 0$, $z = 0$, thus the plot similar to Fig. 1.9.2(b) can not be a solution for Eq. (1.9.3). This means that the CNF characteristic cannot start from the

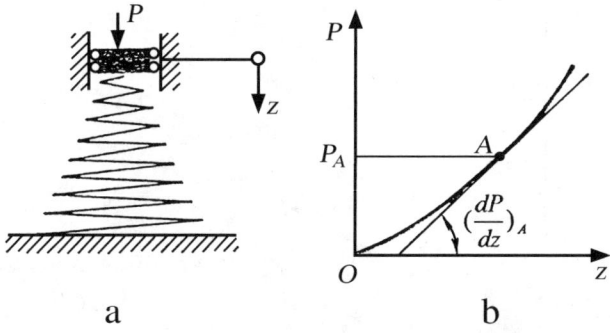

Fig. 1.9.2. (a) Nonlinear (Conical) Spring; and (b) Its Load-Deflection Characteristic.

point $P = 0$. Usually, the CNF characteristic starts from a minimum load P_{min} (see Section 4.5), and for loads below P_{min} the $P(z)$ characteristic is usually linear, so that

$$P = k_o z. \qquad \text{Eq. (1.9.5)}$$

Accordingly, the boundary condition for Eq. (1.9.4) is

$$\frac{dP}{dz} = \frac{P}{z} \quad \text{at} \quad P = P_{min}, \qquad \text{Eq. (1.9.6)}$$

and from Eq. (1.9.4) and Eq. (1.9.6), we arrive at

$$\omega_o^2 \frac{P_{min}}{g} = \frac{P_{min}}{z_{P_{min}}} \qquad \text{Eq. (1.9.7)}$$

or

$$z_{P_{min}} = \Delta_{min} = \frac{g}{\omega_o^2}. \qquad \text{Eq. (1.9.8)}$$

The integration constant C is derived from Eq. (1.9.4) using the boundary condition

$$P = P_{min} \quad \text{at} \quad z = \frac{g}{\omega_o^2} \qquad \text{Eq. (1.9.9)}$$

and

$$P = P_{min} e^{\frac{\omega_o^2}{g} z - 1}. \qquad \text{Eq. (1.9.10)}$$

It is known that subharmonic vibrations cannot develop if damping in the system exceeds a certain critical value that depends on the character

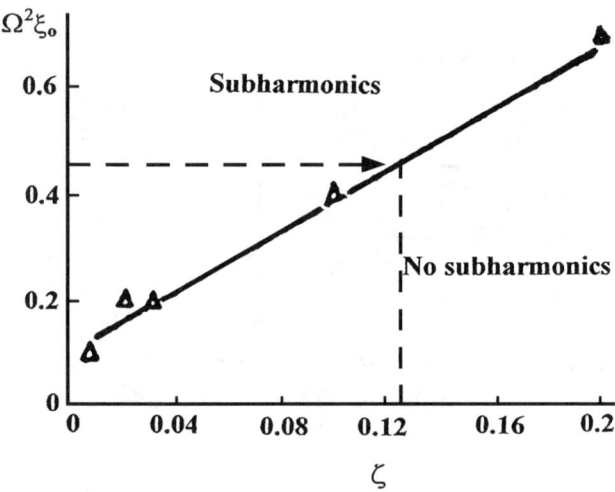

Fig. 1.9.3. Areas of Existence of Subharmonic Oscillations in Vibration Isolation System with CNF Isolators.

and degree of nonlinearity and the excitation amplitude. The critical value of damping for the single-degree-of-freedom system with CNF spring had been computationally determined in [1.15] using a dimensionless equation

$$\ddot{u} + 2\varsigma\dot{u} + e^{u} - 1 = \Omega^{2}\xi_{o}\sin\Omega\tau, \quad\text{Eq. (1.9.11)}$$

where $u = \frac{z}{z_1}$; $\tau = t\sqrt{\frac{g}{z_1}}$; $\Omega = \omega\sqrt{\frac{z_1}{g}}$; $\varsigma = \frac{c}{2m}\sqrt{\frac{z_1}{g}}$; $\xi_o = \frac{z_o}{z_1}$; $g = 9.8$ m/s^2; m is mass of the isolated object; c—damping coefficient; $\zeta = c/c_{cr}$—relative damping (log decrement $\delta = 2\pi\zeta$); z—vertical coordinate of the object; z_1—static deformation of the CNF isolator under the weight mg of the object; z_o, ω—amplitude and frequency of the excitation. The development of subharmonic vibrations in the single-degree-of-freedom system with the nonlinear spring having a CNF characteristic (stiffness is proportional to the weight mg of the object) was found to be determined by the dimensionless amplitude of excitation $\Omega^2\xi_o$ and dimensionless relative damping coefficient ζ. This correlation is shown in Fig. 1.9.3. As an example, for $f = 20$ Hz ($\omega = 125.6$ rad/s); $z_1 = 3$ mm $= 0.003$ m, $z_o = 0.1z_1 = 0.3$ mm $= 0.0003$m, a relatively large vibration amplitude for $f = 20$ Hz; $\xi_o = 0.1$, $\Omega = 2.2$, $\Omega^2\xi_o = 2.2^2 \times 0.1 = 0.485$; and from Fig. 1.9.3, $\zeta = 0.14$ or $\delta = 0.88$.

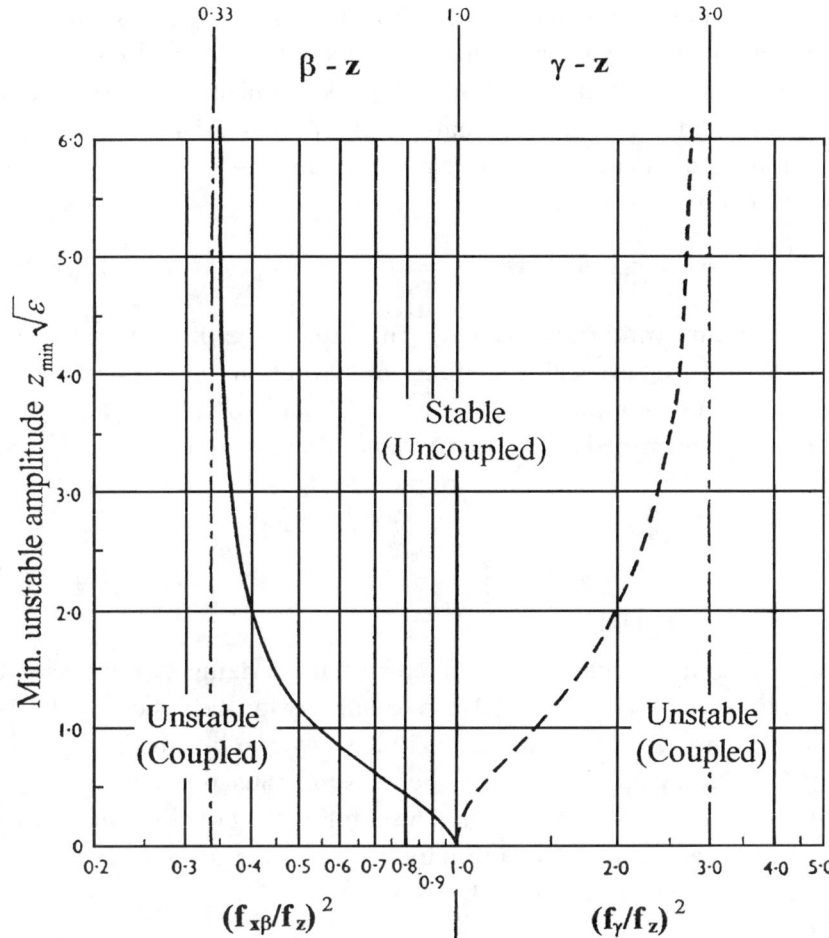

Fig. 1.9.4. Inter-coordinate Coupling Caused by Nonlinearity of Springs; Similar Linear Dynamic System is Uncoupled.

An important effect of nonlinearity in the vibration isolators is the possibility of development of inter-coordinate coupling in the system for which the decoupling conditions discussed in Section 1.3 are satisfied. While a study of inter-coordinate coupling in a six-degrees-of-freedom nonlinear system would be quite cumbersome, some understanding of the effects of the degree of nonlinearity as well as vibration amplitudes on the coupling can be deduced from the results obtained in [1.31]. The inter-coordinate coupling phenomena were studied in two basic two-degrees-of-freedom systems: a planar isolation system in which vertical (z) and angular (β) modes are uncoupled with linear isolators (system $z - \beta$); and a system allowing translational motions of the object along the z-axis and torsional (γ) motions around this axis, uncoupled in the linear system (system $z - \gamma$). The (hardening) nonlinearity of load (F_z)— deflection (z) of isolators for the planar system z—β was assumed to be of a "cubical type"

$$F_z = az + bz^3 \quad \text{or} \quad F_z = a(z + \varepsilon z^3). \qquad \text{Eq. (1.9.12)}$$

The minimum (non-dimensional) amplitude $(z/z_{st})_{min}$ of z-direction vibration that excites rocking (β-direction) vibration in this system, which does not exhibit coupling between the "z" and "β" modes with linear isolators, is determined by the degree of nonlinearity (parameter ϵ), as well as by closeness between natural frequencies f_z and f_β,

$$z_{min} = \frac{1}{\sqrt{\varepsilon}} \sqrt{\frac{4}{3} \frac{1 - n_{\beta z}^2}{3 n_{\beta z}^2 - 1}}, \qquad \text{Eq. (1.9.13)}$$

where $n_{\beta z}$ is a so-called tuning factor characterizing closeness of the natural frequencies, $n_{\beta z} = f_\beta / f_z$. Areas of stability for the system z-γ (tuning factor $n_{\gamma z}$) are similar. The Eq. (1.9.13) is graphically presented in Fig. 1.9.4. When $n_{\beta z} \to 1$, z_{min} is decreasing, thus the strong coupling develops even at small amplitudes. The similar effect of closeness between the "partial" natural frequencies on inter-coordinate coupling in the linear system is shown in Section 1.5.1.

1.10 WAVE EFFECTS IN VIBRATION ISOLATORS

Expressions for transmissibility in single-degree-of-freedom (SDOF) vibration isolation systems were derived in Section 1.4 based on the assumption that the isolator can be represented by a massless spring. In real life, all isolators—both made of metal and of other materials as described in Chapter 3—have some mass. As a result, wave effects may develop at high frequencies of transmitted vibrations whereas dimensions of the isolators become commensurate with multiples of half-wavelengths of the elastic waves passing through the flexible elements of the isolators. Fig. 1.10.1 [1.8] shows an experimentally obtained transmissibility plot for a SDOF vibration isolation system whose flexible element was made of a

Dynamic Properties of Vibration Isolation Systems • 97

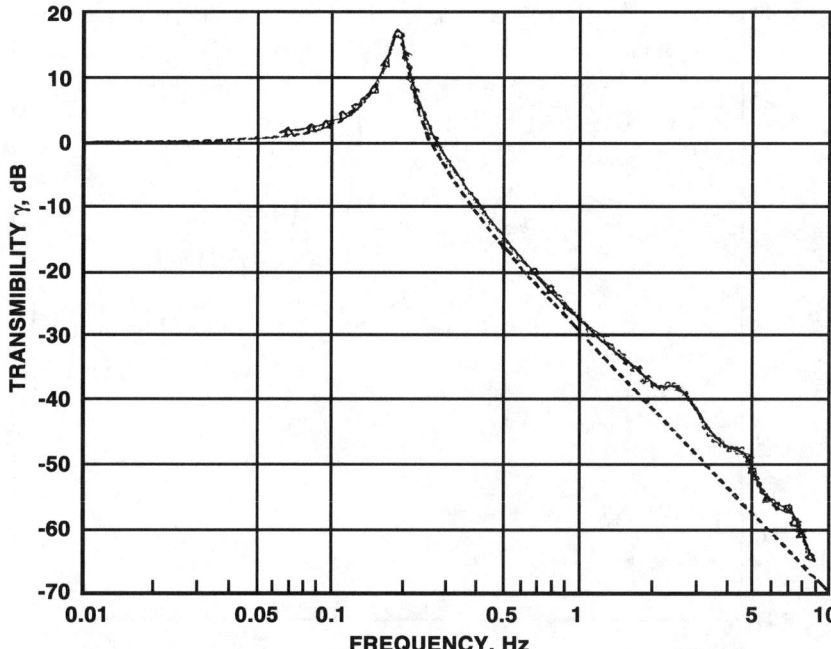

Fig. 1.10.1. Transmissibility in a Broad Frequency Range of an Uniaxial Vibration Isolation System Having Isolator Made from Natural Rubber with 40 Parts by Weight Carbon Black (Solid Line) and Calculated Using Eq. (1.4.18) (Broken Line).

relatively highly damped natural rubber (40 parts by weight of carbon black), $\delta > 0.45$ (solid line), as compared with the transmissibility plot calculated with the "massless isolator" assumption. Fig. 1.10.2 [1.8] shows transmissibility plots for two systems with low damped rubber flexible elements (one in shear and one in compression) and one system with a steel spring, having even lower damping as well as much greater weight/mass. There are visible intense high-frequency peaks resulting in deterioration of transmissibility in the high-frequency range. For the rubber cylinder in compression, the fundamental natural frequency is $f_n = 30$ Hz and the high-frequency peak is observed at $f_n' = 500$ Hz ($f_n'/f_n > 16$), and is ~47 dB (~200 times) below the fundamental peak. For the rubber isolator in shear, $f_n = 26$ Hz, $f_n' = 850$ Hz, $f_n'/f_n = 33$, the difference in resonance peak heights is ~60 dB or 1,000 times. However, for the steel spring there are numerous high-frequency resonances in the shown frequency range. With $f_n = 15$ Hz, first high-frequency peak at

98 • PASSIVE VIBRATION ISOLATION

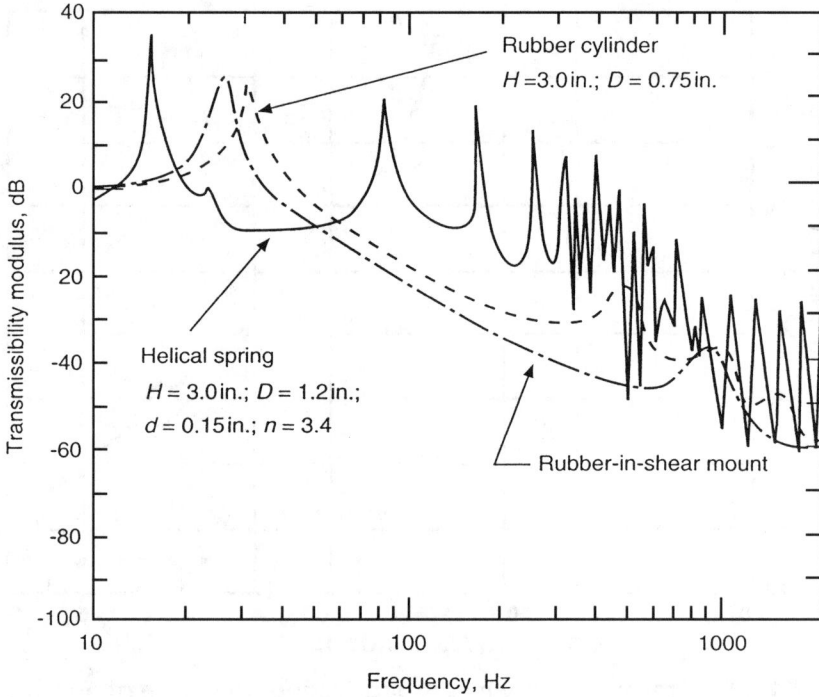

Fig. 1.10.2. Transmissibility in a Broad Frequency Range of a Uniaxial Vibration Isolation System with Isolator Having Rubber Cylinder in Compression; Rubber Element in Shear; and Steel Spring.

$f_n' = 83$ Hz ($f_n'/f_n = 5.5$), and only 16 dB (about 7 times) below the fundamental resonance peak.

With the second approximation assumption that the isolator can be modeled as a cylindrical rod with uniformly distributed mass, it was derived [1.32] that the high-frequency wave resonances in vibration isolators develop at frequencies

$$\omega_i = i\pi\omega_o \sqrt{\gamma} \qquad \text{Eq. (1.10.1)}$$

where ω_o is the fundamental natural frequency, $i = 1, 2, 3\ldots$ is the sequential number of the high-frequency resonance, and

$$\gamma = M/M_f = M/\rho Al \qquad \text{Eq. (1.10.2)}$$

is the mass ratio between the mass M of the object supported by the flexible element, and the total mass M_f of the flexible element (ρ is its density, A—cross sectional area, l—length). The transmissibility at the first high-frequency resonance is, according to [1.32],

$$\mu' = \frac{0.2}{\tan\beta}\frac{M_f}{M} = \frac{0.2}{\gamma\tan\beta},\qquad\text{Eq. (1.10.3)}$$

where $\tan\beta$ is the loss factor of the isolator. It can be seen from Eqs. (1.10.1) and (1.10.3) that both positioning and intensity of the high-frequency resonances caused by standing waves in the isolator are determined by the mass ratio between the isolator and the supported object and by damping in the isolator. Thus, if the high-frequency transmissibility is of importance, steel springs are not the best choice. Rubber has about seven times lower density ρ than steel, and rubber isolators can be made very small for a given natural frequency and load-carrying capacity, especially if the "ideal shape" (streamlined) unbonded flexible elements are used (see Chapters 3 and 4).

Since the wave resonances occur in a broad frequency range, usually much higher than the main resonance frequency, dependence of stiffness and loss factor on frequency for the rubber blend used in the isolator (see Section 3.3.2d) should be considered to obtain more accurate analytical predictions [1.33].

REFERENCES

[1.1] Himelblau, H. and Rubin, S., 1988, "Vibration of a Resiliently Supported Rigid Body," Ch. 3 in *Shock and Vibration Handbook*, edited by C. Harris, 3rd Ed., McGraw-Hill, N.Y.

[1.2] Zavala, P.A.G., Pinto, M.G., Pavanello, R., and Vaqueiro, J., 2001, "Powertrain Mounting Development Based on Computational Simulation and Experimental Verification Method," *Proceedings of the 2001 SAE Noise and Vibration Conference* (*Noise 2001 CD*), SAE Paper 2001-01-1509.

[1.3] Rivin, E.I., 1969, "Some Issues of Vibration Isolation of Machine Tools," in *Dinamika Mashin*, Nauka Publ. House, Moscow [in Russian].

[1.4] Rivin, E.I., 1979, "Principles and Criteria of Vibration Isolation of Machinery," *ASME J. of Mechanical Design*, Vol. 101, pp. 682–692.

[1.5] DenHartog, J.P., 1961, *Mechanics*, Dover Publications, Inc., N.Y.

[1.6] Rivin, E.I., 1999, *Stiffness and Damping in Mechanical Design*, Marcel Dekker, Inc., N.Y.

[1.7] NAVSEA 0900-LP-089-5010, 1977, Navy Resilient Mount Handbook. *A User Guide of Design, Installation and Inspection Information.*

[1.8] Snowdon, J.C., 1979, *Handbook of Vibration and Noise Control*, U.S. Dept. of the Navy, Report TM 79-75.

[1.9] Derby, T.F. and Calcaterra, P.C., 1970, *Response and Optimization of an Isolation System with Relaxation Type Dampers*, NASA Report CR-1542.

[1.10] Crede, C.E. and Ruzicka, J.E., 1988, "Theory of Vibration Isolation," Ch. 30 in *Shock and Vibration Handbook*, edited by C. Harris, 1958, 3rd Ed., McGraw-Hill, N.Y.

[1.11] Ruzicka, J.E. and Cavanaugh, R.D., 1958, "New Method for Vibration Isolation," *Machine Design*, Oct. 16, pp. 114–121.

[1.12] Eliseev, S.V., 1978, "Structural Theory of Vibration Protection Systems," Nauka Publishing House [in Russian].

[1.13] Rivin, E.I., 1967, "Dinamika privoda stankov [Dynamics of Machine Tool Drives]," Mashinostroenie Publishing House, Moscow [in Russian].

[1.14] DeBra, D.B., 1992, "Vibration Isolation of Precision Machine Tools and Instruments," *Annals of the CIRP*, Vol. 41/2, pp. 711–718.

[1.15] Rivin, E.I., 1966, "Vibration Isolation Systems with Constant Natural Frequency Isolators," *Izvestiya VUZov. Mashinostroenie*, No. 3, pp. 62–68 [in Russian].

[1.16] Crede, Ch.E., 1951, "Vibration and Shock Isolation," John Wiley & Sons, N.Y.

[1.17] Banakh, L.Ya., Private communication.

[1.18] Banakh, L. Ya. and Perminov, M.D., 1972, "Study of Complex Dynamic Systems Using Weak Coupling between Subsystems," *Mashinovedenie*, No. 4 [in Russian].

[1.19] Polungian, A.A. and Novokshonov, V.K., 1985, "Analysis of Focal Equi-Frequency Ring-Like Car Engine Suspension," *Izvestiya VUZov. Mashinostroenie*, No. 12, pp. 86–89 [in Russian].

[1.20] Military Standard MIL-STD-810E, 1988, "Environmental Test Methods," U.S. Department of Defence.

[1.21] Miles, J.W. and Thomson, W.T., 1988, "Statistical Concepts in Vibration," Ch. 11 in *Shock and Vibration Handbook*, edited by C. Harris, 3rd Ed., McGraw-Hill, N.Y.

[1.22] Veprik, A.M., Babitsky, V.I., Pundak, N., and Riabzev, S.V., 2000, "Vibration Protection for Linear Split Stirling Cryogenic Cooler of Airborne Infrared Application," *Shock and Vibration*, Vol. 7, No. 6, pp. 363–381.

[1.23] Jacobsen, L.S. and Ayre, R.S., 1958, "Engineering Vibrations," McGraw-Hill, N.Y.

[1.24] Jacobsen, L.S. and Ayre, R.S., 1961, "Transient Response to Step and Pulse Functions," Ch. 7 in *Shock and Vibration Handbook*, edited by C. Harris and Ch. Crede, McGraw-Hill, N.Y.

[1.25] "Application Selection Guide," 1978, *Bulletin C5-178*, Barry Wright Corp.

[1.26] Jarausch, R., 1965, "Hammer, Fundament und Umgebung als Schwingungsystem," *Maschinenmarkt*, Vol. 71, No. 11, pp. 27–38 [in German].

[1.27] Rivin, E.I., 1995, "Vibration Isolation of Precision Equipment," *Precision Engineering*, Vol. 17, pp. 41–56.

[1.28] Rivin, E.I., 1987, "Evaluation of Vibration Isolation Systems for Forging Hammers," in *Meeting Handbook on "Vibration Isolation of Heavy Structures,"* Institute of Acoustics, London, pp. 91–97.

[1.29] Babitsky, V.I. and Veprik, A.M., 1998, "Universal Bumpered Vibration Isolator for Severe Environment," *Journal of Sound and Vibration*, Vol. 218, No. 2, pp. 269–292.

[1.30] Ehrich, F. and Abramson, H.N., 2002, "Nonlinear Vibration," Ch. 4 in *Shock and Vibration Handbook*, edited by C. Harris and A. Piersol, McGraw-Hill, N.Y.

[1.31] Henry, R. and Tobias, S.A., 1959, "Instability and Steady-State Coupled Motions in Vibration Isolating Suspensions," *Journal of Mechanical Engineering Science*, Vol. 1, No. 1.

[1.32] Harrison, M., Sykes, O.A., and Martin, M., 1952, "Wave Effects in Isolation Mounts," *Journal Acoustical Society of America*, Vol. 24, p. 62.

[1.33] Reid, M.C., Oyadiji, S.O., and Tomlinson, G.R., 1988, "Relating Material Properties and Wave Effects in Vibration Isolators," *Proceedings of the 59th Shock and Vibration Symposium*, Sandia National Laboratories, Albuquerque, N.M. October.

[1.34] Soderholm, L.G., 1981, "Load Sensor in Press Mount Indicates Weight Distribution," *Design News*, May 4.

CHAPTER 2

Principles and Criteria of Vibration Isolation

T he vibration isolation of vibration-sensitive/precision objects is rightly considered to be the most important case. However, there are many other situations in which the appropriately applied principles of vibration isolation can be very beneficial. Also, even the objects of the highest precision, while requiring a deep isolation from external disturbances, may themselves be producers of intense dynamic loads and vibrations.

This Chapter gives a systematic treatment of the principal cases wherein vibration isolation is or can be required. The most important case is that of vibration sensitive objects, for which a detailed analysis is given, together with the procedures for selecting parameters of the vibration isolation means. The very important side issues of static rigidity, rocking of the object, and self-generated intense dynamic loads are also addressed.

A comprehensive classification of vibration-producing objects requiring vibration isolation is presented and for each class the practical principles as well as analytical treatments for designing vibration isolation systems are given.

A very important (and the largest) group of so-called "general purpose machines and equipment" was usually neglected. It is demonstrated that a judicial assignment of specifications for vibration isolation systems may result in significant benefits.

A brief treatment of such important but rather special issues as vibration isolating installation of objects on non-rigid structures and of engines in surface vehicles is intended for developing a basic understanding of these contradictory systems and for providing some practical guidelines.

While the approaches described in this Chapter allow solving many practical problems with reasonably positive results, the real-life vibration isolation systems are extremely complex and diverse. An effective experimental approach for optimization of such systems is proposed.

2.1 GENERAL COMMENTS

A major cause for the inadequate performance of vibration isolation systems is often lack of understanding of the goals and principles of vibration isolation of various real-life machines and apparatuses. It is a complex problem to design an optimal vibration isolation system due to diversity of requirements for the vibration isolation of various objects that need vibration isolation. Additional difficulties arise from a variety of working regimes and internal configurations of some objects (such as cutting regimes in machine tools, position variations of heavy tables or gantries on machine tools and CMMs, speed and thus excitation frequency variations of engines and motors, etc.), as well as a variety of environmental conditions (such as dynamic characteristics of floor and other supporting structures, presence of vibration-producing and/or vibration-sensitive equipment in the vicinity, etc.). While development of special requirements for unique objects is warranted, an optimal synthesis of the vibration isolation system for each unit of typical equipment, such as production and measuring equipment, is hardly practical. A more appropriate way of approaching this problem for the production and measuring equipment is to derive more-or-less generic criteria for groups of similar objects.

Four distinct groups of objects can be considered:

1. *Vibration-sensitive machines and devices.* The main goal of vibration isolation for this group of objects is to ensure that relative vibration in the "working" or other critical area (e.g., between tool and workpiece for a precision production machine or between components of an electronic device) does not exceed permissible limits under given external (and, in some cases, also internal) excitations. Typical representatives of this group are precision machine tools, CMMs, photolithography tools, electronic devices.
2. *Vibration-producing machines and devices.* For this group, the main goal of vibration isolation is to reduce intensity of dynamic exertions transmitted from the object to the supporting structure (foundation, floor, etc.) below permissible levels. Typical representatives of this group are forging, stamping, and other impact-generating machines; devices with unbalanced rotors such as fans and compressors; mechanisms generating vibration whose specter is rich in harmonics (e.g., reciprocating mechanisms, vibration shakers).
3. *General-purpose machinery and equipment.* These are neither very sensitive to external vibratory exertions nor do they produce excessive dynamic forces (e.g., machine tools of ordinary precision). The main goals of vibration isolation for this group are: to protect the object from accidental intense external shocks and vibrations; to protect the environment, such as adjacent precision devices, from occasional disturbances caused by the object (such as occurrences of self-excited chatter vibrations or stick-slip vibrations); to reduce noise and general vibration levels (both external and internal); to facilitate installation by eliminating the need to fasten or grout the objects to the floor.

4. *Objects (mostly, machinery and equipment) installed on non-rigid structures.* These objects include machines installed on upper levels of buildings, in ships and surface vehicles, etc. Under such circumstances, both internal and external dynamic excitations may be amplified due to low dynamic stiffness of the supporting load-carrying structures. Thus, even ordinary machines can produce severe vibrations in the floors and other supporting structures, and greater floor vibrations increase the need for protection not only for precision objects but even for ordinary, general-purpose machinery and equipment.

In addition to these groups, there are many special cases characterized by singular requirements, such as mounting internal combustion engines in vehicles, etc.

In all cases of vibration isolation, the normal performance of the object should not be disturbed, e.g., due to static instability or rocking motion of the isolated object or due to loss of required structural rigidity.

The above classification is not absolute. For example, a surface grinder, a CMM, or a photolithography tool are precision objects and need to be protected from floor vibrations. However, start-stop events and reversals of heavy tables or gantries in these machines may lead to intense dynamic loads disturbing the machine itself as well as nearby precision devices. Thus, these machines also have to be treated as vibration-producing objects.

2.1.1 Damping vs. Heat Generation in Vibration Isolators

In this chapter, requirements for parameters of vibration isolation systems for various objects and environments are often formulated as criteria connecting natural frequencies/stiffness constants and damping values. In many applications, performance of the vibration isolation system can be significantly improved if vibration isolators with higher damping are used. In some cases, however, the high-damping vibration isolators are shunned in favor of isolators with lower damping, due to a belief that there is more heat generation in vibration isolators that have higher damping, which may lead to overheating and to accelerated deterioration of the flexible elements made from rubber or other heat-sensitive materials. Below, this notion is shown to be incorrect [2.1].

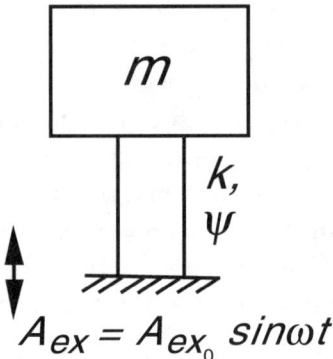

Fig. 2.1.1. Oscillator with a Damped Flexible Element.

Heat is generated in flexible elements as a result of dissipation of mechanical energy in the process of cyclical deformation. Let us consider a simple oscillator, as shown in Fig. 2.1.1, with mass m supported by rubber spring characterized by stiffness k, and by relative energy dissipation during the deformation cycle ψ, see Section 1.4.1. The maximum potential energy of deformation of the rubber element during one cycle of a vibratory process is

$$U = k\frac{A^2}{2}, \quad \text{Eq. (2.1.1)}$$

where A is the amplitude of deformation.

The amount of energy dissipated (transformed into heat) during one deformation cycle is

$$\Delta U = \psi U = \psi k \frac{A^2}{2}. \quad \text{Eq. (2.1.2)}$$

At the resonance, amplitude A_{res} of mass m is

$$A_{res} = A_{ex_0} \frac{2\pi}{\psi}, \quad \text{Eq. (2.1.3)}$$

where A_{ex_0} is the excitation amplitude. Thus, energy dissipation at the resonance (the maximum energy dissipation) is

$$\Delta U = \psi k \frac{A_{res}^2}{2} = 4\pi^2 k A_{ex}^2 \frac{1}{\psi}. \quad \text{Eq. (2.1.4)}$$

Since the excitation amplitude A_{ex_0} is given, *when the degree of damping Ψ is greater, there is less energy dissipation in the rubber element*. This paradoxically-sounding statement is easy to explain: The effect of damping on a mechanical system is, first of all, a change in the system's equilibrium; in this case, it represents a reduction of the vibration amplitudes at resonance. The absolute amount of the energy dissipation is a secondary effect. If an elastic element operates under *forced deformation* (such as in a misalignment-compensating coupling), then the deformation amplitude is given and the heat generation increases proportionally to the degree of damping in the compensating element.

2.2 ISOLATION OF VIBRATION-SENSITIVE OBJECTS

Continuous tightening of machining and measuring tolerances for parts of machines and instruments leads to development of more accurate machine tools and measurement apparatuses. Magnitudes of the tolerances and of dimensional inaccuracies of production and measuring equipment are now often expressed in fractions of micrometers or in nanometers. From the time of World War II, it has been widely accepted that precision machinery and instruments must be isolated from external vibrations whose amplitudes may significantly exceed the magnitudes of allowable machining/processing/measurement deviations. To satisfy this obvious need, a wide inventory of passive vibration isolation mounts and other means was developed and is commercially available [2.2], [2.3], (Chapter 4). Recently, this inventory was complemented by actively controlled isolators that can provide even better vibration protection and/or maintain a constant level of the device mounted on soft isolating mounts, regardless of changes of mass distribution within the machine, e.g. see [2.3]. Since improvements in the vibration isolation efficiency can be achieved by using softer isolating mounts (lower natural frequencies), often extremely soft mounts are used for vibration isolation installations. However, the soft mounting causes large excursions (tilting) of the mounted object caused by shifting masses, etc., as well as large amplitude rocking motions, as well as reduction of effective stiffness of the isolated object. The tilting is a change of leveling when mass distribution within the isolated system alters because of addition/subtraction of components to the system or by movements of massive parts in the system. The rocking motions of the isolated system are attributable to both external excitations (touching of the isolated body, air drafts, etc.) and internal excitations (forces caused by acceleration/deceleration of the moving components inside the object). These rocking motions are undesirable. They are annoying for the operators and may induce displacements in the working zone of the system (e.g., cutting area or measurement area), possibly even exceeding the external vibration-caused displacements from which the system is being isolated. These static and dynamic movements, as well as deterioration of the effective stiffness, can be alleviated by fastening the object to a massive and/or rigid supporting structure ("inertia block"), which is, in turn, mounted on vibration isolators; or by using an appropriate active (servo-controlled) system; or, in some cases, by using vibration isolation systems with motion transformation (Section 1.4.4). However, two former approaches add significantly to the cost and dimensions of the isolated object, and the third one has only a limited effectiveness. In addition, the inertia block impairs mobility of the isolated object, while complex and costly electronic and/or pneumatic active control systems reduce its reliability.

It is shown below that, in many cases, vibration isolation specifications for high-precision objects may not be as stringent as usually assumed. The main reason for this is that sensitivity to external vibrations does not necessarily increase with increasing accuracy requirements, since improvements in accuracy of precision devices are usually accompanied by their better designs. Also, the quality of vibration isolation often can be decoupled from the issues of static and/or dynamic movements by a proper understanding of how the external vibrations influence the object performance, as well as by a better understanding of the dynamics of the vibration isolation system. These statements are illustrated in this section by deriving specific vibration isolation requirements for several types of precision equipment.

2.2.1 Ambient Vibrations

Ambient vibrations affecting vibration-sensitive devices can be transmitted from the supporting structures (floor for stationary equipment, vehicle structures for on-board equipment) or through the air. Vibration isolators are usually intended to reduce the vibrations transmitted from the supporting structures. Floor vibrations are excited by vibratory motions of nearby processing machines, motors, transformers, etc. (e.g., as the result of fast-rotating unbalanced rotors or chatter vibrations in metal-cutting machine tools); by moving trains, trucks, carts, dollies, cranes, etc.; by shocks from impact-acting machines, such as forging hammers and presses that can be located quite far from the precision vibration-sensitive objects; from machines with reversible masses, like surface grinders; from accidental impacts in the process of material handling; from footsteps on non-rigid floors. In many cases vibration of precision equipment is excited by auxiliary devices attached to the equipment units, such as hydraulic pumps in machine tools, vacuum pumps in high-tech production and measuring equipment, etc. For ultra-precision facilities, microseismic motions of the ground should also be considered. Some rather unusual but potentially harmful vibration sources are explosions from military exercises and construction work, "sonic boom" from supersonic aircraft, structural vibrations excited by low-flying airplanes, etc. The character and levels of floor vibrations depend not only on the excitation, but on the vibration-filtering effects of the dynamic system consisting of the soil, foundations, flooring, and other structural components of the building. Natural frequencies of building foundations depend on soil properties and usually are above 10 Hz, and the floors resonate in a vertical direction, usually at 6–30 Hz. Natural frequencies of floors in horizontal directions are higher. Vertical dynamic stiffness of floors is specified by manufacturers of ultra-high-precision photolithography tools (scanners) in the range $1-2.6 \times 10^8$ N/m ($0.6-1.5 \times 10^6$ lb/in) in the 1–10 Hz to 10–250 Hz frequency ranges [2.4].

Floor vibrations usually are made up of a multitude of spectral components with randomly varying amplitudes. A statistical study of an ensemble of floor vibration data measured at various locations on a plant floor showed that floor vibrations represent a non-stationary and non-ergodic random process. It is very difficult to apply exact analytical approaches to such vibratory processes. Often, this fact is not understood properly or is neglected during floor vibration surveys, which are routinely performed by suppliers of vibration isolators or by in-house and outside consultants. A straightforward use of Fast Fourier Transform (FFT) spectrum analyzers in such conditions can give distorted results and smear the important difference between steady or quasi-steady vibrations (when amplitude of a harmonic component is not changing significantly during 3–5 periods of vibratory motion) and transient vibrations with

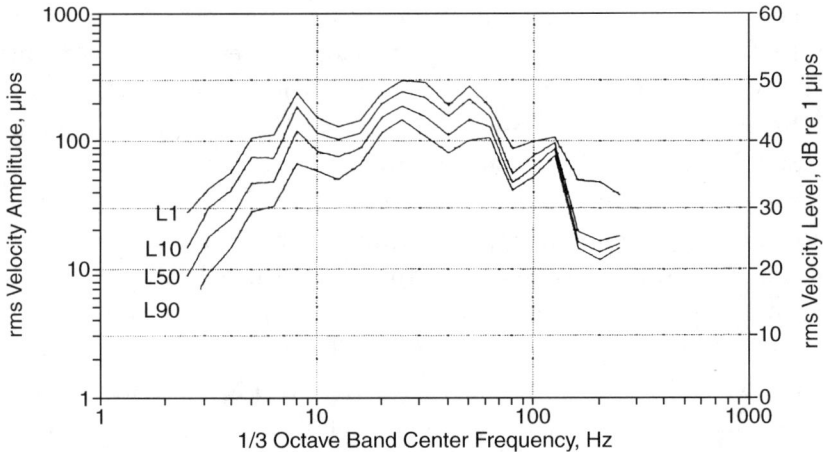

Fig. 2.2.1. Statistical Distribution of Vibration Amplitudes (Second Floor of a Microelectronics Fabrication Facility).

fast-changing amplitudes. Some FFT analyzers have a capability to obtain so-called *"centile spectra"* [2.5]; these can also be obtained by using computer post-processing of FFT analyzer output. A centile spectrum L_n is the envelope of spectral values that are exceeded $n\%$ of the measurement period. Fig. 2.2.1 [2.5] shows one-third octave $L1$, $L10$, $L50$, and $L90$ spectra of vertical vibrations on the second floor of a microelectronics production facility (*"fab"*). The battery of parallel filters was sampled at 1-sec intervals for 45 min. The peak in the 25 Hz band is attributed to the vertical natural frequency of the floor; the peak in the 8 Hz band is due to operation of a 500 rpm air-handling unit. It is obvious that amplitudes of floor vibration existing for only 1% of the time would have much less of an effect on the vibration-sensitive equipment than amplitudes existing for 90% of the time. Thus, if such statistical surveying of the floor vibration had been performed, depending on the circumstances, it seems to be prudent to use L50 or L90 spectra. Unfortunately, high-quality data such as that shown in Fig. 2.2.1 are often not available.

Fig. 2.2.2 shows amplitudes of quasi-steady vertical (a) and horizontal (b) floor vibrations at the various manufacturing plants [2.6]. The broken lines are envelopes of points representing absolute maximums of the amplitudes at various frequencies of vibratory processes that were reasonably stable for 3–5 cycles of the principal harmonic component; the solid lines are the assumed boundaries of the spectra used in analytical

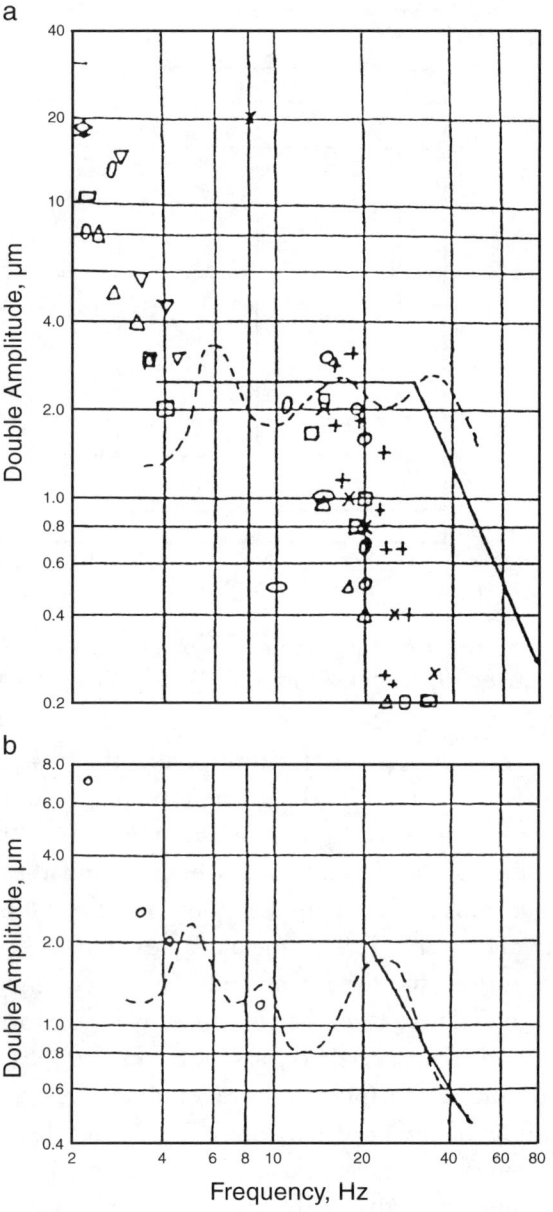

Fig. 2.2.2. Maximum Amplitudes of Quasi-Steady Shop Floor Vibrations: (a) Vertical Vibration; (b) Horizontal Vibration (Broken Lines Indicate Envelopes of Soviet Measurements; Solid Lines, Assumed Boundaries; Individual Marks Pertain to Various U.S. Sites.)

114 • PASSIVE VIBRATION ISOLATION

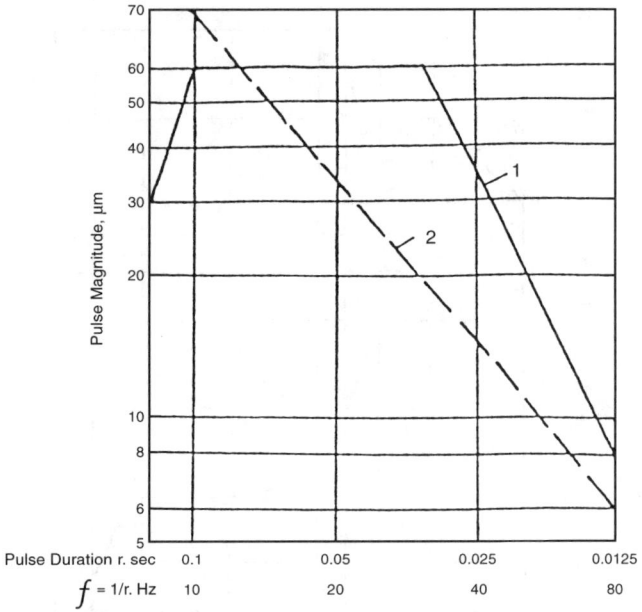

Fig. 2.2.3. Envelopes of Magnitudes of Machine Shop Floor Transient Vibrations Caused by Shocks/Impacts (1—Vertical; 2—Horizontal).

derivations. These lines represent vibrations at the ground floors (solid foundation slabs) measured in the former Soviet Union in the 1960s. Amplitudes of vibrations of upper floors made of thin, prestressed concrete slabs tend to be 1.5–3.0 times higher, and the corresponding frequency ranges tend to be broader. Dots on the plots in Fig. 2.2.2 represent relatively recent (1990s) measurements at U.S. manufacturing plants. The amplitudes for the lower end of the frequency spectrum for the latter are inflated since the measurements were performed using FFT analyzers with windows prescribed for steady-state processes. Thus, these "alleged" low-frequency amplitudes are comparable to amplitudes of transient vibrations of the floor shown in Fig. 2.2.3. The most important for vibration isolation purposes part of the spectrum (5–40 Hz) shows that the old Soviet data give also a conservative upper limit of floor vibrations at the U.S. manufacturing sites equipped with more powerful, faster modern machinery. The upper frequency limit of relatively intense vibrations is much lower for the recent tests at the U.S. sites (~20 Hz) than for the old Soviet tests (30–40 Hz). The envelope of the vertical floor vibratory displacement amplitudes can be considered as a narrow

Principles and Criteria of Vibration Isolation • 115

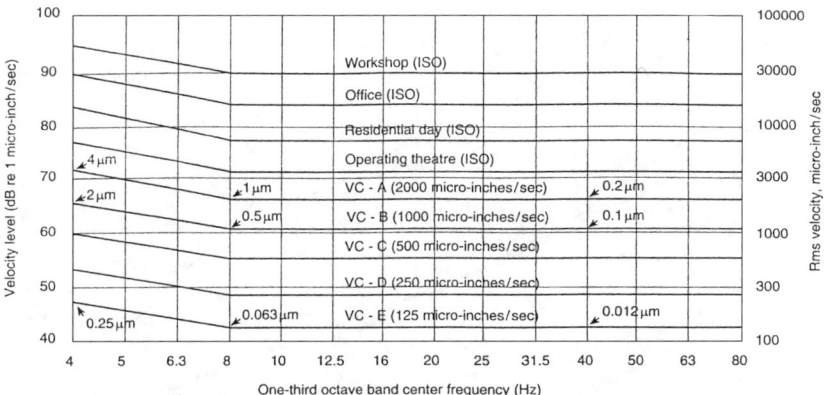

Fig. 2.2.4. BBN Floor Vibration Criteria (VC) for Installation of Precision Equipment.

frequency range (2–20 Hz) noise. The recent floor vibration surveys are usually concerned only with vertical vibrations, while most precision equipment is more sensitive to horizontal vibrations. Fig. 2.2.2(b) shows that the maximum amplitudes of horizontal vibrations are 30–50% lower than the maximum amplitudes of vertical vibrations, Fig. 2.2.2(a).

The intensity of floor vibrations in precision microelectronics manufacturing facilities is much lower since special sites are usually selected for such facilities, and the buildings and their foundations are specially designed. Often such buildings are built on piles 5–15 m (17–50 ft) long driven down to the bedrock. Vibration amplitudes on top of the piles are at least 10 dB (~3 times) smaller than amplitudes on the grade. Caissons 7 m (25 ft) long have shown even more reduction, ~27 dB (~20 times), [2.7]. Vibration spectra of vertical vibrations for three typical clean room floors indicate that maximum levels of vibratory velocity 1.5–5 μm/s (60–200 μin./s) are in 10–60 Hz range, with the corresponding displacement amplitudes not exceeding 0.05 μm (2 μin.) [2.8]. Fig. 2.2.4 shows generic vibration criterion (VC) curves developed by BBN Co., which are frequently used for specifications of vibration levels for newly constructed precision facilities [2.9]. These curves are formulated in terms of vibratory velocity. For the least stringent VC-A, the velocity amplitudes correspond to displacement amplitude 4 μm (160 μin.) at 4 Hz, 1 μm (40 μin.) at 8 Hz, and 0.2 μm (8 μin.) at 40 Hz. For the most stringent criterion VC-E, the respective amplitudes are 0.25 μm (10 μin.), 0.063 μm (2.5 μin.), and 0.0125 μm (0.5 μin.). The designer's selection of a

116 • PASSIVE VIBRATION ISOLATION

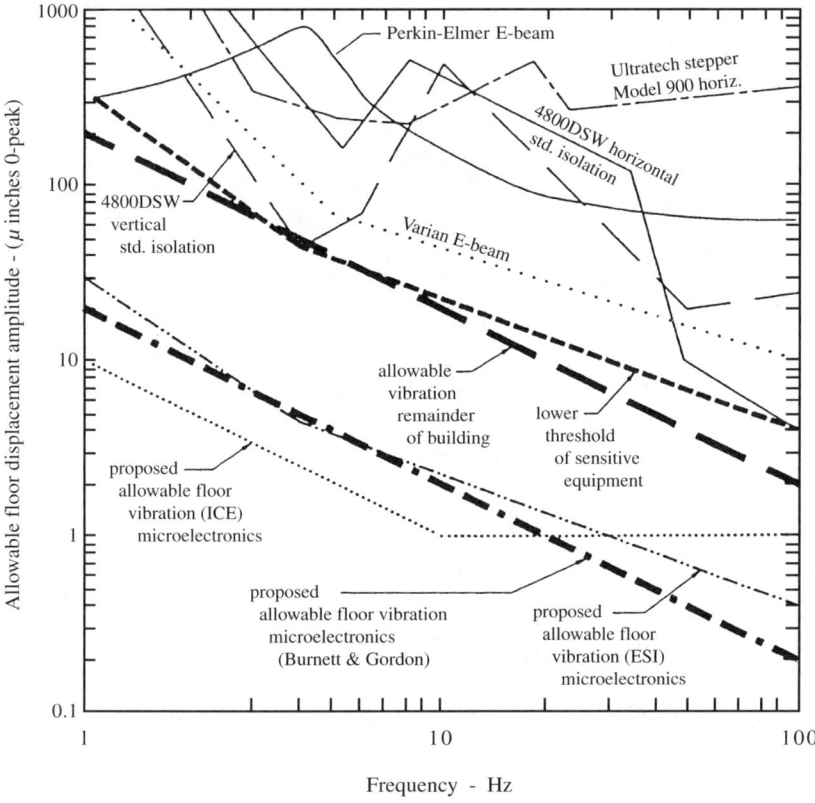

Fig. 2.2.5. Various Floor Vibration Criteria for Microelectronics Production Facilities (Lines Marked with Equipment Models are Manufacturer's Requirements).

particular VC and/or isolation system specifications will make a major impact on the overall cost of the facility. The majority of fab floors built during the last decade have been designed to comply with the VC-D limit, and were proven to be adequate for the state-of-the-art photolithography tools providing for 0.15 μm line width [2.4]. Fig. 2.2.5 [2.7] shows even more conservative requirements for floor vibration levels of microelectronics facilities proposed and promulgated by various consulting groups and companies.

Actual vibration levels on the floors in microelectronics facilities/fabs are usually quite low. Fig. 2.2.6(a) shows vertical vibratory displacements in various locations of a building at the University of Minnesota, which was later razed in order to construct a new microelectronic facility

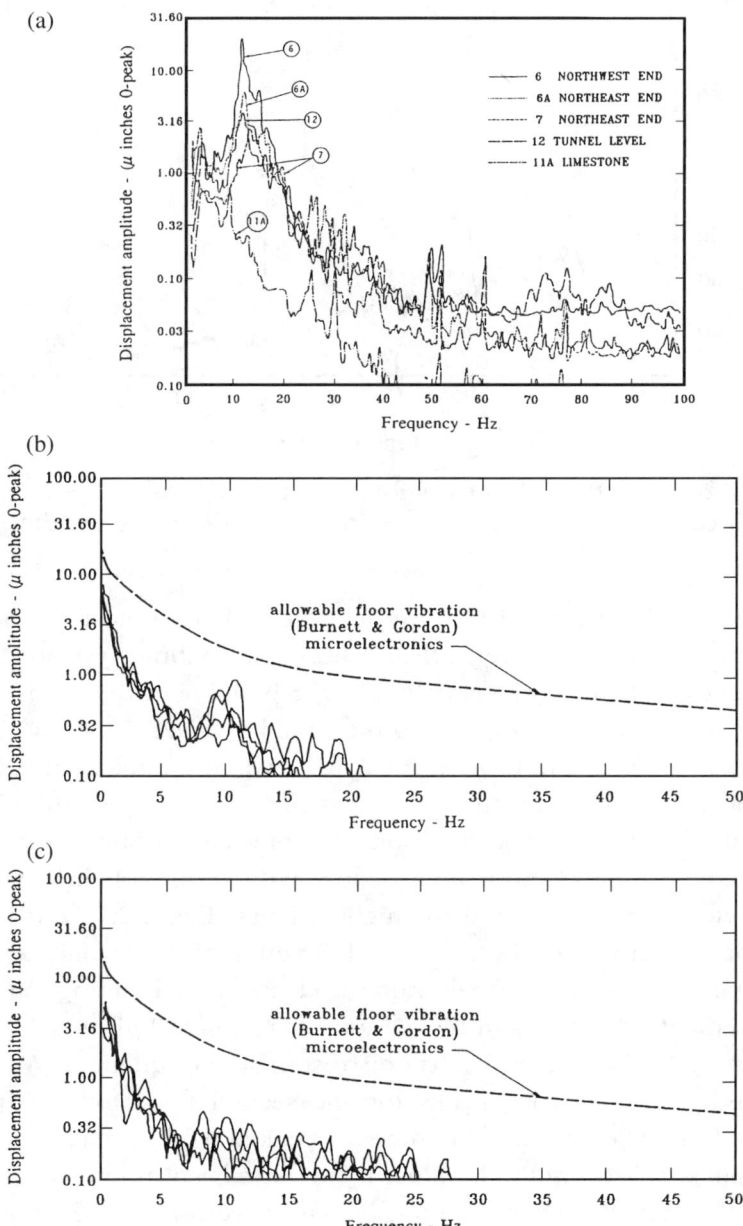

Fig. 2.2.6. Measured Vertical Vibratory Displacement Amplitudes in Various Locations of: (a) University of Minnesota Campus Building; (b) Vertical; and (c) Horizontal Vibratory Displacement Amplitudes of the EE/CS Experimental Microelectronics Fabrication Building Built in Its Place.

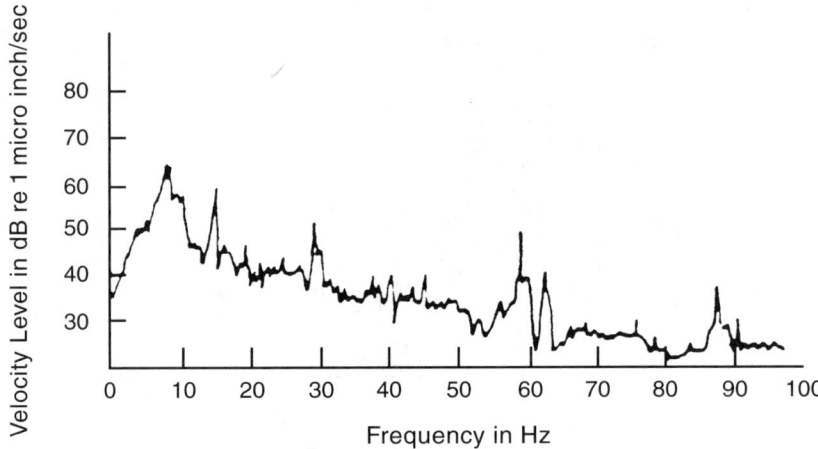

Fig. 2.2.7. A Typical Narrow-Band Vertical Vibration Spectrum at Second-Floor Level in a Steel Frame and Concrete Building.

"EE/CS building"; line 11A shows vibration of the limestone bedrock. Figs. 2.2.6(b), (c) show vertical and horizontal vibratory displacement amplitudes of the newly constructed EE/CS building [2.7]. The maximum displacement amplitudes in the former building were ~0.5 μm (20 μin.) for 12 Hz and 0.08 μm (3 μin.) for 2.5 Hz, while the maximum vertical and horizontal displacement amplitudes (for 1 Hz) in the latter case are only 0.0125 μm (0.5 μin.). Precision electronics and optical instruments (such as scanning electron microscopes with resolution in single-digit nanometers) are often used on higher floors. Fig. 2.2.7 [2.10] shows vibratory velocity levels in the vertical direction at the second-floor level of a specially designed steel frame and concrete building. Vibration displacement amplitudes in this case are 0.25 μm (10 μin.) at 1 Hz, and 0.5 μm (20 μin.) at 8 Hz (floor resonance frequency). Vibration amplitudes at resonance peaks for the second floor for L50 plot in Fig. 2.1.1 are 0.06 μm (2.4 μin.) for 8 Hz peak and 0.03 μm (1.25 μin.) for 25 Hz peak. These amplitude values might have been suppressed as much as two times due to use of the third-octave spectrum (see Section 2.2.3, Fig. 2.2.15); still these values are small even if doubled.

Even with these extremely low amplitudes of floor vibrations, the precision manufacturing and measuring equipment units are installed on low natural frequency pneumatic isolators most of the time. This type of ultra-conservative approach is adopted due to high costs of the equipment and high losses in cases of malfunctions caused by vibration transmission. However, based on analysis summarized in Section 2.2.4 (e.g., Tables 2.2–2.5), it seems that the costs of facilities construction and equipment installation can be significantly reduced without harmful effects by using realistic information for vibration sensitivity of the precision equipment.

In practical cases, notwithstanding these objective data on floor vibration levels at the microelectronics facilities, sometimes customer specifications for design of high-precision microelectronics equipment include requirements for normal operation of the equipment with floor vibration amplitudes up to 25 μm. On the other hand many machine tool manufacturers specify extremely low vibration levels at the sites where even a regular precision machine tool is to be installed, e.g. see Section 2.2.4. The required vibration levels are impractical (too expensive) to achieve and unnecessary, but reduce liability of the manufacturer if the machine does not perform properly, not necessarily due to the floor vibrations.

The most stringent ambient vibration parameters for electronic equipment are specified by military standards, such as [2.11]. According to [2.11], electronic devices must withstand, without loss of performance, a "white noise" excitation with power spectral density of acceleration $S(\omega) = 0.1$ g^2/Hz in the 10–2,000 Hz frequency range.

2.2.2 Detrimental Effects of Vibration

There are various ways in which performance of vibration-sensitive devices can be disrupted by ambient vibrations. Usually precision machines, instruments, and electronic devices comprise assemblies of components. While many connections between components in the electronic devices are made by welding or soldering, many connections in precision machines and measuring instruments, as well as in some electronic devices, are fashioned by frictional joints tightened by bolts or by other means. One of the most important detrimental effects of vibrations is destabilization of dimensional chains. Usually, elements of these chains (e.g., guideways, screws, clamps) are fastened in elastically strained conditions. The forces preventing displacements between the connected elements are static friction forces maintaining the said elastic strains. Vibrations result in variations of the contact pressures in the joints and thus in variations of

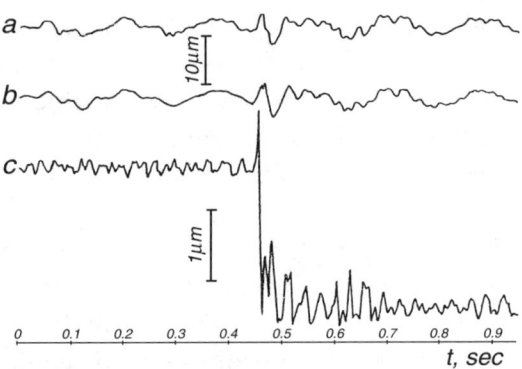

Fig. 2.2.8. Vibratory Displacement of: (a) the Floor; (b) the Bed; and (c) Between the Grinding Wheel and the Workpiece of a Surface Grinder Mounted on Rigid Wedge Mounts, During an Event of Dropping a Heavy Weight on the Floor.

friction forces, leading to relaxation of the elastic strains and to possible "jumps" in the dimensional chains. Such jumps may produce steps in machined surfaces, changes in dimensions within a batch of automatically machined parts, misalignments in optical systems, etc. Fig. 2.2.8 shows vertical vibrations of the floor (a); and the bed (b); and (c) relative vibratory displacement between the grinding wheel and the workpiece being machined on a surface grinder. Dropping a heavy chunk of metal on the floor near the machine resulted in an unrecovered shift between the grinding wheel and the workpiece. Fig. 2.2.9 shows profilograms of tooth profiles after they were ground on a gear grinder using two grinding wheels simulating the generating rack. Fig. 2.2.9(a) shows the tooth profile machined during normal floor vibration conditions (pitch variation ~10 μm), while Fig. 2.2.9(b) shows the profile machined while a heavy chunk of metal was dropped on the floor near the machine (pitch variation ~46 μm). Vibration-caused "jumps" at joints of mounting systems (e.g., wedges, jackscrews) may lead to misalignments and deformations in the machine bed, which, in turn, may result in chatter, reduced accuracy, and increased wear. The strains in the mounting systems can be reduced by using "power-assisted" wedge mounts (see Section 4.1.1).

Ambient (floor) vibrations may also induce relative motions between the critical components of the device, such as between the cutting or measuring tool and the workpiece. These motions may cause waviness in

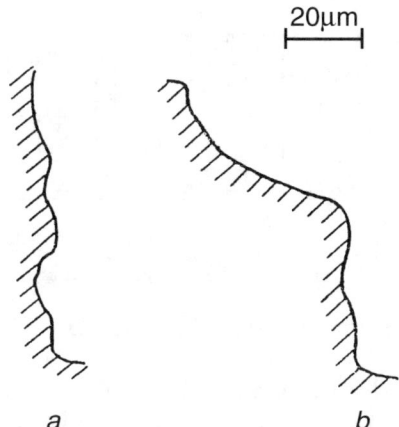

Fig. 2.2.9. Profilograms of Involute Tooth Profiles Ground During (a) the Normal Floor Condition; and (b) an Event of Dropping a Heavy Weight on the Floor.

the machined surface, out-of-roundness, erroneous measurement results, oscillatory misalignments between optical components, line-width variations in microcircuits, signal variations in proximity sensors, and deterioration in surface finish (especially for turning machines, where the waviness during successive revolutions is phase-shifted). Fig. 2.2.10 shows profilograms of parts machined on a surface grinder (with table speed 8 m/min) while the floor was excited with 21 Hz vertical vibration with double amplitude 13 μm. Fig. 2.2.10 (a) shows parts machined while the grinder was installed on rigid wedge mounts; Fig. 2.2.10 (b) shows the part machined while the grinder was installed on Vibrachok (see Chapter 4) vibration isolators. Usually, allowable vibratory displacements in the work zone are 0.1–0.2 of the rated accuracy of the equipment.

Impulsive vibrations can lead to momentary misalignments, to erroneous measurements if the vibrations take place during the measurement process, and to local variations in the depth of cut, which result in dents on the machined surfaces. Vibrations in the measuring zone, between the probe or the light source and the measured surface, can blur the measurement results and effectively reduce resolution of the measuring instrument.

It was observed, e.g. [2.12], that in ultra-precision machine tools—such as diamond-cutting lathes for machining precision optical surfaces—

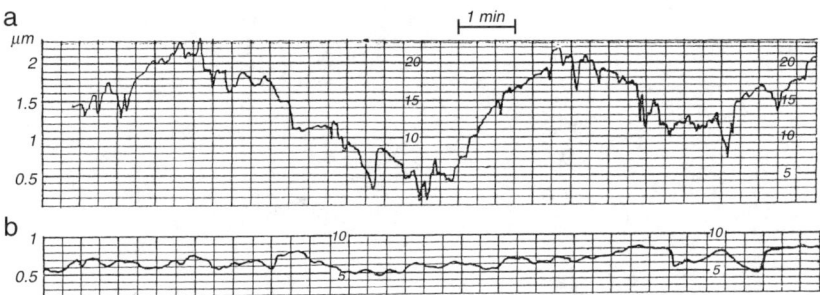

Fig. 2.2.10. Profilograms of Parts Machined on a Surface Grinder While the Floor is Excited by Vertical Vibration $f = 21$ Hz, $2a = 13$ μm, feed rate 8 m/min: Grinder is Installed on (a) Rigid Wedge Mounts; and (b) on Vibrachok Vibration Isolators.

vibratory motion in the work zone may lead to significant variations in cutting force on hard materials at finishing regimes (depth of cut < 2.5 μm on silicon, germanium, zinc selenide, etc.), thereby leading to faster wear of the diamond-cutting tool.

The study described in [2.13] involved surveying of microscope operators to check on problems caused by vibration of the microscope base. It was found that operation of a 400X microscope for hematology analyses is not influenced by horizontal (lateral) vibration with amplitudes of vibratory velocity up to 10 μm/s (400 μin./s). At 12.5–20 μm/s (500–800 μin./s) there were "definite problems," and above 31.5 μm/s (1,260 μin./s) there were "strong vibration-induced task interferences." Below 20 Hz the problem was lateral blurring of the image, above 20 Hz—"out of focus" effects. Similar problems can be observed in other precision sensing and measuring devices in which external vibrations can cause parasitic motions in the working zone, eg. see [2.14].

Environmental vibration may affect the magnet phase stability and, consequently, image quality of the magnetic resonance imaging (MRI) systems widely used for medical diagnostics.

In some cases, absolute vibrations of critical components are objectionable, e.g., vibrations of "quad" magnets in synchrotrons, which can cause deviations of the generated particle beam, e.g., [2.15]. If an electrical component, such as a capacitor (especially in power supplies), inductor, etc., is vibrating near the resonance, the leads can fatigue/break. Cable vibrations can be excited by vibrations in mechanical systems, which can lead to broken wires—e.g., cables in CNC controls of machine tools excited by chatter or forced vibrations in the machine.

2.2.3 Model of Vibration Transmission

Floor vibrations are detrimental to the performance of precision production and measuring equipment because they excite relative vibrations in the working (cutting/measurement) zone or in the joints comprising the dimensional set-up chain. Tests have shown that the effects of floor vibrations on a machine tool or a CMM usually are significant only at the lowest (seldom also at the second) natural frequency of the dynamic system, largely because the higher natural frequencies lie considerably above the frequency range of significant excitation amplitudes. Also, floor/supporting structures vibrations at the higher frequencies usually undergo much greater attenuation by mounting/vibration isolation systems.

For example, the first natural frequency for grinders typically occurs in the 30–70 Hz range, whereas the second natural frequency lies in the 100–150 Hz range. Thus, the higher natural frequencies are in the frequency range of greatly reduced floor amplitudes (e.g., see Fig. 2.2.2). While more intense high-frequency vibrations may be observed on supporting structures of vehicles (spacecraft, aircraft, surface vehicles), the vibration-sensitive on-board devices usually have higher natural frequencies, e.g. see [2.14]. Figs. 2.2.11 to 2.2.13 show vibration-sensitivity curves for precision equipment used in the production of electronic circuits, and Fig. 2.2.14 shows a vibration-sensitivity plot for a high-resolution (X1000) optical microscope [2.8], [2.9]. These plots show what amplitude of vibration at the given frequency results in a relative vibration amplitude in the working zone not exceeding the tolerated amplitude. The minima on these plots represent structural natural frequencies of the devices. It can be seen that the lowest natural frequencies are between 4 and 20 Hz, and many of the higher natural frequencies lie between 30 and 70 Hz.

Fig. 2.2.15 [2.5] repeats the lateral (side-side) sensitivity curve from Fig. 2.2.14. To obtain this plot, the microscope was excited by either a sweep of tonal (narrow frequency band) vibration or by broadband random vibration filtered to one-third octave bands. At each frequency setting, the amplitude was increased until image disturbance became obvious. The solid line shows vibration sensitivity from the tonal sweep

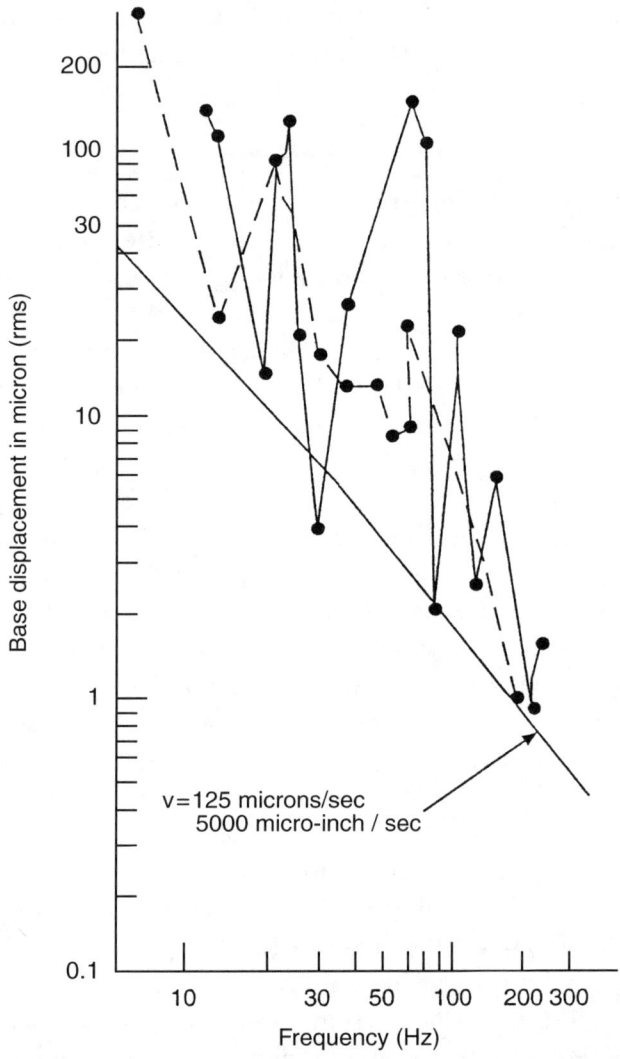

Fig. 2.2.11. Vibration Sensitivity for Projection Aligner Perkin-Elmer Microlign Mod. 341 for 0.1 μm Image Motion (Solid Line—Limit of Vertical Floor Vibration Amplitude, Broken Line—Limit of Horizontal Floor Vibration Amplitude).

and the dots mark the results of the third-octave band excitation. It can be concluded that the sensitivity plot obtained by using wideband (third-octave) excitation substantially distorts the sensitivity data and cannot be used for synthesizing a vibration protection system.

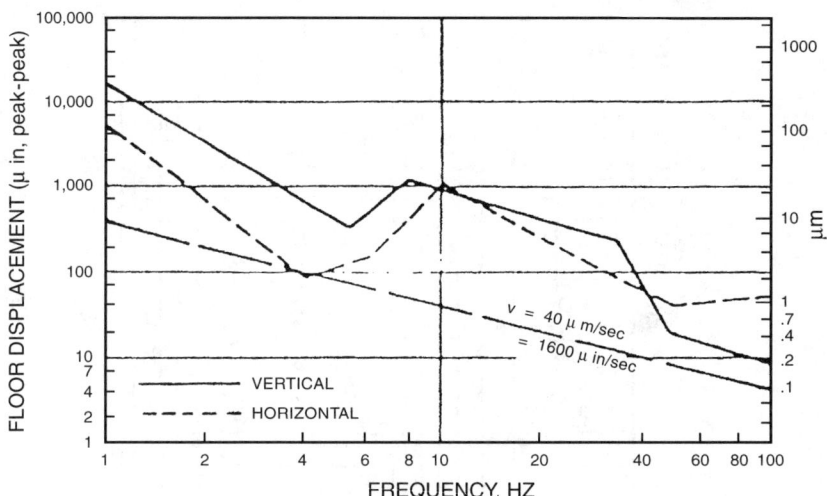

Fig. 2.2.12. Vibration Sensitivity for Mann Wafer Stepper 4800 DSW.

The approximate dynamics of vibration transmission from the floor/supporting structure into the working zone of a precision device can be modeled by three uncoupled two-mass systems (with generalized parameters),—such as the one in Fig. 2.2.16,—one for each coordinate direction of the floor vibrations. Masses and springs in the Fig. 2.2.16 model represent generalized inertias and stiffnesses of the unit, and their values may be different for each coordinate direction. In Fig. 2.2.16, M_b represents mass of the frame (bed) of the object; M_u is the mass of its upper structure (e.g., tool head or measuring head); k_m and c_m represent equivalent stiffness and damping of the internal structural components and joints; and k_v, c_v are stiffness and damping of the mounting devices (e.g., jacks or vibration isolators).

The effect of vibrations on an equipment unit represented by this type of two-mass model may be described in terms of the ratio of the generalized relative displacement amplitude X_{grel} between masses M_u and M_b to the displacement amplitude X_b of the frame M_b. This ratio is given by the following [see Eq. (1.4.11)]:

$$\frac{X_{grel}}{X_b} = \frac{f^2/f_m^2}{\sqrt{\left[1 - \frac{f^2}{f_m^2}\right]^2 + \frac{\delta_m^2}{\pi^2}\frac{f^2}{f_m^2}}} = \mu_{rel}, \qquad \text{Eq. (2.2.1)}$$

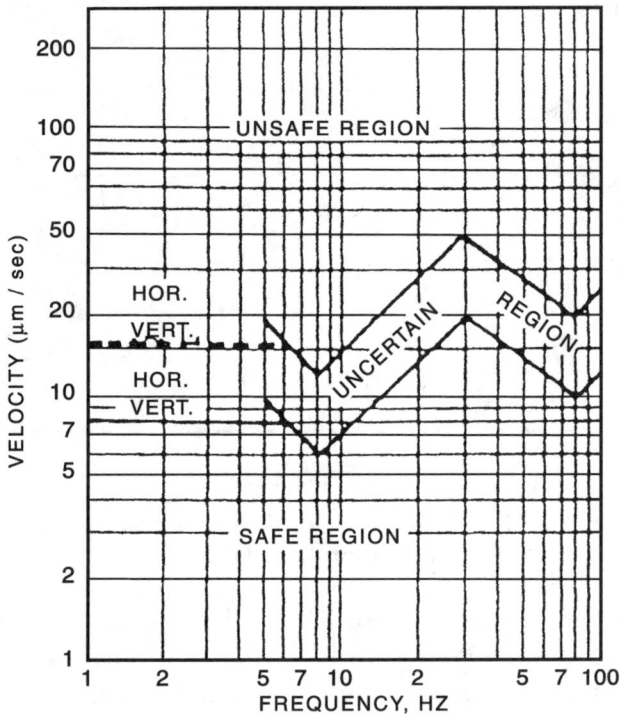

Fig. 2.2.13. Vibration Sensitivity for Philips Electron Beam Pattern Generator Beamwriter EBPG-4.

where f represents frequency of interest, $f_m = 1/2\pi\sqrt{k_m(M_u + M_b)/M_uM_b}$ is the resonance frequency of the partial subsystem representing the upper structure; and δ_m is the log decrement of the upper structure system.

The actual relative displacement in the working area of the equipment unit is proportional to X_{grel},

$$X_{rel} = \gamma \, X_{grel} \qquad \text{Eq. (2.2.2)}$$

and the proportionality constant γ (design constant) depends on the geometry of the device. For example, for an upper structure that can be represented by the model shown in Fig. 2.2.17,

$$\gamma_{zx} = M_u b_1 c / I_{O1} = b_1 c / \left(\rho_y^2 + b^2 + c^2\right) \qquad \text{Eq. (2.2.3)}$$

$$\gamma_{zz} = M_u cd / I_{O1} = cd / \left(\rho_y^2 + b^2 + c^2\right) \qquad \text{Eq. (2.2.4)}$$

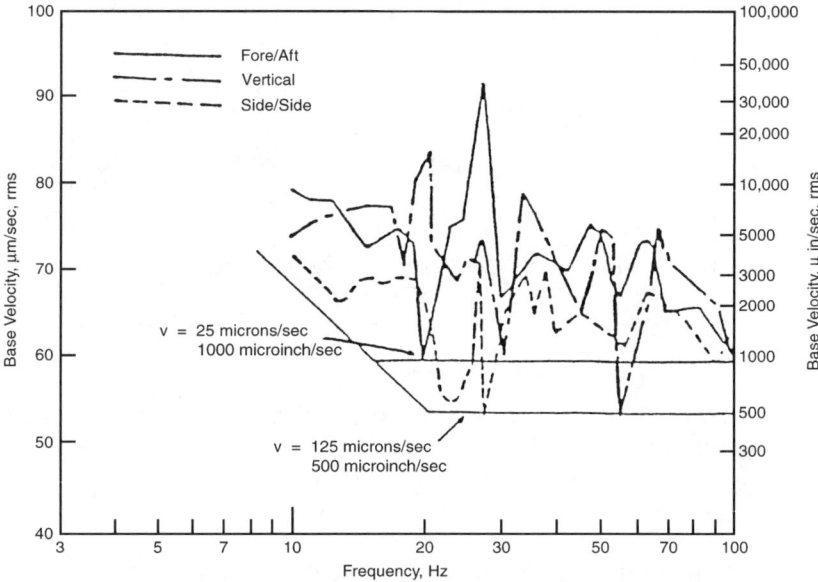

Fig. 2.2.14. Vibration Sensitivity for Nikon 1000X Optical Microscope for Detectable Motion on 1-μm Test Line.

where the first subscript on γ refers to direction of the bed/frame vibration, and the second subscript refers to direction of motion of the upper structure relative to the bed. Here M_u again is mass of the upper structure; I_{O1} its mass moment of inertia about the actual axis of rotation; and ρ_y is the radius of inertia about the principal y-axis. The dimensions $b, b_1, c,$ and d, as indicated in Fig. 2.2.17, are measured from the center of gravity O_1 of the system. Obviously, γ depends upon how the device/machine is set up, and particularly on the position of the upper structure with respect to the bed. In the case of Fig. 2.2.17, the upper structure is connected to the bed by a spring allowing only angular deformation and having infinite vertical stiffness. This is a good approximation for structural designs of numerous machine tools and measuring systems, because guideways usually have a very high translational stiffness perpendicular to the motion direction, but their angular deformations may have significant magnitudes, e.g. see [2.16]. Sensitivity of horizontal displacements in the work zone of such a model to vertical vibrations of the bed (y_{zx}) can be reduced by reducing structural (design) dimensions c and b, by reducing overhang, or by adding counterweights to the back of the

128 • PASSIVE VIBRATION ISOLATION

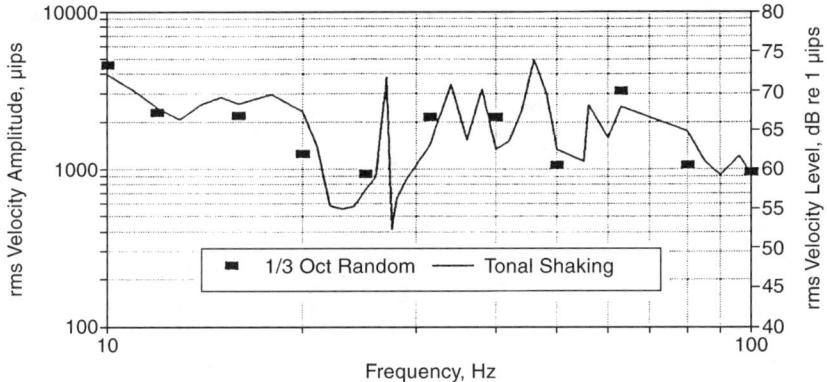

Fig. 2.2.15. Vibration Sensitivity of 1000X Microscope in Lateral (Side–Side) Direction Measured with Narrow-Band and One-Third Octave Excitations.

structure. Sensitivity of vertical displacements in the work zone to vertical vibrations of the bed (y_{zz}) can be reduced by reducing structural (design) dimensions c and d.

Experimentally obtained data for several machine tools are given in Table 2.1[2.6], which shows that the vibration sensitivity can be reduced by proper selection of design parameters, because similar machines in Table 2.1 have very different sensitivities to vibrations of their beds. It can be seen, for example, that in the group of cylindrical grinders γ_{yy} varies from 0.021 (low) to 0.6 (very significant).

It is reported in [2.17] that circuit cards in electronic equipment may have resonant frequencies as low as 12 Hz and up to 80 Hz, while hard disc/floppy disc drive assembly resonates at 29 Hz, which is more or less the same frequency range as for precision machine tools.

The generalized amplitude ratio (transmissibilitry) $X_{grel}/X_b = \mu_{rel}(f)$ from Eq. (2.2.1), which is a measure of the vibration sensitivity of the object as a function of frequency f, does not depend upon the object's mounting, only on its design (structural natural frequency f_m and masses M_b, M_u). To obtain a ratio between amplitudes of the relative vibrations within the working zone of the object and amplitudes of the floor vibrations, X_{rel}/X_b should be multiplied by the ratio of the bed vibration amplitudes to the floor vibration amplitudes (transmissibility of the mounting/isolation system), assuming that the dynamic coupling between the object structure and the vibration isolation system is weak, and that

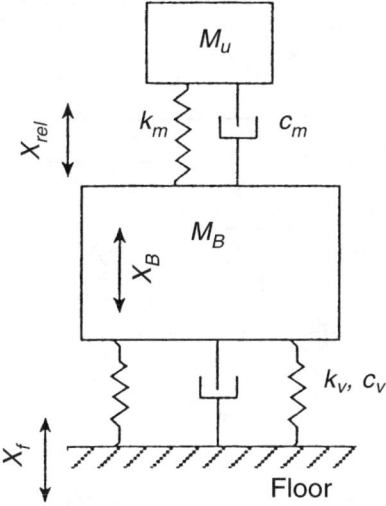

Fig. 2.2.16. Two-mass Dynamic Model for Vibration Sensitivity of Precision Object in One Principal Direction.

these systems can be considered to be independent, (see Section 1.5). This assumption is justified since, in most practical cases, natural frequencies of the isolated bed in the three coordinate directions f_{vx}, f_{vy}, and f_{vz}, are considerably lower than the structural natural frequencies f_m (see Table 2.1), and the mass of the bed M_b is usually much larger than the mass of the upper structure, M_u.

2.2.4 Principles and Criteria of Vibration Isolation

Principles and criteria of vibration isolation of high-precision and/or vibration-sensitive equipment can be considered for three typical cases: The first case deals with protection of a specific unit of equipment from specific steady-state vibrations of the floor or some other supporting structure. Since the steady-state vibrations can be represented by a discrete spectrum, this case is essentially one of isolation of the unit of equipment from sinusoidal vibration of the floor. The second case represents protection of a specific unit of equipment from any combination of typical quasi-steady-state floor vibrations, e.g., within the envelopes represented in Fig. 2.2.2. Although the plots in Fig. 2.2.2 envelop a frequency range of disturbing vibrations, only some spectral components at any particular site may have amplitudes as high as the plot indicates. If a vibration

130 ● **PASSIVE VIBRATION ISOLATION**

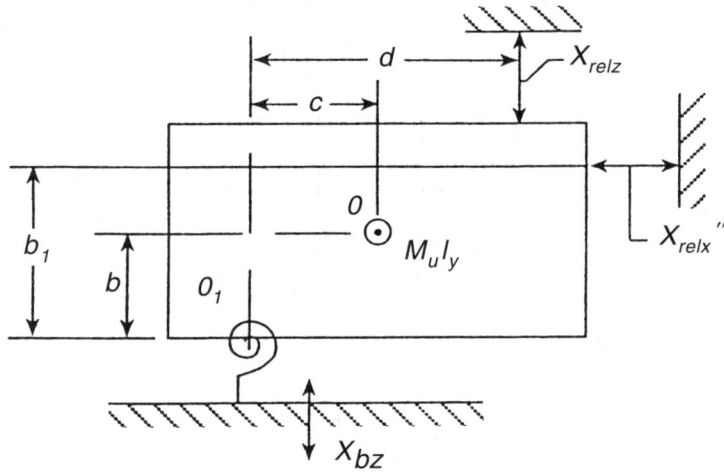

Fig. 2.2.17. Dynamic Model for Influence of Design Parameters on Relative Vibration in Working Zone.

isolation system protecting the unit of equipment from any realization of the floor vibration spectra represented by plots in Figs. 2.2.2(a) and (b) is synthesized, then the unit can be equipped with isolating mounts that would ensure its specified performance for the majority of installation sites. The third case deals with protection of equipment from transient motions of the floor caused by regular or accidental shock excitations. It should be noted that poor understanding of the principles of vibration isolation leads to statements like the following quotation from the manual for OD grinders of a regular precision, similar to ones listed in Table 2.1, produced by one of the large and reputable grinding machine builders: "Subjecting the grinder to greater than 0.00001in. (0.254 μm) peak-to-peak vibrations *at any frequency* [emphasis mine, E.R.] could adversely affect the quality of the finished work part." This requirement precludes installation of the grinders in any facility with a vibration level higher than that specified by VC-E (the most stringent) criterion in Fig. 2.2.4.

2.2.4A Isolation from Steady-State Vibration
In the first case, illustrated by Fig. 2.2.18, the degree of attenuation μ_v of one frequency component f of floor vibrations by the vibration isolation system (system: "frame-isolating mounts") is determined by the frequency

ratio f/f_v where $f_v = 1/2\pi\sqrt{k_v/(M_b + M_u)}$ is the natural frequency of the vibration isolation system in the considered direction. The degree of attenuation (transmissibility) for not-very-high viscous friction and for not-very-high structural (hysteretic) damping are expressed by Eqs. (1.4.6) and (1.4.22), respectively.

Thus, from the two sine waves at frequencies f_1 and f_2 having the same magnitudes, the sine wave having frequency f_1 is attenuated less than the sine wave having higher frequency f_2 (and, thus, higher ratio f_2/f_v) (see Fig. 2.2.18). When there is viscous damping in the isolators, increasing the damping (which reduces the resonance peak, i.e., sensitivity to low-amplitude "background noise" and reduces excursions of the unit caused by accidental impacts and during start/stop regimes) leads to deterioration of isolation at higher frequencies in accordance with Eq. (1.4.6). In case of structural damping, increasing damping would lead to only marginal deterioration of isolation while playing the aforementioned positive roles, in accordance with Eq. (1.4.22).

For the second case the principles involved in isolation of a precision object against floor vibrations are illustrated in Fig. 2.2.19. The bottom-most graph here shows the spectrum of floor displacement amplitudes X_f in a given direction. This graph is a solid line envelope, from that of Fig. 2.2.2 for the considered direction, with its height designated a_o. The low-frequency "ascending" part of Fig. 2.2.2a is caused by measurement errors, so it does not represent the real vibration environment and is omitted. The top-most graph shows the corresponding relative motion in the object's critical (working) area, with Δ_o representing the limit of acceptability of such motions (usually, independent of frequency but determined by the specified accuracy of the object). For precision machine tools and, to a lesser degree, for precision measuring equipment and precision processing equipment for electronics production, the dynamic coupling between the machine structure and the vibration isolation system is relatively weak (see above and Section 1.5). Consequently, the approximate relative motion amplitudes vs. frequency characteristic can be obtained simply by multiplying the X_{rel}/X_b, X_b/X_f and X_f graphs, and the vibration isolation criterion can be expressed as follows

$$a_o \frac{X_b}{X_f} \frac{X_{rel}}{X_b} < \Delta_o \quad \text{or} \quad a_o \frac{X_b}{X_f} \mu(f) < \Delta_o \qquad \text{Eq. (2.2.5)}$$

132 • PASSIVE VIBRATION ISOLATION

Table 2.1. Parameters for the Design of Vibration Isolation Systems for Selected Machine Tools

	γ_{xy}	γ_{zz}	γ_{yy}	γ_{yz}	γ_{xy}
Internal Grinders					
Jotes SOC-100 (M = 1750 kg, D_{max} = 100 mm)	0.025	0.043	0.85	0.25	0.32
Jung (M = 2500 kg, D_{max} = 175 mm)	–	0.48	0.65	0.28	–
Heald 72A (M = 2400 kg, D_{max} = 115 mm)	–	–	0.48	0.4	0.18
Circular Grinders					
Landis (M = 5100 kg, D_{max} = 410 mm)	–	–	0.25	–	0.25
Jotes SWA-25 (M = 3500 kg, D_{max} = 250 mm)	–	–	0.32	–	0.1
Fortuna (M = 3800 kg, D_{max} = 225 mm)	–	–	0.2	–	0.002
Mipsa RUS-450 (M = 2000 kg, D_{max} = 240 mm)	0.26	–	1.5	–	0.45
Mipsa (1948) (M = 2000 kg, D_{max} = 240 mm)	–	–	0.021	–	0.009
Hartex KH620 (M = 2100 kg, D_{max} = 310 mm)	0.12	–	0.35	–	0.08
3A151 (M = 3800 kg, D_{max} = 200 mm)	0.1	–	0.6	–	0.17
3151	0.04	–	0.12	–	0.18
Spline Grinders					
3451G (M = 6500 kg, D_{max} = 125 mm)	–	0.43	–	0.25	–
Surface Grinders					
371M1 (M = 1900 kg, Table 200 × 600 mm)	–	0.52	–	–	–
Jung	–	–	–	0.95	–
3B71 (M = 200 kg, Table 200 × 600 mm)	–	0.024	–	0.45	–
3B722 (M = 6800 kg, Table 320 × 1000 mm)	–	0.28	–	3.5	–
Gear Grinders					
Maag HSS-30	0.12	–	0.45	–	0.55
5831 (M = 4500 kg, m_{max} = 6 mm)	–	–	0.64	–	–
584M (M = 6000 kg, m_{max} = 6 mm)	0.083	–	0.66	–	0.6
Jig Borers					
2B440	0.13	0.39	0.29	0.52	0.39
Diamond Boring Machines					
2A715 (M = 2600 kg, D_{max} = 200 mm)	0.044	0.044	1.7	1.7	–
2706 (M = 2600 kg, D_{max} = 200 mm)	0.23	0.23	0.95	0.95	–
Universal Grinders					
Union (M = 3600 kg, $D_{spindle}$ = 60 mm)	0.15	–	0.14	–	0.045
Lathes					
1K62 (M = 2300 kg, D_{max} = 400 mm)	0.27	–	0.29	–	–
Gustloff A5 (D_{max} = 400 mm)	0.18	–	0.6	–	–
TC-135M (M = 1100 kg, D_{max} = 270 mm)	0.2	–	0.13	–	–

KEY:
$\gamma_{ij} - X_{rel}/X_{relg}$
x_{rel} — actual relative displacement in work area
X_{grel} — generalized relative displacement
 — displacement of upper mass M_u relative to base mass M_B in equivalent 2-mass system (Figure 2.2.16)
$i,j - x, y, z$
i — coordinate direction of motion of bed
j — coordinate direction of motion of upper structure
f_m — resonance frequency of upper system ($M_u - K_m - M_\gamma$ Figure 2.2.16)

γ_{xx}	f_m (Hz)	Δ_o (μm)	Φ_z (Hz)	Φ_y (Hz)	Φ_x (Hz)	f_{vz} (Hz)	f_{vx}/f_{vz}	f_{vx}/f_{vz}	f_{vp} (Hz)
–	70	0.16	45.5	11.5	18.3	32.3	0.4	0.25	32.3
–	70	0.5	25	23.8	–	17.6	–	0.95	15
0.45	60	0.16	–	12.9	12.5	–	–	–	–
–	50	0.16	–	14.8	47.3	–	–	–	–
–	45	0.16	–	12.0	21.3	–	–	–	–
–	40	0.16	–	13.4	13.4	–	–	–	–
–	100	0.16	26.5	11.2	21.3	18.7	0.8	0.42	16.3
–	60	0.16	–	41.3	62.6	–	–	–	–
–	80	0.16	31.5	19.4	41	22.3	1.3	0.61	18.5
–	40	0.32	24.2	10.4	19.5	17.2	0.8	0.43	12.8
–	50	0.32	48	28.9	23.6	34.0	0.49	0.6	32.4
0.11	57	0.64	23.6	33	49.7	16.6	2.1	1.4	14.4
–	60	0.64	22.6	–	–	16.0	–	–	14.1
0.65	50	0.32	–	11	13.3	–	–	–	–
0.11	28	0.32	34.6	8.4	17	24.5	0.49	0.24	17.9
0.21	70	0.32	25.4	7.7	30.7	18.0	1.2	0.3	15.5
–	32	0.5	22.7	12.3	11.1	16.0	0.49	0.54	11
–	45	0.5	–	16.5	–	–	–	–	–
–	35	0.64	33	12.1	12.9	23.2	0.39	0.37	16.7
1.55	45	0.5	17.3	4.9	9.3	12.3	0.53	0.28	10.4
–	90	0.08	43	7.4	–	30.3	–	0.17	30.1
–	100	0.16	28.4	14.4	–	20	–	0.51	17.1
–	40	0.64	28	30.7	52.3	19.7	1.9	1.1	21
–	75	0.64	39	39	–	22.5	–	1.0	29.5
–	70	0.64	44.5	25.8	–	31.4	–	0.58	31.3
–	80	0.32	34.2	44.5	–	24.1	–	1.3	26.4

Δ_o—maximum relative displacement allowed in work area
Φ_z—maximum acceptable value of $f_{vi}\sqrt{\delta_{vi}}$ (vibration isolation criterion)
f_{vi}—resonance frequency of mounted (isolated) system for motion in i direction
δ_{vi}—logarithmic decrement associated with above
f_{vp}—maximum resonance frequency of vertical (z-direction) motion of isolated system for adequate
   vertical isolation of pulse (shock) motions (value of f_{vp} shown in lowest of the two calculated for
   pulses with $a_1 = 50$ μm, $r = 0.05$ s and $a_1 = 30$ μm, $r = 0.025$ s)
M—total mass of machine
D_{max}—greatest overall dimension
m_{max}—maximum module of machined gear ($m = 25.40/DP$ mm where DP—dimetrical pitch of gear)

134 • PASSIVE VIBRATION ISOLATION

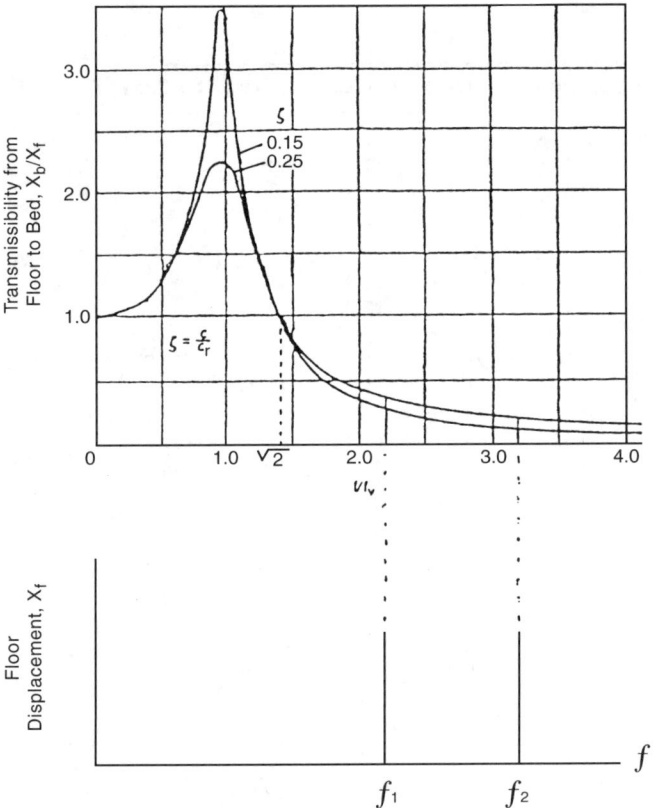

Fig. 2.2.18. Vibration Isolation from Single-Frequency Excitation.

The solid lines (a) in Fig. 2.2.19 correspond to a stiff mounting system under the machine, the dashed lines (b) to a softer isolation system with the same damping as that of the stiffer system, and the dotted lines (c) to a softer isolation system with increased damping. The set of graphs on the left-hand side pertains to viscous damping in the isolators, for which greater damping results in greater vibration transmission at frequencies above $\sqrt{2}f_v$; the graphs on the right-hand side pertain to hysteretic damping in the isolators, which affects the high-frequency vibrations transmissibility into the working area less significantly. It is clear that relative vibrations in the critical area of the object are reduced by using a softer mounting with greater damping. It is important to note that *a resonance of the object on vibration isolators is allowable*, provided that the isolators are selected in a manner such that amplitude of relative

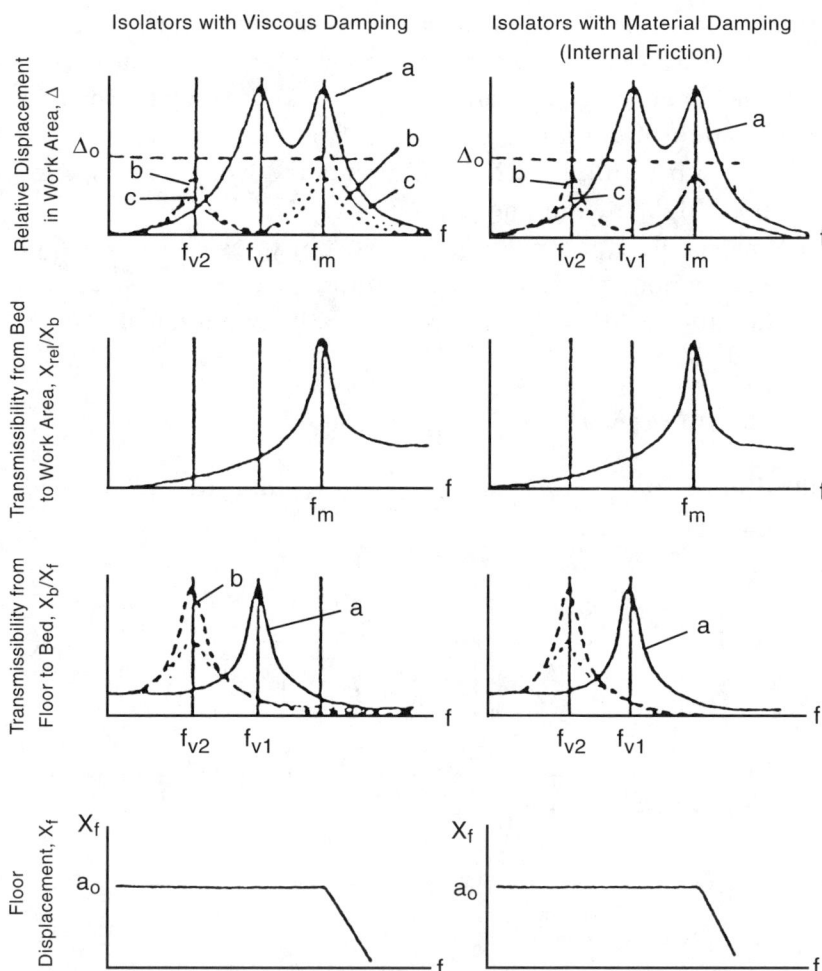

Fig. 2.2.19. Vibration Isolation for Broad Spectrum Excitation.

vibrations X_{rel} at the resonance of the vibration isolation system does not exceed the specified allowable level Δ_o. Accordingly, improvements in isolation can be achieved by reducing f_v or by increasing damping δ_v. While relative displacements X_{rel} at the structural natural frequency f_m increase with the increasing damping in isolators, diminishing amplitudes of floor vibrations at higher frequencies, together with the above-noted features of the transmissibility curve for the isolation system with hysteretic damping, make the resonance at f_m less dangerous than one at f_v.

136 ● PASSIVE VIBRATION ISOLATION

Use of the generalized spectrum of floor vibrations, as shown in Fig. 2.2.19, is a more conservative approach than using actual vibration measurements at the equipment installation site. The latter can change after relocation of old and installation of new equipment, whereas the spectrum assumed in Fig. 2.2.19 represents the worst case of Fig. 2.2.2 based on multiple measurements.

The concept illustrated in Fig. 2.2.19 can be expressed analytically [2.6]. If dynamic coupling between the vibration isolation system and the object structure is neglected, then the relative vibration in the critical zone is as follows:

$$X_{rel}(f) = X_f(f) X_b(f)_{X_f = const} \; X_{rel}(f)_{X_b = const} \qquad \text{Eq. (2.2.6)}$$

Substituting Eq. (1.4.22) for $X_b(f)_{X_f = const}$ into Eq. (2.2.6) yields the following:

$$X_b(f)_{X_f = const} = \mu_v = \sqrt{\frac{1 + (\delta_v/\pi)^2}{\left[1 - (f/f_v)^2\right]^2 + (\delta_v/\pi)^2}} \qquad \text{Eq. (2.2.7)}$$

Substituting into Eq. (2.2.6), Eqs. (2.2.1) and (2.2.2) for $X_{rel}(f)_{X_b = const}$ give the following:

$$\frac{X_{rel}(f)}{X_b} = \gamma \frac{f^2/f_m^2}{\sqrt{\left(1 - f^2/f_m^2\right)^2 + \delta_m^2/\pi^2 \cdot f^2/f_m^2}} = \mu_f \qquad \text{Eq. (2.2.8)}$$

Relative vibrations in the critical area can be computed and compared with the allowable amplitude Δ_o if all the parameters in Eq. (2.2.6) are known. In Eq. (2.2.8) μ_f is the transmissibility from the bed into the critical (working) area at frequency f. It is obvious that if $f \ll f_m$ and $f_1 \ll f_m$ then

$$\mu_{f_1} \approx \mu_f \left(f_1^2/f^2\right) \qquad \text{Eq. (2.2.9)}$$

Since usually $\delta_v \ll \pi$; $\delta_m \ll \pi$; $f_v \ll f_m$, then at the resonance frequency f_v of the vibration isolation system:

$$X_{rel}(f_v) \approx X_f \mu_{f_v} (\pi/\delta_v) \qquad \text{Eq. (2.2.10)}$$

The structural dynamic characteristic of the machine is determined by μ_f, which is measured at one frequency f for which $f \ll f_m$. Then, because it is also assumed that $f_v \ll f_m$, the transmissibility μ_{f_v} can be expressed as follows:

$$\mu_{f_v} \approx \mu_f (f_v^2 / f^2) \qquad \text{Eq. (2.2.11)}$$

Substituting Eq. (2.2.11) into (2.2.10), the relative vibration amplitude in the work zone at f_v (the most dangerous case) is expressed as follows:

$$X_{rel}(f_v) \approx X_f \mu_f (f_v^2 / f^2)(\pi/\delta_v) \qquad \text{Eq. (2.2.12)}$$

If the allowable relative vibration amplitude Δ_o in the work zone is given, then the vibration isolation is effective if

$$X_{rel}(f_v) < \Delta_o . \qquad \text{Eq. (2.2.13)}$$

After substituting Eq. (2.2.12) into Eq. (2.2.13), this can be reformulated as inequality

$$\Phi_p^2 = f_v^2 / \delta_v < \frac{\Delta_o f^2}{\pi X_f \mu_f}, \qquad \text{Eq. (2.2.14)}$$

where

$$\Phi_p = \frac{f_v}{\sqrt{\delta_v}} = \frac{\sqrt{k_v/(M_b + M_u)}}{\delta_v} \qquad \text{Eq. (2.2.15)}$$

is the *vibration isolation criterion* for precision machinery and other vibration-sensitive objects. The criterion Φ_p should be used instead of the natural frequency. It indicates that with the assumptions made above, improvement in isolation can be achieved either by reducing the natural frequency f_v or by increasing the isolation system damping δ_v. Such criteria can be specified for each coordinate direction of the isolation system. The validity of the criterion Φ_p is confirmed by practical data reviewed in Section 4.4.1.

This criterion also allows for a meaningful selection of resilient materials for vibration isolators. Both stiffness and damping of such materials usually depend upon amplitude a and frequency f of vibrations (see Chapter 3). The stiffness can be expressed as

$$k = K_{dyn}(a,f) k_{st}, \qquad \text{Eq. (2.2.16)}$$

where k_{st} is the static stiffness of the isolator and K_{dyn} is the dynamic stiffness coefficient. Substitution of Eq. (2.2.16) into Eq. (2.2.15) yields the following:

$$\Phi_p = \sqrt{\frac{k_{st}}{M_b + M_u}} \sqrt{\frac{K_{dyn}(a,f)}{\delta_v(a,f)}} \qquad \text{Eq. (2.2.17)}$$

It is clear from Eq. (2.2.17) that the static stiffness k_{st} of isolators for a given Φ_p can be increased (thus improving the structural stiffness of the isolated machine and reducing its excursions/rocking) by reducing K_{dyn}/δ_v. Since both K_{dyn} and δ_v depend upon frequency and, especially, amplitude of vibrations, their ratio is different for different materials and for different vibration parameters. Table 4.1 in Chapter 4 gives measured values of this ratio for some materials depending on vibration amplitudes [2.6]. It can be seen from Table 4.1 that selection of materials for vibration isolators for low ambient vibration amplitude conditions should be very different from that for isolators for high amplitude conditions.

Eqs. (2.2.14) and (2.2.15) were derived for generalized ambient vibration environment as shown in Fig. 2.2.2, which can be described as narrow band constant intensity random vibratory displacement. Vibration Criteria (VC) for the precision equipment installation in Fig. 2.2.4 represent "medium band constant magnitude" vibratory velocity starting from 8 Hz; below 8 Hz, these criteria are somewhat between constant amplitude vibratory displacement and constant amplitude vibratory velocity. Military test requirements for the vibration-sensitive devices specified in [2.11] can be described as "broad band white noise" vibratory acceleration, starting from 10 Hz. Because of this, vibration isolation criterion Φ_p is overconservative for VC and MIL specifications. It is much more difficult to provide vibration isolation for the low-frequencies emphasized in Fig. 2.2.2, than for diminishing amplitudes of vibratory displacements at higher frequencies, as emphasized by Fig. 2.2.4 and, especially, by MIL specifications. It can be seen from Tables 2.2, 2.3, 2.4, 2.5 that at high frequencies, requirements for limits shown in Fig. 2.2.2 are much more stringent (much lower natural frequencies required) than for limits shown in Fig. 2.2.4. Thus, the criterion formulated in Eq. (2.2.15) seems to be a conservative one for various applications.

A special comment can be made about vibration isolation of coordinate-measuring machines (CMM). Many CMM designs have high sensitivity to rocking motion. Accordingly, high angular stiffness installations (e.g., see Section 4.7) might be desirable for such units.

2.2.4B Practical Selection of Vibration Isolation Parameters for Precision Objects

Table 2.1 lists acceptable values of Φ_p for the three coordinate directions for various precision machine tools, together with the corresponding permissible relative amplitudes Δ_o, the values of which were chosen to correspond to one-half of the specified precision of the part being machined on a given machine tool. Floor vibration levels as presented in Fig. 2.2.2 (limiting solid lines) were used to compute Φ_p. Values of the natural frequency ratios f_{vx}/f_{vz}, f_{vy}/f_{vz} can be used to determine the required stiffness constant ratios $\eta_{x,y}$ of isolators in the horizontal and vertical directions [see Eq. (1.6.7) in Chapter 1]. For representative precision machine tools, the typical optimal stiffness constant ratios are $\eta_x = 0.5$–1.0; $\eta_y = 1.5$–2.0. Data in Table 2.1 were validated by successful installation of many thousands of machine tools of the listed models, as well as other similar models.

If vibration sensitivity $\mu(f)$ of a precision object is known (for example, from the experimentally obtained plots like ones shown in Figs. 2.2.11–2.2.14), then Eq. (2.2.14) can be used for specifying vibration isolation parameters. For the case of Fig. 2.2.11, $\Delta_o = 0.1$ μm is specified, and for the case of Fig. 2.2.14, $\Delta_o = 1.0$ μm is specified. Assuming $\Delta_o = 0.1$ μm also for the cases in Figs. 2.2.12 and 2.2.13, Eq. (2.2.14) can be used to find the required parameters of isolation mounts for the respective units.

Table 2.2 gives the values of $\mu(f)$ (Δ_o divided by the ordinate of the plot for a given frequency) calculated for critical points from the plots in

Table 2.2. Vibration Isolation Synthesis for Fig. 2.2.11

A. Vertical direction (Y-axis)								
f, Hz	11	12	20	25	30	32	41	70
$\mu(f)$	0.0083	0.010	0.087	0.0091	0.056	0.303	0.05	0.0077
Φ_1, Hz	4.51	12.3	7.0	26.9	13.0	6.3	22.5	128
Φ_2, Hz	12.9	36.6	26.9	116	61	29.7	106	601
B. Horizontal direction (X-axis)								
f, Hz	7	12	22	65	70	100		
$\mu(f)$	0.0033	0.05	0.125	0.071	0.090	0.090		
Φ_1, Hz	13.7	6.05	22.3	49.6	49.2	84		
Φ_2, Hz	23.1	37.5	78	174	172	294		

Table 2.3. Vibration Isolation Synthesis for Fig. 2.2.12

A. Vertical direction (Y-axis)					
f, Hz	5	7.5	35	45	100
$\mu(f)$	0.011	0.0043	0.015	0.25	0.56
Φ_1, Hz	4.9	11.8	32.4	11.6	25.6
Φ_2, Hz	6.7	24.6	152	55	120
B. Horizontal direction (X-axis)					
f, Hz	4	6.5	10	45	100
$\mu(f)$	0.05	0.033	0.0043	0.1	0.5
Φ_1, Hz	2.02	4.03	17.2	24.1	35.6
Φ_2, Hz	2.3	6.2	43	84	125

Fig. 2.2.11 for the vertical and horizontal directions, respectively. The table also contains values of Φ_{p1} calculated for these points using Eq. (2.2.14) and assuming, in accordance with Fig. 2.2.2, that for vertical direction $X_f(f) = const = 3.0$ μm for frequencies 3–30 Hz and $X_f(f) = 3.0(30/f)$ μm for frequencies $f > 30$ Hz; for the horizontal direction, $X_f(f) = const = 2.5$ μm for frequencies 2–20 Hz, and $X_f(f) = 2.5(20/f)$ μm for frequencies $f > 20$ Hz. The values of Φ_{p2} were calculated using floor vibration levels corresponding to line VC-B in Fig. 2.2.4 (both for the vertical and horizontal directions). Values of Φ_{p1} are calculated only for comparison, since high-precision microelectronic production equipment is never used in conventional plant facilities; it is used only in specially designed buildings that comply with some of the VC criteria.

Table 2.2A shows that the lowest value of Φ_{p1} for the vertical direction is 4.51 Hz. If vibration isolators with medium damping $\delta_v = 0.6$ are used, then from Eq. (2.2.15) the required vertical natural frequency $f_v = 4.51\sqrt{0.6} = 3.04$ Hz. However, if isolators made of rubber with high damping $\delta_v = 1.2$ are used, then $f_v = 5.0$ Hz, which can be realized by passive isolators with rubber flexible elements. Much stiffer isolators ($f_{vz} > 14$ Hz) can be used to comply with values of Φ_{p2}, which represent (according to not-very-stringent VC-B) floor conditions at the microelectronics industry installations.

A similar situation is seen in Table 2.2B. However, realization of natural frequencies corresponding to Φ_{p1} (4.7 Hz for $\delta_v = 0.6$, 6.63 Hz for $\delta_v = 1.2$) in horizontal directions with elastomeric isolators does not present any difficulty; even much lower values can be easily realized. Use of Φ_{p2} gives even more latitude in selecting isolator parameters.

Table 2.3 lists similar data for the case presented in Fig. 2.2.12. In this case, the same assumptions as above are made about $X_f(f)$ for computing

Table 2.4. Vibration Isolation Synthesis for Fig. 2.2.13

A. Vertical direction (Y-axis)				
f, Hz	6	8	30	80
$\mu(t)$	0.48	0.83	1.0	5.0
Φ_1, Hz	0.8	0.9	3.08	6.13
Φ_2, Hz	1.2	2.2	14.5	28.8
Φ_3, Hz	3.4	6.2	41	81.5
B. Horizontal direction (X-axis)				
f, Hz	5	8	30	80
$\mu(t)$	0.31	0.83	1.0	5.0
Φ_1, Hz	1.01	0.99	3.38	6.6
Φ_2, Hz	1.4	2.2	14.5	28.2
Φ_3, Hz	4.0	6.2	41	79.8

both Φ_{p1}, and Φ_{p2}. The minimum required Φ_{p1} for the vertical direction is 4.9 Hz, which corresponds to $f_v = 3.8$ Hz for medium-damped isolators ($\delta_v = 0.6$) and $f_v = 5.4$ Hz for highly damped isolators ($\delta_v = 1.2$). The latter value of f_v in the vertical direction can be realized by elastomeric isolators; it can be even 30% higher ($\sim$7 Hz) for the realization of Φ_{p2}. The corresponding numbers for the horizontal direction ($\Phi_{p1} = 2.02$ Hz, $f_v = 1.6$ Hz for $\delta_v = 0.6$, $f_v = 2.2$ Hz for $\delta_v = 1.2$; $\Phi_{p2} = 2.3$ Hz) also can be realized.

For the cases represented in Figs. 2.2.13 and 14, calculations of $\mu(f)$ for critical points requires using vibratory velocity of relative motion in the work zone instead of Δ_o. Calculations for Fig. 2.2.13 were performed for the lowest lines (the boundary of the safe region). The results, shown in Table 2.4, indicate that for frequencies above 5 Hz, the required Φ_{p1} is as low as 0.8 Hz for the vertical direction (which corresponds to $f_v = 0.62$ Hz for $\delta_v = 0.6$ and $f_v = 0.88$ Hz for $\delta_v = 1.2$). These values can be realized by a passive vibration isolation system only with a massive inertia block. Otherwise, an active pneumatic or electronic system with a leveling feature is called for. Since units of such precision are usually installed in special facilities, isolation criteria in this case were also calculated for two classes of precision facilities specified in Fig. 2.2.4: VC-B (Φ_{p2}) and the most stringent VC-E (Φ_{p3}). It can be seen that in the last case the use of passive elastomeric isolators ($f_v = 3.75$ Hz at $\delta_v = 1.2$) is marginally feasible. They become practically feasible if the actual vibration amplitudes on the floor are below VC-E criterion, which is the case for many realized fabs (see Section 2.2.1).

Table 2.5. Vibration Isolation Synthesis for Fig. 2.2.14

A. Vertical direction					
f, Hz	18	22.5	30	43.5	52
$\mu(t)$	0.14	0.18	0.71	0.61	2.7
Φ_1, Hz	5.0	5.6	3.7	7.0	4.4
Φ_2, Hz	18.4	22.7	17.6	33	20.5
B. Fore/aft direction					
f, Hz	20	30	41.5	54	66
$\mu(t)$	0.5	0.31	0.35	0.54	0.87
Φ_1, Hz	3.2	7.4	11.4	13.6	14.5
Φ_2, Hz	11.2	26	40	48	51
C. Side/side direction					
f, Hz	12.5	22	27.5	35	39
$\mu(t)$	0.16	0.93	1.4	0.49	4.0
Φ_1, Hz	3.5	2.6	3.1	7.4	3.1
Φ_2, Hz	9.8	9.0	10.7	26.1	10.7

Table 2.5 lists vibration isolation data for the case of Fig. 2.2.14, assuming relative displacement of 0.1 μm in the work zone. The minimum required Φ_{p1} for the vertical direction is 3.7 Hz, which corresponds to $f_v = 2.9$ Hz for medium-damped isolators, $\delta_v = 0.6$; and $f_v = 4.1$ Hz for $\delta_v = 1.2$. For the horizontal directions, $\Phi_{p1min} = 2.57$, and $f_v = 2$ Hz for $\delta_v = 0.6$, $f_v = 2.8$ for $\delta_v = 1.2$. The natural frequencies for high damping can be realized with passive elastomeric isolators. The minimum value of Φ_{p2} for the vertical direction is 18.4 Hz, which allows use of commercial-off-the-shelf isolators.

Thus, only one of the analyzed four high-precision devices actually requires an active vibration isolation system for protection from the typical spectrum of industrial steady-state floor vibrations. However, as mentioned, such ultra-precision units are never used in general industrial buildings. Accordingly, much stiffer passive isolators could be used in all considered cases, and passive isolators could be used even for the case shown in Fig. 2.2.13 and Table 2.4 if the equipment were located in a building complying with the vibration criterion VC-E. This criterion can be realized through means used by civil engineers, as was shown in [2.8].

2.2.4C *Isolation from Impulsive Vibration*

The third case of vibration isolation involves isolation from short duration (impulsive or shock) motions of the floor or supporting structure. Such excitations tend to be less troublesome because of their short durations. Accordingly, the permissible value of the peak relative displacement Δ_{op} in response to floor shocks can be, approximately, taken to be three times the value of Δ_o permissible for steady-state vibrations. An isolation system for an object subjected to shock excitation can be considered using the modified shock spectrum approach described in Section 1.8 to obtain the following:

$$\Delta_p = A_{rel}\, a_b\, \gamma = A_1 A_{rel} \gamma a_f \qquad \text{Eq. (2.2.18)}$$

where a_b denotes the peak displacement of the bed caused by shock motion of the floor with displacement magnitude a_f; A_1 represents the shock spectrum value corresponding to a versed-sine pulse in Fig. 1.8.2(c) acting on the vibration isolation system and A_{rel} represents the shock spectrum value for relative displacement in the work area corresponding to a versed-sine motion of the bed (Fig. 1.8.3).

If γ and Δ_{op} are known, as well as the pulse magnitude a_f and duration τ_f (e.g., from Fig. 2.2.3), then the values of the natural frequency f_{vp} of the isolation system that are necessary for adequate isolation may be determined from the foregoing expression by trial and error. Values of f_{vp} for the z-direction that have been determined in this manner are given in Table 2.1 for a number of precision machine tools. We may note that these values are very close to the natural frequency values f_v that are necessary to provide sufficient isolation of steady-state vertical vibration according to the Φ_{pz} criterion and $\delta_v = 0.5$. Since this correlation holds for practically all machine tools in Table 2.1 that have very different and diverse design structures, it can be extrapolated to other precision equipment—such as those whose sensitivities to floor vibrations are presented in Figs. 2.2.11–2.2.14—provided the impulsive excitation from the floor is as depicted by the plots in Fig. 2.2.3. However, modern ultra-high-precision equipment is always installed far from sources of intensive impulsive loads, and 30–60 μm displacement magnitudes of floor motions, as shown in Fig. 2.2.3, would never occur in the controlled environment facilities for high-precision equipment. Accordingly, required parameters of the vibration isolation systems determined in Tables 2.2, 2.3, 2.4, 2.5 would be adequate for isolation of impulsive vibrations as well as steady vibrations of the floor. It was shown (e.g., Section 1.8, Fig. 1.8.3) that isolator damping has less effect on transmission of shocks than of steady motions. Thus, use of increased

144 ● **PASSIVE VIBRATION ISOLATION**

damping may often permit use of stiffer isolators for dealing with steady excitations, which usually are the most important. However, for cases where impulsive floor motions are particularly severe, stiffer isolators should not be used, even if they are highly damped.

2.2.5 Vibration Isolation Systems

The required parameters of vibration isolation systems for various units of precision equipment, as presented in Tables 2.1–2.5, are calculated based on the assumption that the unit is mounted in such a way that there is no dynamic coupling between the vertical and horizontal vibrations of the unit. However, as was shown in Chapter 1, use of conventional constant-stiffness vibration isolators results in strong coupling of vibratory processes in various directions. This coupling can be caused by substantially non-uniform mass distribution in typical equipment units, by difficulties in determining the C.G. positions and calculating weight distribution between the mounting points, by large differences in nominal stiffness constants between sequential models of vibration isolators in a given isolator design line, and by large deviations of the stiffness constant values from the nominal values due to manufacturing tolerances.

The undesirable coupling problem can be alleviated considerably by use of constant natural-frequency (CNF) isolators, in which stiffness in both the vertical and horizontal directions is proportional to the weight load on the isolator (see Section 1.3). As a result, use of such isolators results in a system with a very low degree of coupling between the vertical and horizontal/rocking vibrations without the need to determine the C.G. position, to calculate weight load distribution between the mounting points, etc. In addition, such isolators have a significantly reduced sensitivity to manufacturing tolerances (see Chapter 4).

The effect of using CNF isolators is illustrated by Fig. 2.2.20, which shows the frequency response of the relative motion between the grinding wheel and the table of the surface grinder 3B71 (see Table 2.1) caused by vertical floor vibrations of 5 μm amplitude in 7–38 Hz range. It can be seen that use of CNF isolators that provide a natural frequency in the vertical (z) direction of 20 Hz resulted in much better isolation than the best constant-stiffness isolators with the substantially lower 15 Hz natural frequency in the z-direction. The difference between vertical

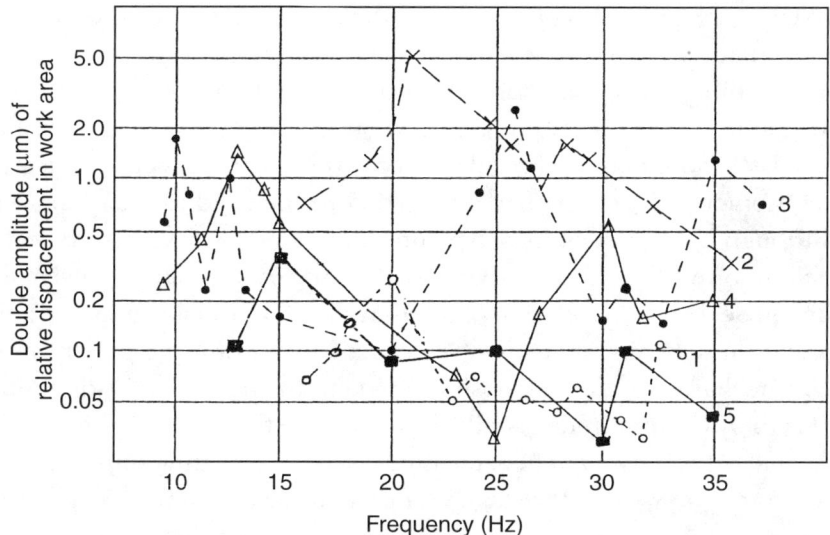

Fig. 2.2.20. Amplitudes of Relative Motion in Working Area (Between Grinding Wheel and Table) of 3B71 Surface Grinder Excited by 5-μm Amplitude Floor Vibration for the Machine Installed on Various Isolators (1—Rubber-Metal CNF Isolators, $f_{vz} = 20$ Hz; 2—Steel Wedges $f_{vz} = 27$ Hz; 3—Wire-Mesh Isolators, $f_{vz} = 25$ Hz; 4—Plastic Pads, $f_{vz} = 30$ Hz; 5—Rubber-Metal Isolators, $f_{vz} = 15$ Hz.).

natural frequencies of 15 and 20 Hz is equivalent to approximately doubling the isolator stiffness for $f_{vz} = 20$ Hz as compared with $f_{vz} = 15$ Hz. Such difference can be very significant for stiffening machine frames (see Section 2.2.6), for simplifying leveling, and for reducing rocking motions due to reversals of the heavy table. Production models of wide load range CNF isolators for vertical natural frequencies of 10, 15, 20, 30 Hz have been developed (see Chapter 4). Use of streamlined elastomeric elements (see Section 3.3.2, Section 4.5) enables for relatively easy realization of various natural frequency CNF isolators, as low as 1–2 Hz.

2.2.6 Side Issues for Vibration Isolated Precision Equipment

Vibration isolation using passive isolators results in a reduced stiffness of the connection between the isolated equipment and the rigid supporting structure (foundation). This can lead to translational and, especially undesirable, rocking motions of the object if it contains internal massive moving units (e.g., tables in surface grinders, gantries in CMMs, stages in photolithography tools). These motions are a result of inertia forces that are caused by accelerations/decelerations of the massive units and by reaction forces acting on their driving motors.

Installing a machine or apparatus on vibration isolators that have reduced stiffness may also result in reduction of the effective rigidity of the equipment frame not structurally tied to the rigid foundation.

Although these issues may be important for any device installed on flexible mounts, they are especially important for vibration-sensitive precision devices. It is noted in [2.18] that the greatest vibration excitation in precision optical systems is often caused by vibratory inputs from the "stage mechanism" used to position targets in the X- and Y-directions. While vibration amplitude usually decays before the critical imaging event takes place, it is important to determine how much settling time is allowed between the stopping of the stage motion and the beginning of the critical imaging.

2.2.6A Reduction of Mobility of Isolated Objects Caused by Internal Dynamic Forces

The rocking or translational motion of isolated precision machines may become objectionable if it does not settle until the next cycle of operation (such as the next grinding pass on a surface grinder, next measurement event on CMM, next exposure in a photolithography scanner or stepper, or next staging act on a microscope). Although these motions usually have low frequency, they may be very intense and thus may induce unacceptable displacements in the working zone. Two typical cases are: (a) moving heavy structural components (e.g., in surface grinders), and (b) moving relatively small components with high accelerations (e.g., wafer and reticle stages in photolithography tools).

In Case (a), the rocking may be reduced by installing the machine rigidly on massive foundation blocks and by placing vibration isolation means under these blocks. This approach tends to be expensive and time-consuming for installation and, especially, for relocation of the object/machine. If, as is often the case, the direction of maximum vibration sensitivity is at a right angle to that of the internal excitation, use of anisotropic isolators with the optimized stiffness ratios can be very beneficial (see Chapter 4.7). A higher stiffness of the isolators in the direction of internal excitation reduces amplitudes of the rocking motion and increases its frequency. The latter leads to faster decaying of the rocking motion. Another effective approach to reduce rocking motion by reducing time of its settling is to increase damping of the vibration isolators, especially in the direction of the rocking motion. Yet another approach is to use increased distances between isolators in the direction of the internal excitations, thus increasing the angular stiffness of the system per Eq. (1.2.7). This decreases the angular motion component (associated with large amplitudes of motion) and increases the translational component, thus significantly reducing excursions of the isolated object. The increased distances between isolators can be achieved by installing isolators under a plate or rails attached to the frame/bed of the isolated object. Plates 25–100 mm thick are often used for enhancing frame stiffness of precision machine tools and other precision objects; it is relatively easy to design positioning of mounting holes in the plate in order to reduce rocking motions. An effect similar to that of increasing effective angular stiffness of the isolation system can be achieved by using vibration isolators with high angular stiffness, per Eq. (1.2.7). While conventional isolators have very low angular stiffness, use of judiciously designed flexible elements, e.g., employing thin-layered rubber-metal laminates (see Section 3.3.2C and Section 4.7) can provide the desired characteristics. While a complete decoupling can be achieved by locating isolators in the plane containing C.G. of the isolated object (see Chapter 1), this type of arrangement can be achieved, practically, only in isolation systems wherein the object is installed on a massive and bulky inertia block. However, reduction of the distance between the plane in which the isolators are located and the C.G. plane, by raising the former, is beneficial since it reduces the coupling between translational and angular axes in the system, and increases the translational component at the expense of the more objectionable angular component of the rocking motion.

To reduce the influence of internal dynamics for Case (b), all of the above techniques can be used to achieve a radical reduction of disturbances in the working zone caused by the reaction forces of the driving motors and by acceleration/deceleration of the moving stages. In addition, two structural design techniques are gaining popularity. One technique is introduction of a so-called "reaction" or "force" frame carrying the driving motor, e.g., see Fig. 2.2.21 [2.19]. The force frame is rigidly connected to the supporting structure (support frame in

148 • PASSIVE VIBRATION ISOLATION

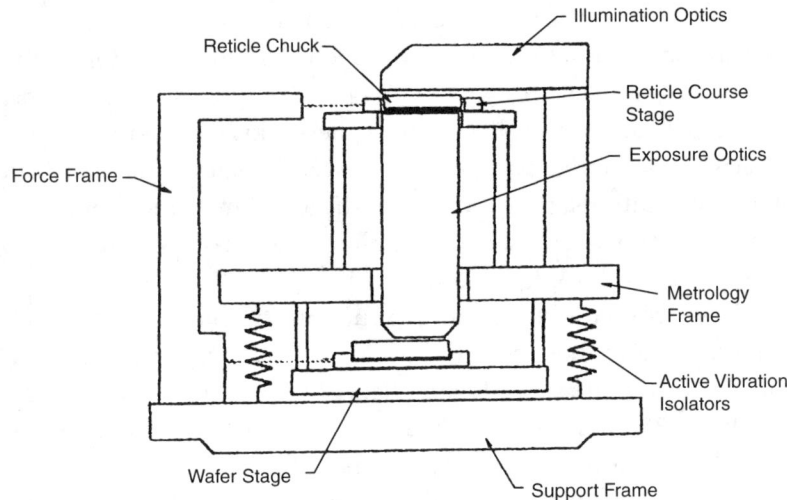

Fig. 2.2.21. Architecture of ASML PAS 5500 Scanner (Actuator Reaction Forces Transmitted to Force Frame).

Fig. 2.2.21) attached to the floor, while the metrology frame, e.g. carrying vibration-sensitive wafer stage and reticle stage (chuck in Fig. 2.2.21), is isolated from the reaction frame for protection from external vibrations. When the wafer stage and/or the reticle stage is accelerated/decelerated by the motor attached to the reaction frame, the reaction forces from the motor excite the reaction frame and the floor, amplitudes of floor vibrations are relatively low due to extremely high specified stiffness of the floors in the microelectronics fabs, see Section 2.2.1 and [2.4]. Transmission of the floor/support frame vibration is attenuated by vibration isolators of the metrology frame and thus do not degrade performance. Due to very low friction of the connections (air bearings) supporting the wafer and the reticle stages, the stages accelerations do not generate vibrations in the metrology frame. The price to pay is a heavier and more complex structure. Fig. 2.2.22 [2.19] shows a schematic of another approach wherein the relatively heavy frame carrying the reticle stage is driven by a powerful long stroke (LoS) actuator in the y-direction, and is fine-positioned by smaller and slower short stroke (SS) actuators in six degrees of mobility. Tilting of the machine due to C. G. shifts generated by the LoS actuator are prevented from developing by use of two counterweights (balance masses), generating equal but oppositely directed C. G. shifts thus canceling the undesirable effects.

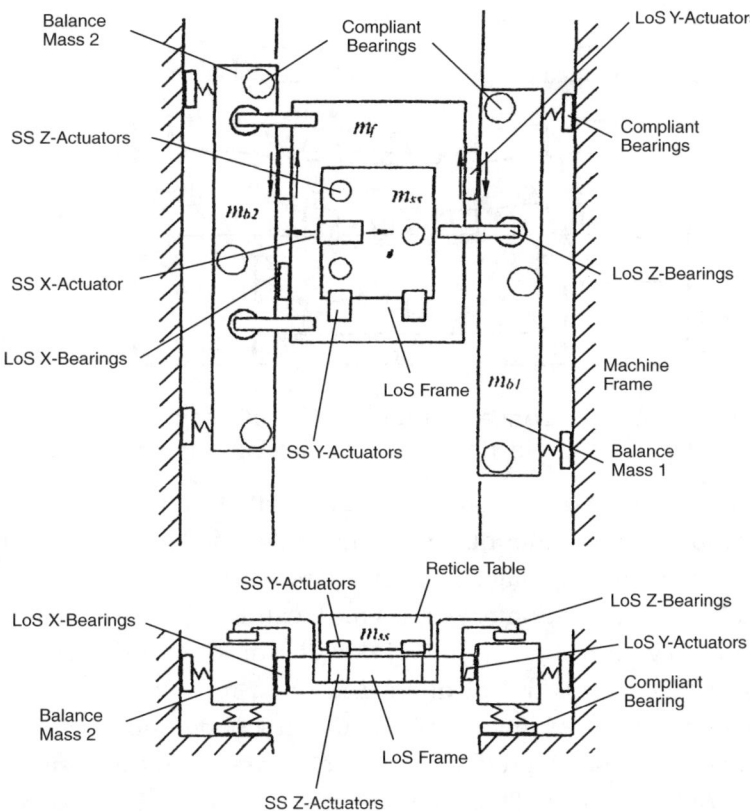

Fig. 2.2.22. Balanced Reticle Stage for a Precision Lithography Scanner.

2.2.6B Influence of Vibration Isolation on Effective Stiffness of Isolated Object

The stiffness, number, and locations of supporting elements have a great influence on deformations/stiffness of the supported system. Even a "flimsy" structure may have a decent effective stiffness if it is judiciously supported. Smaller precision devices usually do not require the addition of a foundation for enhancing rigidity of their frames, whereas larger devices (e.g., machine tools weighing over ten tons) usually do, unless they are specially designed to be mounted on three supporting points (kinematic mounting) e.g., see [2.16]. It should be noted, that the structural natural frequencies f_n of smaller objects are higher than those of similar but larger objects proportionally to the scaling factor $1/L$, where L is a reference linear dimension. In many cases, and particularly for machines of

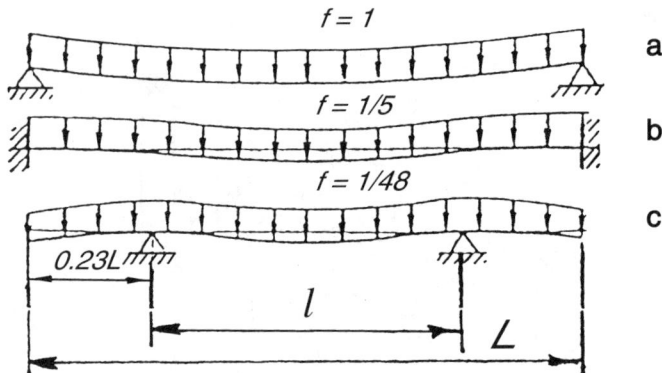

Fig. 2.2.23. Influence of Type and Locations of Supports on Maximum Deflection of a Double-Supported Beam.

intermediate weight (5–10 tons, or 10–20,000 lbs), judicious placement of the vibration isolation mounts may reduce static deflections of the frame, thus effectively making it act as if it were more rigid.

One of the optimization principles for locating supports is *balancing of deformations* within the system, so that the maximum deformations in various critical points are of about the same magnitudes. For simple-supported beams providing simplistic unidirectional models for real-life structures, the balancing effect can be achieved by placing supports at so-called *Bessel's points*. This approach is illustrated in Fig. 2.2.23, which shows a beam loaded by a uniformly distributed load (e.g., by the weight load). Replacement of simple supports in Fig. 2.2.23(a) with built-in ends [Fig. 2.2.23(b)] reduces the maximum deformation by a factor of 5. However, placing just simple supports at the Bessel's points located for this structure and this loading pattern at about 0.23L from the beam's ends ($L/l = 1.86$) reduces the maximum deformation by 48 times [Fig. 2.2.23(c)]. Instead of one point where the maximum deformation occurs in cases of Figs. 2.2.23(a),(b), in the case of Fig. 2.2.23(c) there are three points where the deformation is maximum: the midspan and the two ends [2.16]. It is obvious that a similar effect would occur if compliant supports were used instead of the rigid supports in Fig. 2.2.23. Since the propensity for deformation is reduced in scheme presented in Fig. 2.2.23(c), the influence of the support compliance will also be reduced. The same concept can be applied in cases of supporting more complex structures, such as plates or complex frames [2.16], as illustrated in the *"Example"*.

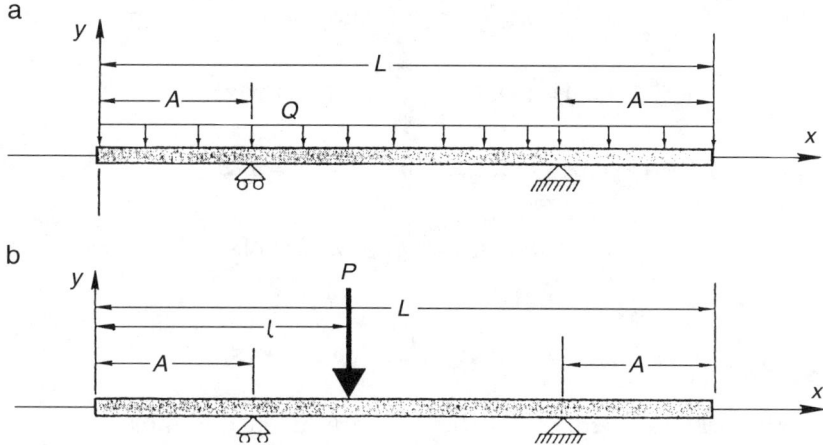

Fig. 2.2.24. A Double-Supported Beam with Different Loading Conditions.

Influence of the number and locations of the mounting elements (isolators) on relative deformations between the structural components (e.g., between the grinding wheel and the part being machined in grinding machine, see "*Example*") is due to changing reaction forces acting from the mounts on the structure. These reaction forces cause changes in the structural deformation patterns. It is important to distinguish between the two cases of deformations caused by the mount reactions. In the first case, the supported structure is a single part or unit in which the mass distribution does not change noticeably in time. In the second, there are changes in the mass distribution after the structure is installed on its supports/mounts. Some examples include machine tools designed for machining parts with significantly different weights and/or having moving massive parts (columns, tables, links, etc.) inside the structure.

Positioning of the Bessel's points depends on the loading schematic of the structure and the deformation parameter (slope or deflection). Fig. 2.2.24(a) shows a simply supported beam uniformly loaded with intensity Q [similar to that of Fig. 2.2.23(c)]. Fig. 2.2.24(b) presents the same beam loaded by a concentrated (moving) load P. For the case of Fig. 2.2.24(a), the maximum slope within (at $x = L/3$) and outside (at $x = 0$ and $x = L$) supports is equalized as

$$\left| y'_{x=0} \right| = \left| y'_{x=L/3} \right| = 0.001543 \frac{QL^3}{EI} \quad \text{at } A = 0.222L, \qquad \text{Eq. (2.2.19)}$$

and the maximum deflections (at $x = 0$ and $x = L/2$) are equalized as

$$y_{x=0} = y_{x=L/2} = 0.0002698 \frac{QL^4}{EI} \quad \text{at } A = 0.223L. \qquad \text{Eq. (2.2.20)}$$

Similarly, for the case of Fig. 2.2.24(b) when $l = L/2$,

$$y'_{x=0} = y'_{x=0.441L} = 0.03727 \frac{PL^2}{EI} \quad \text{at } A = 0.119L \qquad \text{Eq. (2.2.21)}$$

$$|y_{x=0}| = |y_{x=L/2}| = 0.006811 \frac{PL^3}{EI} \quad \text{at } A = 0.156L \qquad \text{Eq. (2.2.22)}$$

While the optimal location of the supports from the ends of the beam varies only from $0.119L$ to $0.223L$, support placement is quite critical. If the supports for all cases were placed at the average value at $0.18L$ rather than at the given optimal locations, the peak slope would increase by 160% and peak deflection would increase by 280%.

Fig. 2.2.25 shows results of experimental optimization of the location of the supporting mounts for a boring mill with a moving column. The column weight $W_c = 42,500$ N (9,400 lbs) was imitated by two dead-weights $P_1 = P_2 = 21,250$ N, which were sequentially placed on the actual machine bed in two locations, at the ends of the rated travel distance of the column. The deadweights were placed symmetrically between the flat and V-shaped guideways, thus assuring that the deformations of the bed were strictly bending (no torsion). Deflections of the bed were measured by gages with 0.5 μm resolution relative to a reference base (not shown). The optimization procedure was performed by shifting the support B by 150 mm after each loading event. After moving the support B by 150 mm from its initial position [Fig. 2.2.25(b)], maximum deflection between the supports decreased by 4 μm (15%), by 300 mm – by 10 μm (40%) [Fig. 2.2.25(c)]. Further shifting of the support [Fig. 2.2.25(d)] causes a significant increase of upward deflection on the outside of the bed (at C), thus the case in Fig. 2.2.25(c) was assumed to be the optimal one.

Structural frames are finish-machined while installed on mounts that are resting on some rigid supporting surface (plate). If the number of the mounts is three (*"kinematic support"*), then their reaction forces do not depend on variations of height or stiffness of the mounts, and do not significantly depend on variations in the geometry of supporting surfaces.

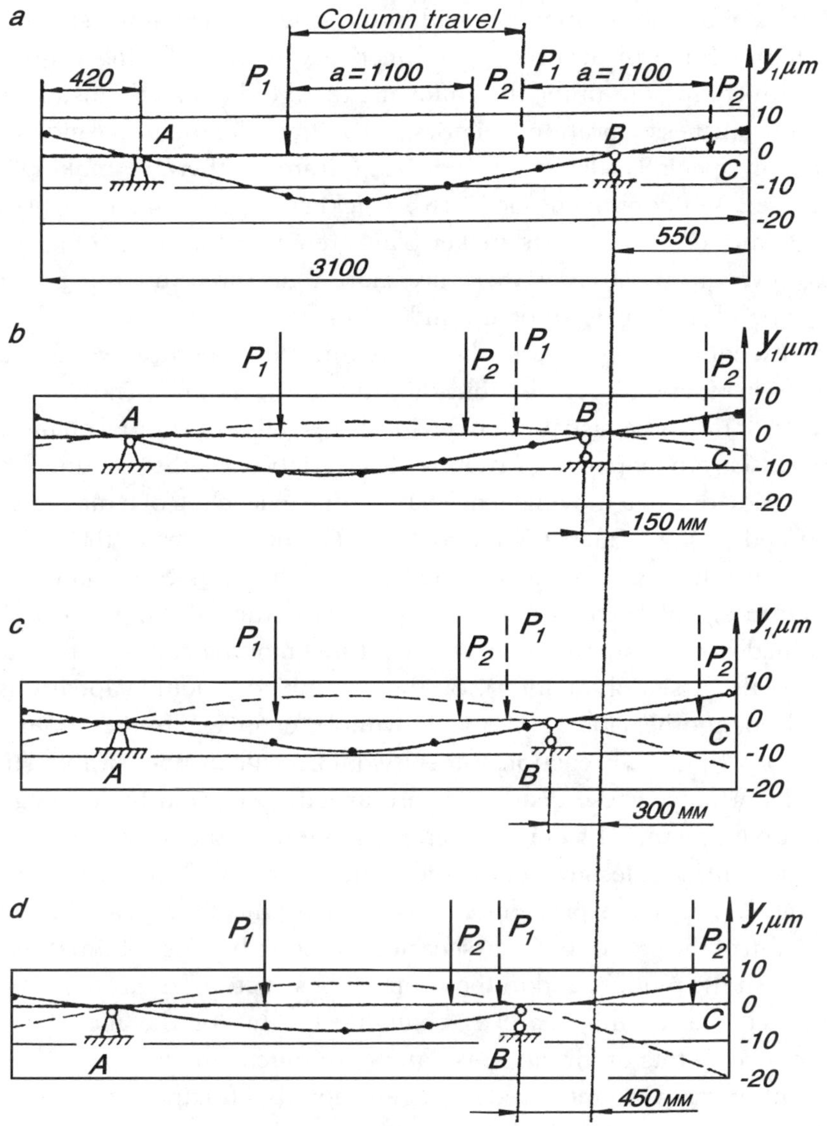

Fig. 2.2.25. Optimization of Placement of Rigid Supports Under the Frame of a Horizontal Boring Mill; All Dimensions in mm.

Such structure would have the same effective stiffness installed on vibration isolators as if it were installed on rigid mounts. The reaction forces are fully determined by the position of the C.G. of the part/structure and by coordinates of the mounts relative to the C.G. As a result, relative positions of all the components within the structure would be stable to a very high degree of accuracy. An effective way of stiffening frames of precision equipment units, thus allowing them to be directly installed on even low stiffness vibration isolators, is by attaching them to rigid (steel) plates or specially designed rigidizing frames, thereby making the structure a self-contained one [2.16]. Thickness of such plates is usually 25–100 mm (1–4 in.), with thicker plates required for larger objects.

However, in most cases there are more than three mounting points. This type of system is statically indeterminate and its relocation, even without changes in mass distribution within the structure, would cause changes in reaction force distribution between the mounts. These changes in the reaction force distribution obviously result in substantial changes in 3-D stress/strain map of the structure. The variations in the reaction forces distribution due to a moving heavy part or due to leveling of a machine are illustrated by Figs. 1.3.2, 1.3.4, and 4.4.2. The differences in the reaction force distribution can be easily translated into differences in moments of these forces, which are also applied to the structure. If the part is finish-machined while mounted on more than three mounts and then relocated (e.g., to the assembly station), or the assembled machine/apparatus is moved to another location, the inevitable changes in the mounting conditions (thus the reaction forces) would result in undesirable structural distortions. These distortions are usually corrected by leveling the structure (adjusting the height of each mount until the distortions of the reference surface designated on the structure are within an acceptable tolerance). The same procedure must be periodically repeated at each relocation of the part or of the assembled structure because of time-related changes in the mounts and in the supporting structure. It has to be noted that since the leveling involves changing heights of the mounts, and stiffness constants of vibration isolators are much smaller than stiffness constants of rigid mounts (wedges, screw mounts, etc), the reaction forces of the isolators are much more robust than those of rigid mounts. In addition, the isolators do not experience "strain relaxation" processes, which are typical for the rigid mounts (see Section 2.2.2 and *"Example"*).

Although increasing the number of supporting mounts for a frame of a given size reduces distances l between the mounts and, theoretically, reduces deformations that are proportional to l^3, the system becomes more sensitive to small deviations of the supporting surfaces geometry, and to conditions and height variations of the mounts, especially those of

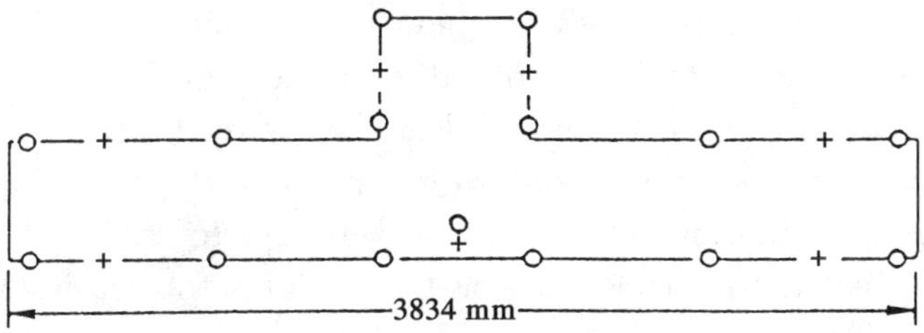

Principles and Criteria of Vibration Isolation • 155

	Longitudinal Transverse Axis Rotation			Longitudinal Axis Rotation		
Type of Installation	Wheelhead position			Table position		
	Front	Rear	Diff.	Left	Right	Diff.
On 15 rigid wedge-type mounts	-3.0	-0.5	2.5	0	-1.0	1.0
On 7 isolators (f_{vz} - 20 Hz)	+3.3	+0.8	2.5	+0.8	+0.3	0.5

Fig. 2.2.26. Mount Locations Under Bed of Schaudt AR-1500 Precision OD grinder (o—as Recommended by Manufacturer; +—Optimized Installation as Determined by Rigidity Evaluation) and Angular Deformations in Working Zone in 10^{-5} rad Units Caused by Travel of Heavy Structural Units of the Grinder.

rigid mounts. These variations may significantly affect reaction forces and, since the structure becomes more rigid with the increasing number of mounting elements, even a small deviation may result in large reaction force fluctuations (and thus deformations) or, conversely, may result in "switching off" of some rigid mounts. Since the proper selection of the number and location of the mounting points is very important, both to reduce rocking and to increase the effective rigidity of the structure of precision vibration-sensitive devices, this problem generally deserves special attention. Designers rarely give this issue the consideration it deserves.

Example. The manufacturer recommended installation of a large (3.8 m/13 ft long) precision cylindrical (OD) grinder on 15 rigid wedge mounts placed in the locations indicated by circles in Fig. 2.2.26. The main problem with this installation was strain relaxation in the wedges caused by ambient shocks and vibrations, with the resulting changes in the wedge heights (see Section 2.2.2). The ensuing repetitive loss of alignment between the critical units of the grinder as well as development of excessive positioning errors when a very heavy table carrying the spindle head was traveling along the bed, or a heavy grinding wheel head was traveling in the transverse direction, led to deterioration of geometry of the ground parts. The surface finish quality was also very sensitive to ambient vibrations. Since constructing a special vibration isolated foundation was undesirable, it was suggested that the wedge mounts be replaced with vibration isolators providing vertical natural frequency $f_{vz} \approx 20$ Hz. However, this replacement resulted in unacceptable relative displacements between the grinding wheel and the workpiece when the wheel head or the heavy workpiece table were traveling along their guideways. These displacements are caused by distortions of the frame parts separated from the rigid floor plate by complaint vibration isolators. It was found that the use of 7 isolators instead of 15, placed as indicated by + in Fig. 2.2.26, resulted in a significantly greater effective rigidity [2.6], even exceeding rigidity of the original installation, thus making it possible to install the grinder on the isolation mounts without using a stiffening foundation block. The table in Fig. 2.2.26 shows deformations (relative angular displacements) between the grinding wheel and the machined part during longitudinal travel of the heavy table and during transverse travel of the grinding wheel head. It is remarkable that the deformations were *smaller* when this large machine was installed on 7 resilient mounts than when it was installed on 15 rigid mounts. The machine was producing quality parts for several years without changing the mounts.

2.2.7 Vibration Protection of Civil Engineering Structures

While the principles of vibration isolation are generic, in many cases specifics such as, but not only, the size and weight of the civil engineering structures require somewhat different approaches to the design of their isolation systems. Some of these approaches are described in the specialized literature, such as [2.20]. This section gives only a brief overview of the issue.

Civil engineering structures (buildings) often require protection from external vibration sources. Two major sources of vibratory excitation are ground-born vibrations caused by nearby road and rail traffic and by earthquakes.

The road/rail vibration may cause personal discomfort of the building occupants as well as unacceptable disturbances for vibration-sensitive devices installed in the building. In some cases, a building may be rendered incompatible with its intended use (e.g., as a hospital or a precision manufacturing facility). Vibration isolation of buildings allows for a more effective utilization of land in areas known for high levels of vibration, such as those adjacent to surface and underground railways and busy roads. This is especially so in areas where land value is high, where the higher costs of the vibration-isolated buildings are easily justified.

Buildings to be constructed in earthquake-prone areas have to be designed to withstand the full horizontal acceleration load produced by a "standard" design earthquake for the particular area. While it significantly inflates costs of the building, the occupants and various devices in the building still experience the full force of the earth movement, thus making this method of earthquake protection far from perfect. The mounting of a building on flexible vibration isolators results in the building "riding" the earthquake, thus reducing dynamic loads acting on both the building and its occupants. In such cases, both costs of construction might be reduced and the psychological trauma of the occupants as well as losses caused by destruction of vibration-sensitive devices are also reduced.

Two basic techniques of vibration isolation of buildings are: mounting them on steel springs (see Section 3.1.1 and Section 4.6.2) or on rubber-metal laminated pads (see Section 3.3.2C). The steel springs provide significant flexibility in the vertical and two principal horizontal directions, as well as in three angular directions about the vertical and horizontal axes. Rubber-metal laminated pads provide the required flexibility in principal horizontal directions, while having high vertical stiffness and high angular stiffnesses about the principal horizontal axes. A judgment on these two isolation means can be rendered by analyzing both dynamic and static requirements for vibration isolation systems of buildings.

While typical horizontal natural frequencies of multistory buildings are in the range of $f_{n_{x,y}} = 1-5$ Hz (bending-like mode shapes), their vertical natural frequencies are much higher, $f_{n_z} = 5-15$ Hz (compression mode shapes). As it was shown above in Section 2.2.4, lower structural natural frequencies result in higher vibration sensitivity, thus requiring lower natural frequencies of the vibration isolation system.

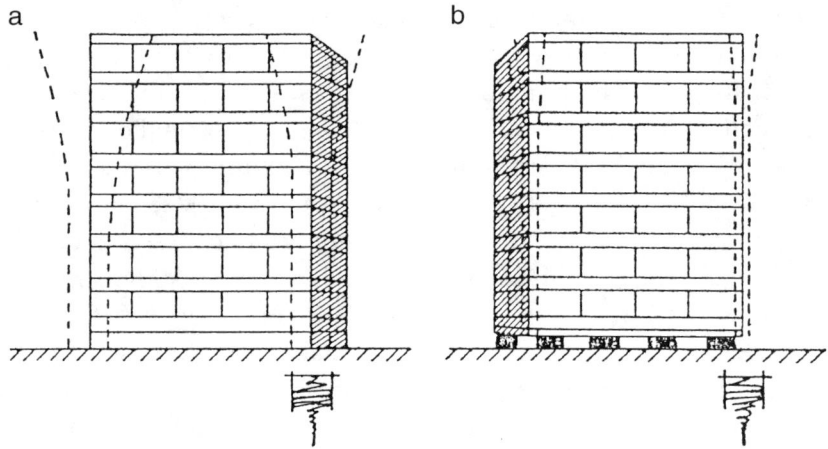

Fig. 2.2.27. Response of Buildings to an Earthquake.

Thus, the natural frequencies of horizontal-rocking modes of the isolated building should be much lower than the vertical natural frequency. In addition, the earthquake ground motions are the most intense in horizontal directions in the frequency range $f_{ex} = 1-10$ Hz, peaking at about 4 Hz, with a potential for exciting resonance-like conditions for the horizontal vibratory modes of multistory buildings. Accordingly, the desirable lowest horizontal-rocking natural frequency of the earthquake-protected building should be $f_{x\beta,y\alpha} = 0.5$ Hz. If the steel springs are used as vibration isolators, the vertical natural frequency of such a system would be in the range of $f_z = 1-1.5$ Hz, and the system would be characterized by large amplitude rocking vibrations [see Eqs. (1.6.3) and (1.6.4)], that may cause disastrous effects at the top floors of a high-rise building. Reduction of the rocking component of the horizontal-rocking vibration can be effected (for the same $f_{x\beta,y\alpha}$) by increasing f_z and/or the angular stiffness of the isolators. Rubber-metal laminated pads (Section 3.3.2c and Section 4.7), provide for both increased f_z and high angular stiffness, thus resulting in quasi-horizontal vibration modes with only minor rocking components, as shown in Fig. 2.2.27 [2.21]. These types of systems provide for attenuation of the most critical horizontal seismic excitation, practically not influencing the less dangerous vertical vibrations.

The means for protecting buildings from road- and rail-induced vibration are similar to the means employed for earthquake protection, but

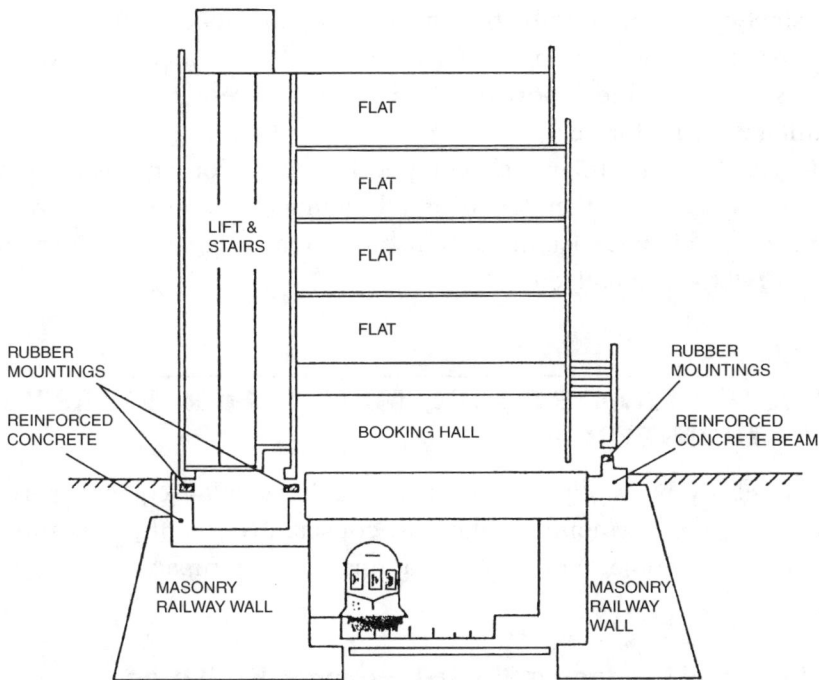

Fig. 2.2.28. Vibration Isolated Residential Building.

with significantly higher natural frequencies. Fig. 2.2.28 [2.21] presents a schematic of a luxury residential building in London built on top of the underground metro station. It is installed on 13 rubber-metal laminates providing vertical natural frequency $f_z = 7$ Hz and $f_{x,y} = 2.5$ Hz. While the former is adequate (low enough) for protection from the train-induced excitations, the latter is adequate (high enough) to prevent wind-induced swaying of the building. The cost of the isolation system is only about 5% of the total cost of the building.

In addition, the "static issues" must also be considered. While buildings are much stiffer in the vertical direction (compression) than in the horizontal directions (bending/shear), their weight has to be supported by a rigid base (foundation) plate/block preventing structural deformations of the building. If the base plate rests on springs that have relatively low vertical stiffness, the building has to be self-contained and to have a very high stiffness (thus, a greater weight) since it is not supported by the ground. On the other hand, if the rubber-metal laminate pads that have high vertical stiffness are used, the base plate can be reduced in size and cost since it is supported by the ground via stiff rubber-metal laminated pads. Much higher damping of rubber-metal laminated pads vs. steel springs in the critical horizontal direction represents another major advantage of the former.

However, springs might be a competitive means for vibration isolation of single-story and other low-height buildings, since their natural frequencies are higher, rocking motions are not as critical, size of the base plate is relatively small, etc.

2.3 ISOLATION REQUIREMENTS FOR VIBRATION PRODUCING OBJECTS

Main cases in this group are objects producing single-frequency periodic excitations; polyharmonic excitations; conservative (confined within the structure of the object) impact excitations; inertial impacts.

2.3.1 Objects Producing Single-Frequency Excitation

Two cases of vibration isolation installations are considered below. The first case (a unidirectional excitation), not encountered very often but very important notwithstanding, is characterized by unidirectional excitation and by isolators providing isolation mainly in one direction. In the second case (a multidirectional excitation), the single-frequency force may have several coordinate components (e.g., produced by an unbalanced rotor) and the isolators may have commensurate stiffness constants in several directions.

2.3.1A Objects Producing Unidirectional Excitation

A typical representative of this group is vibratory (most often, electrodynamic) shaker for performance testing of components and devices under intensive vibrations. Shakers are characterized by the maximum amplitude of harmonic (sinusoidal) force they can apply to the table holding the test object over the specified frequency range. The reaction force amplitude acting on the supporting structure (floor) can significantly exceed the rated force amplitude [2.22] and thus become a hazard for vibration-sensitive equipment and for humans at a large distance. Although most of the tests are performed with vertical vibrations, the shaker body can tilt in the vertical plane, thus changing the direction of the vibration vector. The tests can also be performed by application of vibration with a complex spectral content (including shocks) within the specified amplitude-frequency range. However, analysis of vibration isolation is usually performed for single-frequency excitation.

Since the frequency range is usually quite broad and often starts from low frequencies (e.g., 2.0 Hz – to −7,000 Hz for Data Physics S-100 shaker in [2.22]), shakers are usually installed on low vertical natural frequency ($f_v = 1-3$ Hz) pneumatic isolators whose stiffness in the horizontal direction is much higher (about 2:1) than that in the vertical direction. Due to such low vertical natural frequency the shaker structure is "floating", and this reactive motion takes some energy from the useful vibratory motion of the vibrating table carrying the object being tested, thus potentially reducing efficiency. The efficiency can be improved if the shaker is installed on the isolator not directly, but if it is first attached to an inertia block. The case described in [2.22] involves the shaker with rated peak amplitude of vibratory force 650 N (145 lb) and weighing 80 kg (176 lb). Fig. 2.3.1 shows the reaction force transmitted to the floor for the shaker rigidly attached to the floor (line 1); installed on pneumatic vibration isolators at $f_v = 2.5$ Hz (line 2); and the shaker rigidly attached to a concrete slab weighing 1,000 kg (2,200 lb) which is, in its turn, installed on $f_v = 0.5$ Hz isolators with damping $\eta = 0.01$. The "direct" isolation reduces vibration transmission in the important frequency range 7–20 Hz (where most of floor structure resonances are located) by 1-to-1.5 order of magnitude. The transmissibility, of course, deteriorates around $f_v = 2.5$ Hz, in the 1–3.3 Hz frequency range. The isolation system with the inertia block 12.5 times heavier than the shaker, and with five times lower natural frequency, reduces transmissibility in the same frequency range about 35 times more [hardly necessary since it results in the excitation force in this frequency range ~1.5 N (0.3 lb)].

Use of the inertia mass (slab) is also important for improving efficiency of the shaker. The reduction of the useful stroke of the shaker is determined by the factor $K = M_B/(M_B + M_D + M_T + M_C)$ [2.22], where $M_B = 80$ kg is mass of the body, $M_D + M_T = 25$ kg is mass of the test object together with the vibrating table of the shaker, and $M_C = 1.3$ kg is mass of

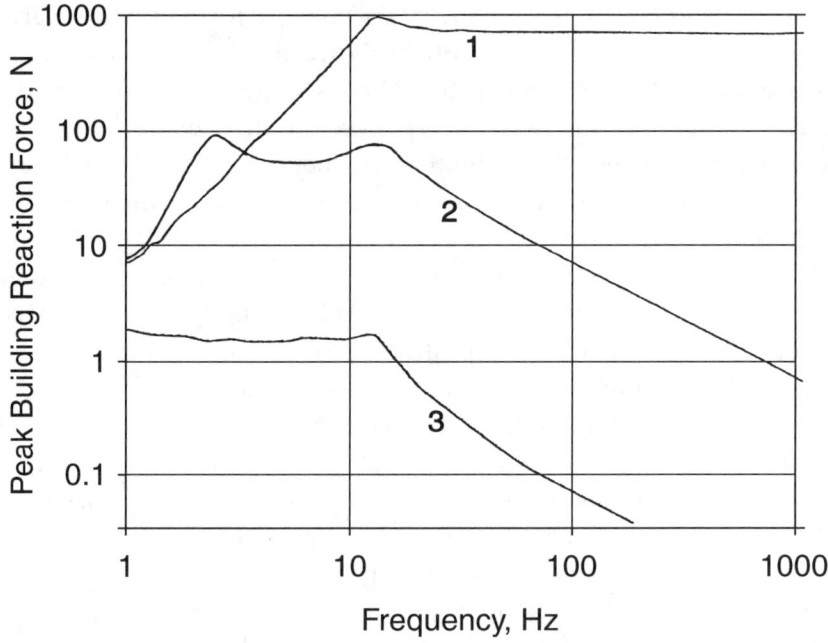

Fig. 2.3.1. Floor Reaction Forces for Three Different Shaker Mounting Configurations [1—Attached to Floor; 2—Installed on Isolators ($f_v = 2.5$ Hz); 3—Attached to Concrete Slab Supported by Isolators ($f_v = 0.5$ Hz)].

the moving coil. $K = 0.75$ for the direct installation on isolators, and $K = 0.98$ for the inertia block installation, since the mass of the inertia block (slab) is, effectively, increasing M_B. Thus, the inertia block in the case of low natural frequency isolation not only helps reduce transmissibility, but noticeably improves dynamic efficiency of the system. The efficiency can also be increased by increasing f_v or by using the "virtual" inertia block provided by the isolators with motion transformation (see Section 1.4.4).

2.3.1B Objects Producing Multidirectional Excitation

Frequency ratio $f/f_{vz} \geq 4–5$ is usually recommended for isolation of objects generating a *single frequency (f) dynamic force* (e.g. created by rotating unbalanced parts), where f_{vz} is the vertical natural frequency of the isolation system. Thus, about 15 to 25 times reduction of transmission of the vertical component of the dynamic force can be achieved per Eqs. (1.4.6) and (1.4.18), although with modern balancing techniques this is often unnecessary. Consequently f_{vz} can become rather low—e.g., $f_{vz} = 6-7$ Hz for rotors with rotational speed of 1,300 rpm, 3–4 Hz at 900 rpm—and therefore steel-spring isolators are frequently being used. These isolators have low damping and poor performance in a high-frequency range, $f > \sim 10 f_{vz}$, thus high-frequency excitations due to inaccuracies of bearings, clearances etc., are not attenuated and may even be amplified.

The frequency ratio can be reduced and specified according to the real circumstances, considering transmissibility in all modes of isolation systems. The critical cases are ones wherein the greatest components of the dynamic force generated within the object are vertical. Horizontal force components are easier to isolate since reasonably low natural frequencies of horizontal (and/or rocking) vibrations are not associated with large excursions of the isolated object, unless the vertical natural frequency is low. Typical cases associated with generation of intense vertical components of the dynamic force are objects containing horizontal (or inclined) unbalanced rotors.

For a horizontal (y-axis) unbalanced rotor, there are two principal force components: z-component (F_z) and x-component (F_x). Transmissibility of the vibration isolation system is determined for F_z by the natural frequency f_{vz}, and for F_x, by the natural frequency $f_{x\beta}$ of the lower rocking mode (see Section 1.6). If the torque T acting on the rotor varies with the frequency f, then transmissibility for the *oscillatory torque* T_y is determined by the natural frequencies $f_{x\beta}$ and $f_{\beta x}$ of the rocking modes (mostly, by higher rocking mode $f_{\beta x}$). The oscillatory torque is not always present, but may be pronounced in internal combustion engine installations. For a vertical (z-axis) rotor, the correspondence is as follows: $F_x - f_{x\beta}$; $F_y - f_{y\alpha}$ (lower rocking modes); oscillatory torque, f_γ. Modal coupling is not considered.

The effectiveness of vibration isolation depends on attenuation of *all* components of the dynamic force and torque. Orthogonal components of the dynamic forces are phase-shifted by $\pi/2$. If for a z-axis rotor F_o is the centrifugal force amplitude, then

$$F_x = F_o \sin 2\pi ft; \quad F_y = F_o \sin (2\pi ft - \pi/2); \quad F'_x = F'_{xo} \sin 2\pi ft;$$

$$F'_y = F'_{yo} \sin (2\pi ft - \pi/2); \qquad \text{Eq. (2.3.1)}$$

164 ● PASSIVE VIBRATION ISOLATION

$$\mu_{F_x} = F'_{x_o}/F_o; \quad \mu_{F_y} = F'_{y_o}/F_o; \quad \mu_F = \frac{\sqrt{(F'_{x_o} \sin 2\pi ft)^2 + (F'_{y_o} \cos 2\pi ft)^2}}{F_o}.$$

Eq. (2.3.2)

Here (') denotes components on the output side of vibration isolators; μ_{F_x}, μ_{F_y} are transmissibilities along the respective axes; and μ_F is the overall force transmissibility. The amplitude of the trigonometrical expression $(A\sin\varphi)^2 + (B\cos\varphi)^2$ is equal to $max(A^2, B^2)$, therefore the overall transmissibility

$$\mu_F = \frac{\max(F'_{x_o}, F'_{y_o})}{F} = \max(\mu_{F_x}, \mu_{F_y}).$$

Eq. (2.3.3)

The oscillatory torque for a horizontal or for a vertical rotor has only one component, so the transmissibility expression is a straightforward one.

Thus, effectiveness of vibration isolation is determined by transmissibility in only two modes—the highest one for force and one for torque. In many cases, only one significant factor (force or torque) is present, thus transmissibility in just one mode can be considered. If both factors are present, then the *"effective transmissibility"* μ, or its weighted average, can be used. For the z-axis and y-axis rotors, respectively,

$$\mu_z = 1/2(W_F \mu_F + W_T \mu_T) = 1/2[W_F \max(\mu_{F_x}, \mu_{F_y}) + W_T \max(\mu_{T_x}, \mu_{T_y})$$

Eq. (2.3.4)

$$\mu_y = 1/2[W_F \max(\mu_{F_z}, \mu_{F_x}) + W_T \max(\mu_{T_z}, \mu_{T_y})],$$

Eq. (2.3.5)

where weighting factors $W_F + W_T = 2$. In general, transmission of torque is less dangerous. With widely spaced isolators, the forces creating the output torque are small and with narrow spaced isolators, action of the torque on the supporting structure is local. Thus, in many cases $W_F = 2\,W_T$ can be assumed. Eqs. (2.3.4) and (2.3.5) give conservative values since reactions of the isolators, induced by forces and by torques, should be added up as vector quantities. Typical for vibration isolation systems for machinery are the following ratios: $\eta_{x,y} = k_z/k_{x,y} = 4-6; f_{x\beta}, f_{y\alpha} = 0.4-0.45 f_z$; $f_{\beta y}, f_{\alpha x} \cong 1.2\ f_z$; $f_\gamma \cong 1.15\ f_z$. For these values, for a medium damping characterized by $\delta_y = 0.6$, and for the required $\mu = 0.20$ (five times attenuation, which is adequate in most cases), the needed frequency ratios are: from Eqs. (2.3.4) and (1.4.6) for z-axis rotors $f/f_z > \sim 2.0$; from Eqs. (2.3.5) and (1.4.6) for x-axis rotors $f/f_z > \sim 2.6$. If the torque oscillations are insignificant and can be neglected ($W_T = 0$), then $f/f_z > \sim 0.9$ for vertical (z-axis) rotors and $f/f_z > \sim 2.0$ for x-axis rotors. If a greater attenuation is required, these frequency ratios should be, obviously, increased, and their values can be found following the above procedure. However, in the former case the formal conclusion should be modified since the closeness of f_z to f may promote intense vertical vibrations due to inevitable inter-coordinate coupling. Thus, it is prudent to adopt $f/f_z > \sim 1.5$ condition for vertical rotors.

This approach was validated by experiments on a vertical-axis hammer crusher ($f = 12$Hz; $f_{x\beta} = f_{y\alpha} = 1.66$ Hz: $f_{\beta x} = 3.5$ Hz; $f_{\alpha y} = 4.0$ Hz, and $\mu_{F_x} = 0.02$; $\mu_{T_x} = 0.093$; $\mu_{T_y} = 0.125$). Using Eq. (2.3.4) with $W_F = 2W_F = 1.33$, $\mu = 0.042$. By measuring floor vibrations with and without isolators, $\mu = 0.0355$.

From Eqs. (2.3.4), (2.3.2), and (1.4.6), necessary natural frequencies could be easily determined for a required attenuation μ. In cases when the exciting force is generated by unbalanced rotors, the most intensive forces transmitted to the supporting structures may develop during transient periods—during acceleration of the rotor to its working rpm or, often even more dangerous, during deceleration (stopping) of the rotor. The latter condition is more dangerous due to usually slower rates of deceleration than of acceleration, unless a brake system is used, and a resulting possibility of quasi-resonance conditions. Table 2.6 lists design parameters and measured dynamic loads transmitted to the supporting structures for six (A–F) vibratory screens for mining industry. The transmitted dynamic loads during the transient regimes exceed the steady-state dynamic loads up to five- or six-fold due to exciting quasi-resonance conditions on the low natural frequency of the isolation system. Although the transient period is relatively short and in most cases the short duration/high-intensity dynamic forces are not critically dangerous, it is desirable to reduce their amplitudes. This can be achieved by design means (faster braking) and by increasing damping in the isolators, thus slowing down the resonance amplification process in the vibration isolation system. In Table 2.6: W—weight of machine (screen) with vibrating box, kg; n—main rotor speed, rev/s; a_{ac}—average acceleration, rev/s^2; a_{dec}—average deceleration, rev/s^2; k_v—total vertical stiffness of isolators, 10^5 N/m; k_h—total horizontal stiffness of isolators, 10^5 N/m; f_v—vertical natural frequency, Hz; F_{wv}, F_{wh}—amplitude of vertical, horizontal component of dynamic force transmitted through isolators in

Table 2.6. Dynamic Forces from Vibratory Screens in Working and Transient Conditions

Model	W	n	a_{ac}	a_{dec}	k_v	k_h	f_v	F_{wv}	F_{wh}	F_{tv}	F_{th}
A	1,740	25	18	1.6	8.4	35	4.5	2,760	8,380	16,560	50,000
B	2,135	21.5	18	1.6	8.4	35	4.1	3,300	10,100	19,800	60,600
C	2,395	21.5	6.8	1.3	8.4	35	3.9	2,940	9,020	17,640	54,120
D	3,080	16.5	6.8	1.3	10.9	45.5	4.0	4,530	14,350	27,180	86,100
E	2,750	20	–	–	11.7	–	5.0	3,500	–	21,000	–
F	1,800	12	6.7	1.1	–	–	–	–	4,200	–	17,000

working regime, N; F_{tv}, F_{th}—maximum amplitude of vertical, horizontal component of dynamic force transmitted through isolators during deceleration, N.

2.3.2 Objects Producing Polyharmonic Excitations

Approach to isolation in the case of *polyharmonic dynamic force* depends on factors such as the spectral content of excitation, the absolute values of the lowest frequencies of dangerous components, the dynamic characteristics of structures and equipment to be protected, the mode of operation of the vibration producing objects to be isolated, etc. In some cases, especially with an inertia block under the machine, when highly damped isolators (with internal damping that does not significantly impair post-resonant transmissibility) are used, resonance (of not the most dangerous spectral component) could be a permissible working regime. Also, it is very important to pay attention to the spectral composition of dynamic forces and their directions (vertical or horizontal).

Table 2.7 gives experimentally determined amplitudes of spectral components of dynamic forces generated by swinging frame (x-direction, F_x) and by flying shuttle (y-direction, F_y) for two textile looms of different designs: (a) having 220 strokes per minute (3.7 strokes per second); and (b) having 185 spm (3.1 sps). While in some cases the first harmonic is the most intensive one, in other cases the 3rd, 4th, and even 8th harmonic has the greatest force amplitude. While the first harmonics have very low frequencies that require extremely low stiffness values for isolators, it has to be noted that the dynamic forces are acting in horizontal directions, thus relatively stable isolators with high vertical stiffness and low horizontal stiffness can be used. Such stiffness patterns

Table 2.7. Amplitudes F_i of Spectral Hatmonics of Dynamic Forces Generated by Looms

Harmonic #, i	a			b		
	f_i, Hz	F_{x_i}, N	F_{y_i}, N	f_i, Hz	F_{x_i}, N	F_{y_i}, N
1	3.7	4,670	108	3.1	501	330
2	7.4	1,630	241	6.2	–	400
3	11.1	1,800	596	9.3	1,390	–
4	14.8	705	242	12.4	3,000	410
5	18.5	593	279	15.5	1,370	500
6	22.5	387	230	18.6	664	320
7	25.9	263	77	21.7	–	–
8	–	–	–	24.8	–	600
9	–	–	–	27.9	627	480
10	37	267	82	31	254	290

of the isolators also result in translational motion rather than rocking vibrations of the isolated looms, thus reducing maximum amplitudes of the loom displacements. Also, it should be taken into consideration that low (especially, the 1st) harmonics are not very dangerous, while the higher harmonics are easier to isolate. Even if all of the harmonics are not isolated (one or two can even be amplified), the vibratory climate in the shop can be noticeably improved. Due to specifics of textile plants, there may be hundreds of looms in one shop, thus building foundations for individual (or even for clustered) looms is prohibitively expensive. However, the looms can be connected by relatively light and inexpensive rigid frames extending their effective "dynamic" footprint, thus resulting in purely translational rather than rocking motions even if they are installed on isolators with low vertical stiffness.

Usually, the first harmonic of a polyharmonic excitation is determined by the basic kinematics of the vibration producing object, while higher harmonics can be significantly influenced by dimensional scatter, variations of clearances in joints, shape variations, etc. As a result, amplitudes of the higher harmonics may vary for different copies of the same model of a machine or mechanism. For example, Fig. 2.3.2 [2.23] shows spectral components of vertical dynamic force F_z and dynamic moments in the X-Z (M_y) and Y-Z (M_x) planes measured at the mounting points of two copies of high-quality induction motors of the same model ($n = 1,500$ rpm) for cylindrical grinders. The most pronounced are harmonics with the rpm frequency (25 Hz) and with frequencies 100, 200, and 300 Hz, which are multiples of the electric supply frequency $f_{el} = 50$ Hz.

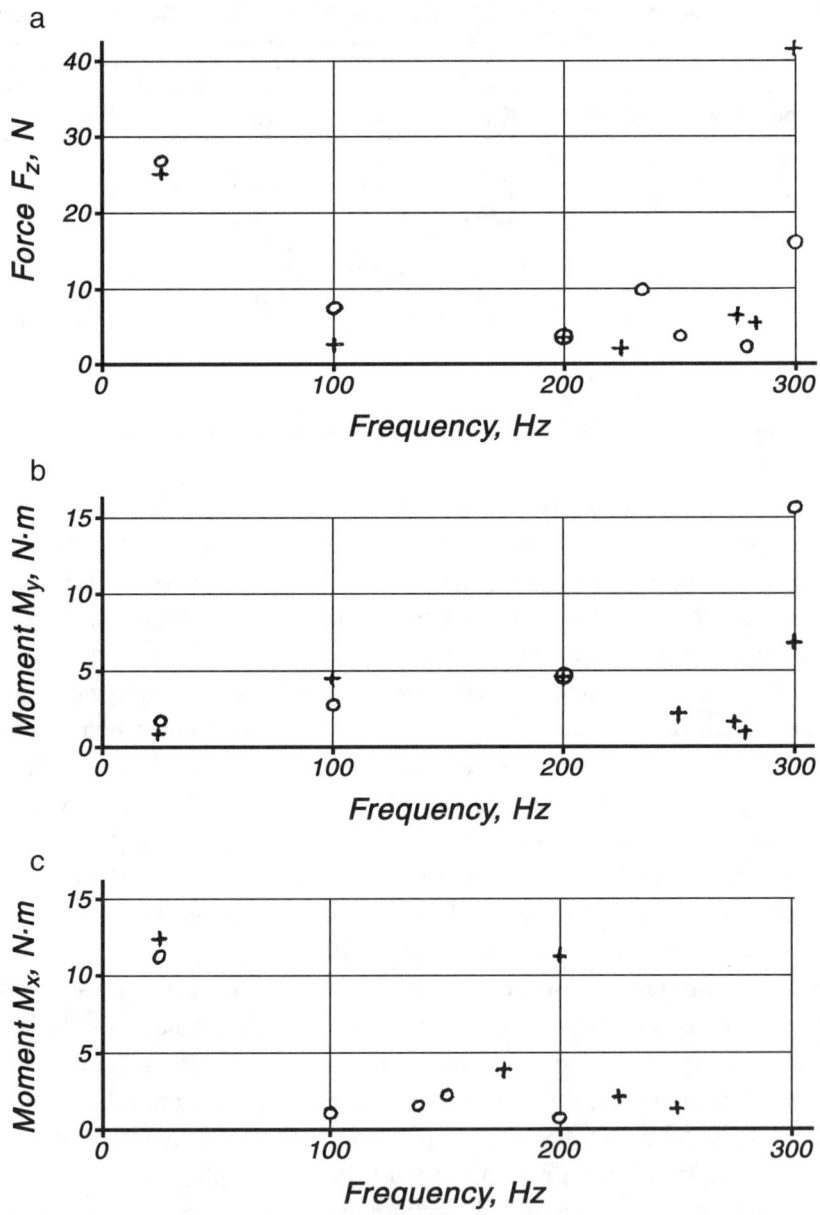

Fig. 2.3.2. Spectral Components of Reaction Force (F_z) and Reaction Moment (M_x and M_y) Variations at Mounting Points for Two Copies of Induction Motors of the Same Model.

While the 25 Hz amplitudes are quite similar for both specimens, amplitudes of other spectral components, both listed above and others, differ significantly. This phenomenon should be considered when designing vibration isolation installations for objects producing polyharmonic dynamic excitations.

2.3.3 Objects Producing Conservative Impact Excitations [2.24]

Typical objects that produce dynamic forces of the *"conservative pulse"* type are stamping presses. Reaction moments generated by the production pulse loads excite the lower rocking mode in the $x-z$ plane. Extensive experimental studies established that pulses generated by stamping presses are shaped close to the *"versed-sine"* pulse (see Section 1.8). Typical working conditions of the most widely used crank presses can be represented by pulses whose duration $\tau = \sim 0.1 t_c$ for automatic operation and $\tau = \sim 0.05 t_c$ for manual operation (single strokes), where t_c is the duration of one working cycle (one full revolution of the crank). Since the automatic regime usually corresponds to a higher number of strokes per minute (spm), and manual operation to lower spm, the nominal pulse duration for synthesis of the vibration isolation system can be assumed to be calculated for n_{max}, the maximum rate of spm that is usually realized in the automatic mode. Thus, the nominal pulse duration is

$$\tau = 0.1(60/n_{\max}) = 6/n_{\max} \qquad \text{Eq. (2.3.6)}$$

When the press is rigidly attached to the floor, the natural frequencies of the lower rocking modes are $f_{x\beta}$, $f_{y\alpha}$ = 15–25 Hz, close to the fundamental vertical natural frequency of the floor. Thus, for widely used presses with n_{max} = 60–200 spm, dynamic loads to the floor may be amplified by a transmissibility factor 1.2–1.5, in accordance with the "shock spectrum" for the versed-sine pulse given in Section 1.8. Sometimes, presses are fastened to inertia (foundation) blocks whose mass is about equal to the press mass. This reduces the dynamic pulse transmission to the environment by $\sim 50\%$, thus the transmissibility factor is about $\mu = 0.6$–0.75. Considering that installation of the press on vibration isolators would attenuate the most objectionable high frequency components of the pulse much more than the magnitude of the pulse, the required degree of attenuation of the pulse magnitude by vibration isolation can be assumed as $\mu = 0.5$. From this, and assuming damping in

170 • PASSIVE VIBRATION ISOLATION

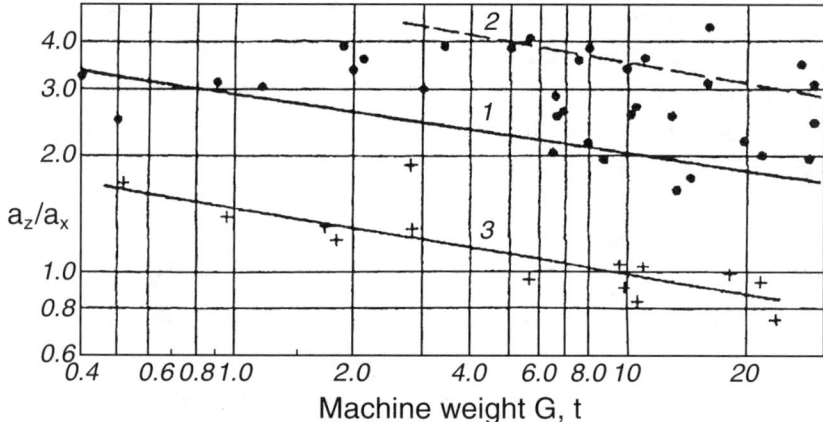

Fig. 2.3.3. Average Ratios of Mounting Points Location for Forming Machines vs. Machine Weight (●—Vertical Mechanical Punch Presses; +—Horizontal Cold Headers).

the isolators $\delta_v = 0.6$, from the shock spectrum plot in Fig. 1.8.2(b) obtain the required pulse duration/natural period ratio $\tau/T_v \approx 0.2$, or

$$T_v = \tau/0.2 = (6/0.2)(1/n_{max}) = 30/n_{max}(\sec); \; f_v = f_{x\beta} = n_{max}/30(Hz).$$
Eq. (2.3.7)

In some cases, presses are installed on vibration isolators instead of rigid mounts just to simplify the installation/relocation procedures and to eliminate anchoring the machines, as well as to reduce noise transmission. In such cases, no attenuation of the pulse magnitude is required or $\mu = 1$ (while the high-frequency vibration and noise producing spectral components of the pulse are still attenuated). In such cases, from Fig. 1.8.2(b) $\tau/T_v \approx 0.4$, or

$$T'_v = \tau/0.4 = 15/n_{max}(\sec); \; f'_{x\beta} = n_{max}/15 \, (Hz). \qquad \text{Eq. (2.3.8)}$$

The main parameter of vibration isolation systems is vertical natural frequency f_{vz}, which is the basis for selecting isolators. It can be easily determined from known $f_{x\beta}$ using Eq. (1.6.6) if a_x/a_z and η_x are known. Here a_z is the C.G. elevation over the supporting surface (floor), a_x is an average distance in the x-direction between the projection of the C.G. on the supporting surface and the mounting points, and η_x is the stiffness ratio of the employed vibration isolation mounts. Fig. 2.3.3 shows values of a_z/a_x for a multitude of commercially available stamping presses as a

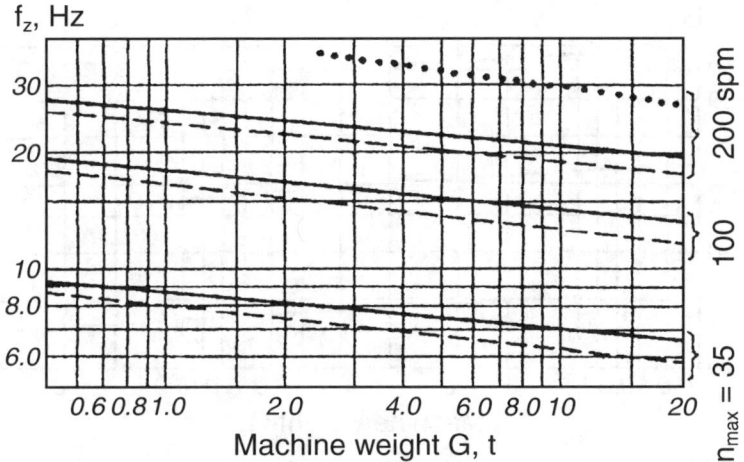

Fig. 2.3.4. Recommended Natural Frequencies for Vibration Isolation of Vertical Punch Presses.

function of press size (weight) W. Line 1 represents the prevailing correlation of a_z/a_x vs. W, while line 2 represents the designs with high C.G. (large a_z) and/or small footprint (small a_x), less suitable for installation on vibration isolators (see below).

Transmission of dynamic loads to the floor from low spm presses usually is not dangerous for the supporting structures. With higher spm, dynamic loading of the supporting structures increase. There might even be a closeness to resonance between the higher harmonics of the pulse train and natural frequencies (usually 15–40 Hz) of the supporting structures. Thus, for isolation of low spm mechanical presses ($n_{max} \leq$ ~100 spm), Eq. (2.3.8) can be adopted, while for high-speed presses ($n_{max} \geq$ ~ 200 spm), Eq. (2.3.7) can be used. Intermediate values of $f_{x\beta}$ can be obtained by interpolation for intermediate values of n_{max}. Fig. 2.3.4 shows recommended vertical natural frequencies f_{vz} of press installations based on the above considerations for presses represented by line 1 in Fig. 2.3.3, and for two typical stiffness ratios of vibration isolators: $\eta = 4.0$ (solid line) and $\eta = 2.5$ (broken line).

For high C.G./narrow base presses (line 2 in Fig. 2.3.3) the required f_{vz} are much higher (dotted line in Fig. 2.3.4), and still an intense rocking of the machine can be observed. This situation can be corrected by extending a_x by mounting the press on a frame or rails. The best results can be achieved if the press designers were considering vibration

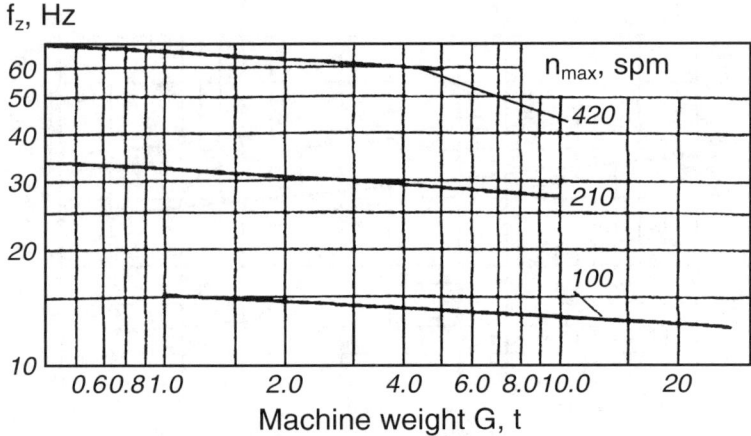

Fig. 2.3.5. Recommended Natural Frequencies for Vibration Isolation of Horizontal Cold Headers.

isolation installation during the press design process and were taking (simple) measures aimed for reduction of the a_z/a_x ratio.

While stamping presses generate dynamic loads in the vertical plane, some production machines (cold heading machines, injection molding machines, etc.) generate intensive dynamic loads of the conservative nature in a horizontal direction. Usually, such machines have heavy reversing masses generating large horizontal inertia forces in addition to the conservative pulses associated with the production process. Such forces may result in very intense rocking of the machine on vibration isolators, somewhat reduced by usually large values of a_x. Reduction of the undesirable rocking can be achieved by increasing stiffness of the isolators. Because of this, Eq. (2.3.8) should be applied. Fig. 2.3.5 shows values of f_{vz} obtained for cold heading machines using Eq. (2.3.8); statistics of a_z/a_x vs. W is shown by line 3 in Fig. 2.3.3, and $\eta_x = 2.5$.

2.3.4 Objects Generating Pulses of Inertial Nature [2.25]

Machines generating pulses of inertial nature, such as forging hammers, mold conditioning (jolting) machines, impact testers, etc., are the most hazardous industrial sources of vibration. Forging (drop) hammers combine high process efficiency with a positive effect on the mechanical properties of the produced parts and with a relatively simple and inexpensive design. However, applications of forging hammers are

Principles and Criteria of Vibration Isolation • 173

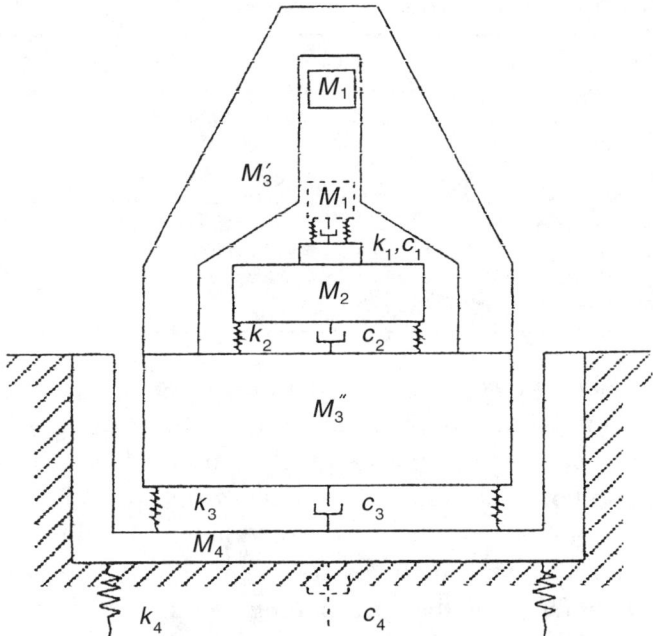

Fig. 2.3.6. Dynamic Model of Forging Hammer Isolation System.

diminishing, to a large extent due to the unacceptable intensity of vibrations transmitted to the environment and high costs of (and large room required for) the means for abating (isolating) these vibrations. Usually a hammer is installed on a massive foundation block, which is supported by a multitude of compliant steel springs. The foundation block plays a dual role. On the one hand, it provides for a steady system even with low natural frequencies of the isolation system (f_{vz} = 2–7 Hz), and on the other hand, it improves efficiency of transmission of the impact energy to the billet (see Section 2.3.1A for the similar role of the inertia mass for shakers). Some damping is provided by powerful viscous friction units or by large rubber cubes, but damping is usually low (δ = 0.1~0.3). The first stage of vibration isolation is provided by oak beams or reinforced rubber belting pads placed under the anvil. Analytical expressions for transmissibility in this complicated system are derived below, using a modified "shock spectrum" technique (Section 1.8). These expressions open the possibilities for reducing costs and size of vibration isolation means for forging hammers and other objects that generate powerful inertial pulses without losing their performance.

Table 2.8. Approximate Natural Frequencies of Soils

Soil type	f_{n4}, Hz	Soil type	f_{n4}, Hz
Peat	7.5	Very light soft clay	12
Peat, 2 m thick, overlaying sand	12.5	Stiff clay	19
Light waterlogged sand	15	Tertiary clay, moist	21.8
Fairly dense medium sand	24.1	Gravelly sand with clay lenses	19.4
Fine sand with 30% medium sand	24.3	Silt and sand mixed	23.3
Uniform coarse sand	26.2	Limestone	30
Very dense mixed grain sand	26.7	Granite	40
Dense pea gravel	28.1		

Fig. 2.3.6 shows a schematic model for an isolated forging hammer; m_1 represents mass of dropping tup with the upper die; m_2 is the anvil, $m_3 = m_3' + m_3''$ represents frame m_3' attached to foundation block m_3''; m_4 is the foundation box. Stiffness and damping, respectively, of the die-billet subsystem are k_1, c_1; of the first isolation stage between anvil and foundation block k_2, c_2; of main isolators k_3, c_3; and effective stiffness and damping coefficients for the surrounding soil are k_4, c_4. Usually, load-deflection characteristics of soils are nonlinear of the stiffening type, close to the CNF-type characteristics. Thus, the system k_4, c_4 can be characterized by the (partial) natural frequency f_{n_4} determined by the soil type, Table 2.8 [2.26]. Damping of most soil types is quite high, $\delta \approx 0.5$–0.6 for medium vibration amplitudes ($a = 2$–3 μm) and much less for lower amplitudes, similar to the properties of wire-mesh materials described in Section 3.3.1A. The natural frequencies of other partial subsystems for typical vibration isolated hammers are: for $m_2 - k_2 - m_3 - \sim 50$ Hz; for $m_3 - k_3 - m_4 -$ between 2 and 7 Hz; for the foundation box-soil subsystem $m_4 - k_4 -$ usually 15–25 Hz (see Table 2.8). Due to substantial differences between these frequencies, dynamic coupling between the subsystems is not very significant (see Section 1.5), and they are considered as independent in the first approximation below.

The impact pulse in forging hammers can be approximated as a half-sine pulse whose magnitude P_o and duration τ_p depend on the billet material, its temperature, design, and size of the die. The worst case is that of the impact of the upper die against the lower die without billet material inside. For this case, the pulse parameters were measured in [2.27] for different weights of the dropping parts (Table 2.9).

Due to the pulse character of excitation, it is natural to use the shock spectrum method (Section 1.8). For a load pulse P_o acting on a two-mass

Table 2.9. Impact Pulse Parameters for Forging Hammers [2.27]

Dropping Mass (tup + upper die), m_1, t	1	2	3	5	5.5	10	15	20
Pulse duration, τ_{p1}, ms	1.0	1.26	1.44	1.69	1.75	2.07	2.45	2.64
Load pulse magnitude, P_o, kN	1,925	2,830	3,700	5,130	5,500	7,500	10,890	12,400

system, as in Fig. 2.3.7, the peak load in the flexible connection k is, per Eq. (1.8.2),

$$P_{k\max} = \frac{m_b}{m_a + m_b} A P_o, \qquad \text{Eq. (2.3.9)}$$

where A is the shock spectrum transmissibility. Since the first period of the response to the pulse excitation can be approximated as a versed-sine pulse, and the subsystems in the model in Fig. 2.3.6 are considered as dynamically independent, the first period of the response of one subsystem can be considered as a new versed-sine pulse acting on the next subsystem in the chain.

The forging load pulse (duration τ_{p1}, magnitude P_o) acts on anvil m_2 when tup m_1 drops. According to Eq. (2.3.9), the maximum force in isolator k_2, c_2 is

$$P_{2\max} = \frac{m_3}{(m_1 + m_2) + m_3} A_1 P_0, \qquad \text{Eq. (2.3.10)}$$

where A_1 is the value of the shock spectrum for the "half-sine" pulse as a function of τ_{p1}/T_{n2}, from Fig. 1.8.2(b) and $T_{n2} = 2\pi\sqrt{[(m_1 + m_2)m_3/k_2(m_1 + m_2 + m_3)]}$ is the natural period of the system $(m_1 + m_2) - k_2 - m_3$. Stiffness k_1 is usually large, thus masses m_1 and m_2 are assumed to act as a single mass. For the typical range of $\tau_{p1}/T_{n2} = 0.05$–0.15, from Fig. 1.8.3 $\tau_{p2} = (0.5$–$0.6)T_{n2} \approx 0.55 T_{n2}$. If a "versed-sine" pulse of duration τ_{p2} and magnitude P_{2max} acts on m_3, the maximum force magnitude in the main vibration isolator k_3, c_3 is

$$P_{3\max} = \frac{m_4}{m_3 + m_4} A_2 P_{2\max} = \frac{m_4}{m_3 + m_4} \frac{m_3}{m_1 + m_2 + m_3} A_1 A_2 P_o.$$

$$\text{Eq. (2.3.11)}$$

176 • PASSIVE VIBRATION ISOLATION

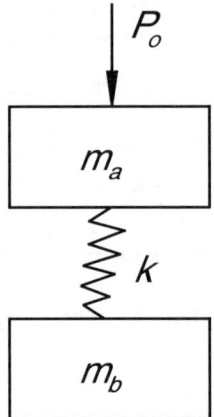

Fig. 2.3.7. Two-Mass Oscillatory System.

Here A_2 is the value of the shock spectrum for the versed-sine pulse from Fig. 1.8.2(b), a function of τ_{p2}/T_{n3}, where $T_{n3} = \frac{1}{f_{n3}} = 2\pi\sqrt{[m_3 m_4 / k_3(m_3 + m_4)]}$ is the natural period of the subsystem $m_3 - k_3 - m_4$. The fundamental natural frequency of the isolated hammer is

$$f_{vz} = \frac{1}{2\pi}\sqrt{\frac{k_3(m_1 + m_2 + m_3 + m_4)}{(m_1 + m_2 + m_3)m_4}} = f_{n3}\sqrt{\frac{(m_1 + m_2 + m_3 + m_4)m_3}{(m_1 + m_2 + m_3)(m_3 + m_4)}}$$

Eq. (2.3.12)

Since usually $\tau_{p1} \ll T_{n2}$, $\tau_{p2} \ll T_{n3}$, analytical expressions for A_1, A_2 from Eq. (1.8.4) can be substituted into Eq. (2.3.11), thus

$$\frac{P_{3\max}}{P_o} = \frac{m_4}{m_3 + m_4} \frac{m_3}{m_1 + m_2 + m_3} 4\frac{\tau_{p1}}{T_{n2}}\left(\cos\frac{\pi\tau_{p1}}{T_{n2}}\right)\frac{\pi\tau_{p2}}{T_{n3}} \quad \text{Eq. (2.3.13)}$$

Substituting Eq. (2.3.12), as well as $\tau_{p2} \approx 0.55 T_{n2}$ into Eq. (2.3.13), arrive at

$$\frac{P_{3\max}}{P_o} \cong 6.9\sqrt{\frac{m_3 m_4^2}{(m_3 + m_4)(m_1 + m_2 + m_3)(m_1 + m_2 + m_3 + m_4)}}\,\tau_{p1} f_{vz}$$

Eq. (2.3.14)

Eq. (2.3.14) describes the effectiveness of the vibration isolation system. It shows that for the short pulse excitation the *transmitted force* ($P_{3\max}$)

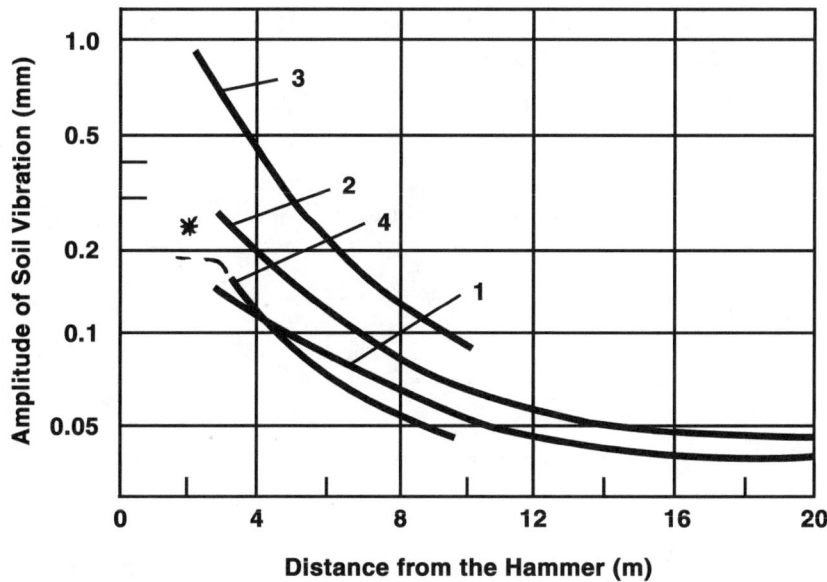

Fig. 2.3.8. Vibration Displacement Amplitudes of Soil Measured in the Vicinity of Forging Hammers [1—Vibration Isolated Hammer, $f_n = 6$ Hz, Dropping Mass 2 t; 2—Unisolated Hammer, $f_n = 20$ Hz, Dropping Mass 0.75 t; 3—Unisolated Hammer, $f_n = 18$ Hz, Dropping Mass 2 t; 4—Vibration Isolated Hammer, $f_n = 4.5$ Hz, Dropping Mass 5 t; *—Computed Amplitude for Case "4" Using Eq. (2.3.16)].

is proportional to the natural frequency of the isolation system, while in isolation cases of steady (e.g., sinusoidal) vibrations the *transmitted force is proportional to the second power of the natural frequency*. However, in the latter case the vibration transmitted to the supporting structure or to the soil is of the same frequency as the excitation, while for the pulse excitation the transmitted vibration has a less hazardous low-frequency equal to the natural frequency of the main isolation system. Within the assumed approximations, the effectiveness does not directly depend on stiffness k_2 of the auxiliary isolation system.

Force $P_{3\max}$ is acting on the foundation box. Its vibration amplitude is approximately equal to its static deflection under the load $P_{3\max}$, since the latter changes with frequency $f_{vz} = 2\text{--}7$ Hz, substantially lower than the natural frequency $f_{n4} = 15\text{--}20$ Hz of the foundation box—soil ($m_4 - k_4$)

subsystem (see Table 2.8). Since soil stiffness k_4 is nonlinear, the effective stiffness k_4 can be calculated as

$$k_4 = 4\pi^2 f_{n4}^2 m_4. \qquad \text{Eq. (2.3.15)}$$

Thus, the amplitude of soil vibration in the first approximation is

$$a_s = \frac{P_{3\,\max}}{k_4} = \frac{P_{3\,\max}}{4\pi^2 f_{n4}^2 m_4}$$

$$= 0.17 \sqrt{\frac{m_3 m_4^2}{(m_3 + m_4)(m_1 + m_2 + m_3)(m_1 + m_2 + m_3 + m_4)} \cdot \frac{\tau_{p1} f_{vz}}{f_{n4}^2}}.$$

Eq. (2.3.16)

Example. For hammer with $m_1 = 5{,}000$ kg $= 5$ t, $m_2 = 20\, m_1$, $m_3 = 60\, m_1$, $m_4 = 40\, m_1$, $\tau_{p1} = 0.0017$ s, $f_{vz} = 4.5$ Hz, $P_o = 5.1 \times 10^6$ N, from Eq. (2.3.14) $P_{3\max} = 0.0165\, P_{o\max} = 0.083 \times 10^6$ N. From Eq. (2.3.16), displacement amplitude of the foundation box $a_s = 2.3 \times 10^{-4}$ m (* in Fig. 2.3.8). The measured amplitude of the foundation box vibration was $a_s = 1.8 \times 10^{-4}$ m at ~3 m from the hammer (line 4 in Fig. 2.3.8). The discrepancy ~24% is due to several reasons: less severe impact conditions during the test (a billet was forged instead of rigid dies coimpacting); vibration decay in the soil; damping c_2, c_3, c_4 as well as nonlinearity of soil behavior were not considered in the analysis.

The damping of anvil isolators typically amounts to the value of log decrement $\delta_2 = 0.2-0.3$; for the main isolators (steel springs with parallel dampers, usually rubber cubes), usually $\delta_3 = 0.1-0.3$. Since fundamental natural frequencies of industrial and residential buildings are in the same range as the natural frequencies of the isolated foundations, quasi-resonant amplification of building vibrations may occur due to slow decay of soil vibrations resulting from one hammer blow. This phenomenon was observed in [2.28] on a building 400 m from an isolated hammer ($f_{vz} = 3.9$ Hz).

The increased damping of both anvil and main isolators would reduce, albeit not very significantly, the shock spectrum values in accordance with Eq. (1.8.5). For δ between 1.0 and 1.5 that can be realized with some rubber and wire-mesh materials (see Chapter 3), A_1 and A_2 in Eq. (2.3.11) can be reduced by 1.15 to 1.2 times, thus allowing increasing f_{vz} by 1.3–1.4 times without any detriment to the effectiveness of isolation.

In addition, the increased damping in the main isolators accelerates decaying of the foundation vibrations and thus reduces adverse effects of forging hammer-induced vibrations on humans, on precision equipment, and on structures. As can be seen from Fig. 1.8.2(e), increase of δ of the excitation pulse "decaying sine wave" from 0.3 to 1.25 reduces the relative vibrations (i.e., the hazard from the hammer operation) by 2.55 times for the most dangerous quasi-resonant case, and for the case when frequency of the excitation pulse f is $2f_s$ (natural frequency of the excited structure), by about 1.25 times.

Considering both effects of the damping increase, the principal parameter of the vibration isolation system for an object generating inertial pulses, such as forging hammers, becomes not its main natural frequency f_{vz}, but a shock isolation criterion

$$\Phi_{sh} \cong \frac{f_{vz}}{\delta^{0.25}}. \qquad \text{Eq. (2.3.17)}$$

Thus, increasing damping in both main and anvil isolators from $\delta = 0.3$ to 1.25 allows an increase in the natural frequency of the vibration isolation system by a factor of ~1.5, from $f_{vz} = 2$–7 Hz to $f_{vz} = 3$–10 Hz without reducing isolation effectiveness. This is a very significant change since stiffer springs and/or reduced mass m_3 of the foundation block can be used without degrading the system performance and steadiness, but significantly cutting costs. The higher natural frequencies can be realized by using elastomeric springs combining stiffness element k_3 and damping element c_3 in Fig. 2.3.6 in one unit, thus further reducing costs.

These conclusions were validated by testing a 0.75 t hammer installed on the foundation block supported by several layers of rubber isolating carpets KB-2 (see Chapter 4), thus realizing $f_n = 7$ Hz and $\delta = 0.8$. The system performance was found to be satisfactory, similar to the performance of spring-suspended foundations having significantly lower natural frequencies.

Further reduction of costs and dimensions of vibration isolation systems for objects producing inertial pulses can be achieved by using vibration isolation foundations with motion transformers (Section 1.4.4).

2.4 GENERAL PURPOSE MACHINERY AND EQUIPMENT

For this group, which is the largest and the most diversified one, selection of isolation system parameters is influenced by factors such as dynamic stability of the production process, vibration level of the machine, and dynamic loads in the joints (especially bearings). In many cases, installation on vibration isolators is used to remove the need for anchoring the object to the floor in order to prevent "walking" of the vibration-generating machine, since the vibrations reduce the effective friction between the machine and the floor. Vibration isolation is used also to reduce noise levels in the shop and in adjacent rooms.

2.4.1. Influence of Mounting Conditions on Dynamic Stability (Chatter) [2.29]

The dynamic stability of working processes in production machinery and equipment, especially of cutting processes, can be disturbed by several mechanisms, all of which, in effect, introduce negative damping into the working area (e.g., see [2.30]). Thus, when the total damping balance in the system comprising the machine structure and the production process becomes negative, chatter (a self-excited vibratory process) develops. Fastening the machine bed with stiff mounts to a rigid foundation improves the dynamic stiffness of the machine structure and, consequently, its chatter resistance. On the other hand, compliant mounts, such as vibration isolators, may not have this type of effect.

The influence of natural frequency and damping of the isolation system on the dynamic stability of the production process can be analyzed using correlations developed in Section 1.5. For a metal-cutting machine tool, influence of its vibration isolation system on the dynamic stability can be estimated by using the dynamic model in Fig. 1.5.1(a), where $m_1 = m_u$ is the effective mass of the upper (tool-carrying) unit of the machine; $m_2 = m_b$ is the effective mass of the bed/frame; $k_{12} = k_m$ is the effective structural stiffness coefficient of the machine; $k_2 = k_v$ is stiffness of the isolators; and $m_3 = \infty$ is mass of the foundation. This system has two natural frequencies: f_1 and f_2. The lower natural frequency f_1 is close to the partial natural frequency f_v of the isolation subsystem $m_b - k_v$; the higher natural frequency f_2 is close to the partial natural frequency f_m of the structural subsystem $m_u - k_m - m_b$. The damping constant $\Delta_2 = 2\pi f_2 (c/c_{cr})_2$ of the higher mode (responsible for the chatter resistance) of the full system $m_u - k_m - m_b - k_v$ can be expressed (according to Section 1.5) in terms of damping constants of the partial subsystems $\Delta_m = 2\pi f_m (c/c_{cr})_m$ and $\Delta_v = 2\pi f_v (c/c_{cr})_v$, and the coupling coefficient $\sigma = 2\gamma \frac{f_m f_v}{f_m^2 - f_v^2}$, see (1.5.5), where elastic coupling coefficient $\gamma = \sqrt{m_b/m_b + m_u}$; and $f_m = \frac{1}{2\pi}\sqrt{k_m(m_b + m_u)/m_b m_u}$, $f_v = \frac{1}{2\pi}\sqrt{k_v/m_b}$ are partial natural frequencies of the respective subsystems. Per Eq. (1.5.12)

$$\Delta_2 = \frac{1}{2}\frac{\left(1+\sqrt{1+\sigma^2}\right)\Delta_m - \left(1-\sqrt{1+\sigma^2}\right)\Delta_v}{\sqrt{1+\sigma^2}}. \quad \text{Eq. (2.4.1)}$$

Since for typical machine tools $f_m \gg f_v$, e.g. see Table 2.1, then

$$\sigma^2 = 4\gamma^2 \frac{f_m^2 f_v^2}{(f_m^2 - f_v^2)^2} \approx 4\gamma^2 \frac{f_m^2 f_v^2}{f_m^4} = 4 \frac{m_u}{m_b + m_u}\frac{f_v^2}{f_m^2} < 1 \quad \text{Eq. (2.4.2)}$$

and

$$\Delta_2 \approx \frac{1}{2}\left[\left(1 + \frac{1}{1+\sigma^2/2}\right)\Delta_m + \frac{\sigma^2}{2+\sigma^2}\Delta_v\right]. \quad \text{Eq. (2.4.3)}$$

At the stability/chatter threshold, the stabilizing effect of the structural damping is canceled by the destabilizing effect of the cutting process, so that $\Delta_m = 0$ and the only damping in the system is provided by the mounting elements. Thus from Eq. (2.4.3) at the threshold of stability

$$\Delta_2 = 2\pi f_1 \left(\frac{c}{c_{cr}}\right)_1 \approx \frac{1}{2}\frac{\sigma^2}{2+\sigma^2}\Delta_v \approx \frac{\sigma^2}{4}\Delta_v = \frac{\sigma^2}{4}\left[2\pi f_v\left(\frac{c}{c_{cr}}\right)_v\right], \quad \text{Eq. (2.4.4)}$$

or

$$\left(\frac{c}{c_{cr}}\right)_1 \approx \frac{\sigma^2}{4}\frac{f_v}{f_1}\left(\frac{c}{c_{cr}}\right)_v. \quad \text{Eq. (2.4.5)}$$

Substituting Eq. (2.4.2) into Eq. (2.4.5), using identity $(c/c_{cr})_v = \delta_v/2\pi$, and considering that $f_1 \approx f_m$, the final expression becomes

$$\left(\frac{c}{c_{cr}}\right)_1 \cong \frac{1}{2\pi}\frac{m_u}{m_b + m_u}\frac{f_v^3}{f_m^3}\delta_v. \quad \text{Eq. (2.4.6)}$$

Table 2.10. Mounting Systems for Lathe Chatter Tests

Mounts	f_{vz}, Hz	δ_{vz}	f_{vy}, Hz	δ_{vy}	Φ_{dsz}
A	32	0.39	13	0.09	12,800
B	30	0.57	10	0.67	15,400
C	24	0.5	7.8	0.5	6,900
D	20	0.9	6.3	0.88	7,200
E	17	0.42	5.4	0.44	2,060
F	12	0.38	3.2	0.3	660

The meaning of Eq. (2.4.6) is as follows: if dynamic stability of the machine tool is sensitive to mounting conditions, the effect of the mounting system is determined by the magnitude of the dynamic stability criterion

$$\Phi_{ds} = f_v^3 \delta_v. \qquad \text{Eq. (2.4.7)}$$

This criterion states that a small reduction in the natural frequency of the mounting or vibration isolation system (i.e., stiffness of the attachment to the foundation) can be compensated only by a substantial increase in the damping of the mounting system.

It has to be understood that the mounting elements or vibration isolators influence chatter resistance of machine tools only when the structural stiffness k_m and/or natural frequency f_m and/or structural damping δ_m are low, or other structural and/or process features are not optimal. Machine tools rigid by design are usually not noticeably influenced by the mounting conditions. It is possible that the model in Fig. 2.2.8, and thus the machine sensitivity to the mounting conditions, may change as a result of change in its internal configuration (position of the tool carriage in a lathe, position of the gantry on a plano-milling machine, etc.).

Eq. (2.4.7) was validated by extensive tests performed on a medium-size lathe [2.29]. Chatter resistance, which was characterized by the maximum depth of cut without chatter (t_{lim}), was measured with different sets of mounts (Table 2.10). Mounts A are all-metal jack mounts; all other mounts (B–F) are constant natural frequency (CNF) isolators with rubber flexible elements, providing various natural frequencies and degrees of damping of the installation/vibration isolation system. Fig. 2.4.1 shows that the values of the maximum stable depth of cut t_{lim} for different mounting conditions are closely correlated with values Φ_{ds} calculated for the vertical natural frequency of the mounting

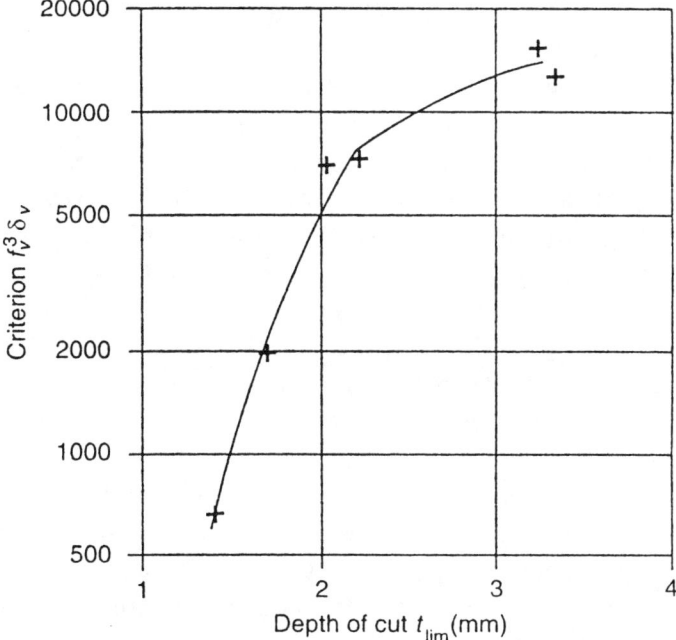

Fig. 2.4.1. Maximum Stable Depth of Cut t_{lim} vs. Criterion $f_v^3 \delta_v$.

system. It can be seen that the stable depth of cut t_{lim} for a lathe installed on high-damping vibration isolators (B) is about the same as t_{lim} for the rigidly-mounted lathe (A).

2.4.2 Vibration Levels of General Purpose Machines

In some cases intensive vibrations with natural frequencies of the mounting system interfere with the normal performance of general purpose machines. A real machine has a multitude of vibration sources acting in different directions and with different frequencies; dynamic characteristics of the floors and other supporting building structures may also affect the vibration levels. While non-excessive vibration levels may not be detrimental for the performance of general purpose machines, they may cause annoyance and discomfort for the operators. Though only an experimental selection of isolators is reliable, crude analytical criteria proposed in this section proved to be useful in the development of optimal isolators.

The most common source of vibrations are centrifugal forces generated by rotating components that are not perfectly balanced, such as gears, pulleys, flanges, etc. The centrifugal force amplitude for an unbalanced rotating part is proportional to the second power f^2 of its rotational frequency f. However, tolerances on the unbalance become stricter for the components rotating with higher f. As a result, the relation between f and the allowable amplitudes F_{cf} of centrifugal forces for specific parts and components must be modified. Survey of industrial standards shows that the allowable F_{cf} is proportional to $\sim f^{0.5}$ for rotors of electric motors; to $\sim f^{-0.5}$ for grinding wheels; to $\sim f^{0.5} - \sim f^{-0.5}$ for unbalanced blanks for machining on lathes, such as forgings, castings, etc. As a first approximation, it can be assumed that the rated amplitudes of centrifugal forces do not depend on frequency. The same is often true for other sources of machine vibrations. For example, the most pronounced spectral amplitudes of vibratory displacement for several types of machine tools, determined during idling and averaged over an ensemble, can be considered as frequency-independent in a rather broad frequency range, about 10–100 Hz. Thus, resonances in the "machine on its mounts" system are very probable both for rigid and for soft mounting elements.

Assuming amplitudes F_o of exciting forces to be independent of frequency, and mass m_b of the machine bed/frame to constitute the bulk of the machine mass, the resonance amplitude of the machine bed/frame on its mounts is

$$(x_b)_{res} = \frac{F_o}{k_v} \frac{\pi}{\delta_v} = \frac{F_o}{4\pi f_v^2 m_b \delta_v} \qquad \text{Eq. (2.4.8)}$$

where k_v, δ_v are stiffness and damping of the mounts; and f_v is the natural frequency of the machine on its mounts. The amplitude of relative vibrations in the working zone is then

$$(x_{rel})_{f_v} = \frac{F_o}{4\pi f_v^2 m_b \delta_v} \cdot \frac{1}{\frac{m_b}{m_b + m_u} \frac{f_m^2}{f_v^2} - 1} \qquad \text{Eq. (2.4.9)}$$

If $f_v \ll f_m$, which is usually true for installation on vibration isolators, then

$$(x_{rel})_{f_v} \cong \frac{m_b + m_u}{m_b^2} \frac{F_o}{4\pi f_m^2 \delta_v} \qquad \text{Eq. (2.4.10)}$$

Thus the maximum amplitudes of relative displacement in the working area that are based on the above assumptions do not depend on the natural frequency of the isolation system, but only on its damping, which could be considered as a vibration level criterion,

$$\Phi_{vl} = \delta_v.$$ Eq. (2.4.11)

In case of stiff mounts, when f_m and f_v are commensurate, the relative displacement would substantially increase. Fastening the machine to a massive foundation block is equivalent to increasing m_b in Eq. (2.3.9) with a corresponding reduction of relative vibrations.

Often the vibratory velocity level is considered as an indicator of the machine vibration status. An equally hazardous or annoying action of vibration on personnel corresponds to an equal vibratory velocity level at frequencies higher than 2–8 Hz (International Standard ISO 2631). From Eq. (2.4.8), the maximum velocity amplitude at resonance is

$$(v_o)_{f_v} = 2\pi f_v (x_{m_v})_{f_v} = \frac{F_o}{2m_b f_v \delta_v}$$ Eq. (2.4.12)

Thus, a comparison of alternative mounting systems by the vibratory velocity level can be performed by the criterion

$$\Phi_{vv} = f_v \delta_v.$$ Eq. (2.4.13)

Thus, to reduce the vibratory velocity level of an isolated object, increase in the natural frequency and in damping have similar effects.

2.4.3 Influence of Vibration Isolation on Bearing Loads

With the proliferation of installation of industrial machinery and equipment on vibration isolators, it was noticed that in many cases the useful life of bearings inside the machines installed on vibration isolators was lengthened as compared with rigidly installed similar machines. A substantial component of dynamic loads acting on bearings is the centrifugal force due to imbalance in the rotor supported by the bearings.

Influence of vibration isolation on dynamic loads of rotor bearings caused by imbalance of the rotor can be studied on a simple model in Fig. 2.4.2 [2.31]. While the main effect of vibration isolation is reduction of transmission of the unbalanced force $F(t)$ to the foundation as characterized by the force transmissibility coefficient, μ_F [e.g., see Eq. (1.4.4)], the secondary effect of vibration isolation is change of dynamic load $R(t) = R_o \cos\omega\, t$ in the bearing. The bearing is represented in the model in Fig. 2.3.2 by its stiffness k_1 and damping c_1 (e.g., see [2.16]), and influence

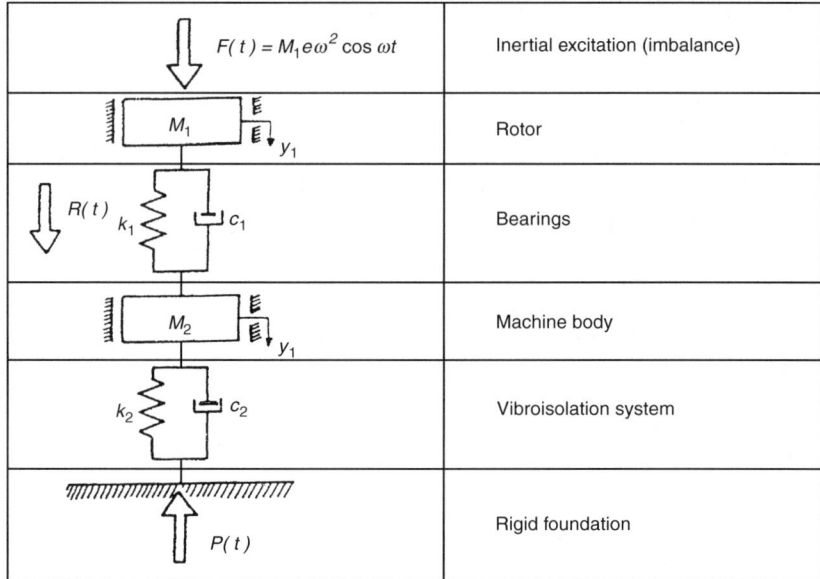

Fig. 2.4.2. Model of Dynamic Loading for Rotor Bearings.

of the isolator k_2, c_2 in the model on the dynamic load amplitude R_o can be expressed as *influence function*

$$W = \frac{(R_o)_v}{(R_o)_r} \qquad \text{Eq. (2.4.14)}$$

where $(R_o)_v$ is the dynamic load amplitude in the bearings of the isolated object comprising the rotor; and $(R_o)_r$ is same for the rigidly installed object ($k_2 = \infty$). It is shown in [2.31] that in a broad frequency range, both $\mu_F < 1$, and $W < 1$, thus vibration isolation is accompanied by reduction of dynamic loads in bearings. Fig. 2.4.3 shows μ_F and W vs. frequency ratio f/f_v, where $f_v = 1/2\pi\sqrt{k_2/(M_1 + M_2)}$, for a case of the light rotor ($M_1/M_2 = 0.1$) [Fig. 2.4.3(a)], and for the case of a heavy rotor ($M_1/M_2 = 1.0$) [Fig. 2.4.3(b)]. Here M_1 is mass of the rotor; M_2 is mass of the machine; k_1 is stiffness of the bearing; and k_2 is stiffness of the isolators. The stiffness ratio is (for both cases) $k_1/k_2 = 10$ and damping ratios for both partial subsystems ($\zeta_1 = c_1/2\sqrt{k_1 M_1 M_2/(M_1 + M_2)}$ for the rotor subsystem $M_1 - k_1 - M_2$ and $\zeta_2 = c_2/2\sqrt{k_2 M_2}$ for the vibration isolation subsystem $M_2 - k_2$) are $\zeta_1 = \zeta_2 = 0.005$. It can be seen that the frequency ranges in which both $\mu_F < 1$ and $W < 1$ are quite broad. This indicates the simultaneous presence of both the vibration isolation

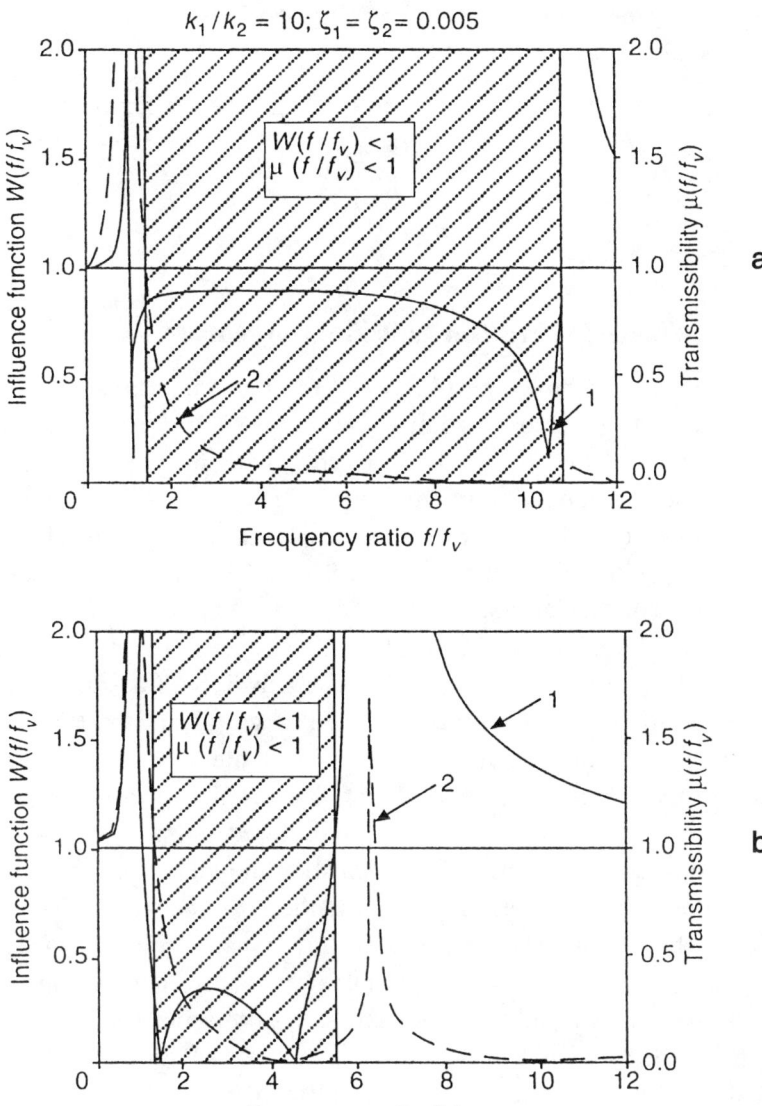

Fig. 2.4.3. Frequency Dependence of Influence Function W (1) and Transmissibility μ_F (2): (a) Light Rotor ($M_1/M_2 = 0.1$); and (b) Heavy Rotor ($M_1/M_2 = 1.0$).

effect and the effect of reduction of dynamic loads on bearings. While this frequency range is broadening for lighter rotors, the reduction of the dynamic load is more significant for heavier rotors. For $\zeta_1 = \zeta_2 = 0$, close-form expression for lower f_l and upper f_u frequency limits of the area where $W < 1$ is [2.31]

$$\frac{f_l}{f_v}, \frac{f_u}{f_v} = \sqrt{\frac{1}{2}(1+\mu)\left[1 + \frac{2+\mu k_1}{2\mu} \frac{k_1}{k_2} \mp \sqrt{\left(1 + \frac{2+\mu k_1}{2\mu} \frac{k_1}{k_2}\right)^2 - \frac{4 k_1}{\mu k_2}}\right]}.$$

Eq. (2.4.15)

2.4.4 Influence of Vibration Isolation on Noise

It is obvious that vibration isolation is reducing transmission of high-frequency vibrations across structures and thus should result in reduction of noise in spaces adjacent to a device with moving mechanisms (such as a general purpose machine, a machine tool of ordinary precision, etc.), but separated from this device by structural partitions. An example of such transmission reduction is illustrated in Section 2.6. The fate of the near-field noise at the device itself is less obvious. As shown in Section 2.6, noise level at the device can stay the same or even increase. There are published observations that noise levels measured at a stamping press were reduced when the press was installed on vibration isolation mounts.

A noticeable reduction in noise can be achieved if the majority of production machinery in a shop is installed on vibration isolators, most probably due to reduction of the floor-radiated noise. Fig. 2.4.4 [2.32] shows comparison of the noise level in a shop in which ten lathes operate. Initially, these lathes were rigidly attached to the floor (by cement grouting), then they were reinstalled on constant natural frequency rubber-metal isolators, $f_{vz} = \sim 20$ Hz. For the machines operating with the same regimes, the overall (A-weighted) sound pressure level reduction is 2 dBA (87.5 dBA for vibration isolated machines vs. 89.5 dBA for rigidly installed machines). An even more significant 4–5 dBA reduction was measured in the most annoying high-frequency range (2,000–3,000 Hz).

2.5 VIBRATION ISOLATED OBJECTS INSTALLED ON NON-RIGID SUPPORTING STRUCTURES

A majority of stationary (e.g., production and measuring) equipment are installed by means of vibration isolators attached to rigid supporting

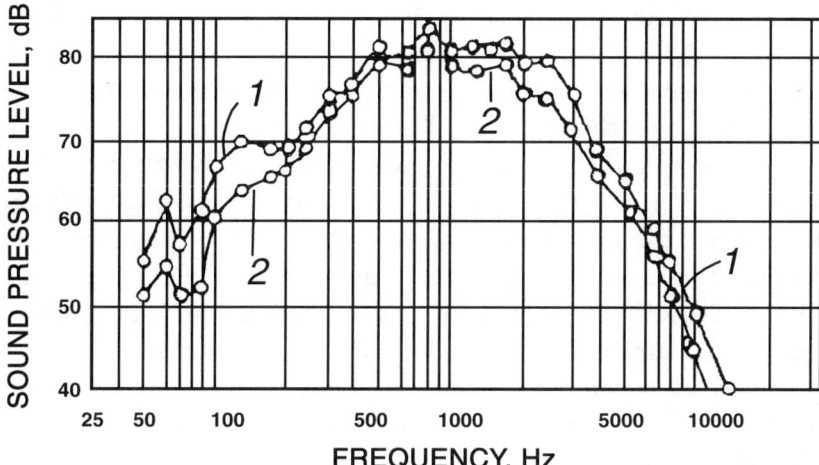

Fig. 2.4.4. Noise Spectra for Machine Shop with 10 Operating Lathes (1—Lathes Rigidly Attached to Floor; 2—Lathes are Installed on Rubber-Metal CNF Isolators, $f_{vz} = 20$ Hz).

structures, such as foundation plates and individual foundation blocks. However, some stationary equipment requiring vibration isolation, both vibration-sensitive and vibration producing, is installed on non-rigid structures, such as floors over the basement or on higher floors of multi-story buildings. The installation of production equipment on higher floors is rather widely practiced in Europe, especially in the East European countries and recently in Asian countries with high real estate prices. This problem also becomes critically important for objects installed in vehicles, such as car engines, machinery on surface and under-surface ships, aircraft, etc.

This problem can be (and has been many times) analyzed by using a model comprising dynamic characteristics of the isolated object, of the supporting structure, and of the vibration isolators, e.g. [2.33]. Usually, for this type of analysis the dynamic descriptions characterizing each of these subsystems are experimentally determined *impedances* for each subsystem. It seems that such treatment is useful only in the high (acoustical) frequency range wherein the supporting structure, the isolated object, and the isolators (considered as distributed parameters system with corresponding high-frequency resonances; e.g., see Section 1.10) are dynamically interacting. In most cases, the mass of the isolated object is substantially greater than the *"effective mass"* of the supporting structure significantly involved in the vibratory process associated with the object. On the other hand, the effective stiffness of the supporting structure in the vicinity of the object is, usually, substantially greater than stiffness values of the isolators. For production machines which are installed on floors other than foundation plates, mass of the machine is usually 1.5–2.5 times greater than the associated mass of the supporting floor structure that is usually made of thin prestressed-concrete plates. Often, objects such as precision machine tools installed on higher floors, are attached to stiffening plates thus further increasing their masses. The stiffness of the supporting structure is significantly greater than stiffness of the vibration isolators, especially in the horizontal directions, which are usually the most important for isolation of precision machinery. A similar situation exists for engine mounting systems in various vehicles. For example, mass of the engine is three-to-five times greater than the associated mass of the engine cradle, while stiffness of the cradle is much greater than stiffness of the engine mounts. As a result, the influence of the supporting structure on the object natural frequencies is not very significant (usually less than 10–15%), and the most important influence of the non-rigid supporting structure is modification (usually reduction) of effective damping in the isolation system. This statement can be illustrated by analysis in [2.34] in which the vibration isolation system "car engine on engine mounts" was analyzed for the case of rigid foundation and for the case of actual supporting conditions on the low-stiffness vehicle sub-frame. Fig. 2.5.1 shows force transmissibility for both cases: rigid foundation and actual "flexible" supporting structure. The plots are very similar; the resonance peaks of transmissibility are at the same frequencies (representing engine rpm) in both cases. Some peaks have increased height (reduction of damping), some have reduced height (increase of damping), and one peak is the same for both cases. Similar results are obtained in [2.35].

Since the change in damping is the main consequence of non-rigidity of the supporting structure in the low-frequency range, a simplified analysis can be used. The simplified model for an object installed on a non-rigid supporting structure is the model in Fig. 1.5.1(a), where m_1 is mass of the isolated object, k_{12}, c_{12} are stiffness and damping of the isolators, m_2, k_{23}, c_{23} are effective mass, stiffness, and damping of the supporting structure, and $m_3 = \infty$. The techniques presented in Section 1.5 allow for complete analysis of damping distribution in this system and

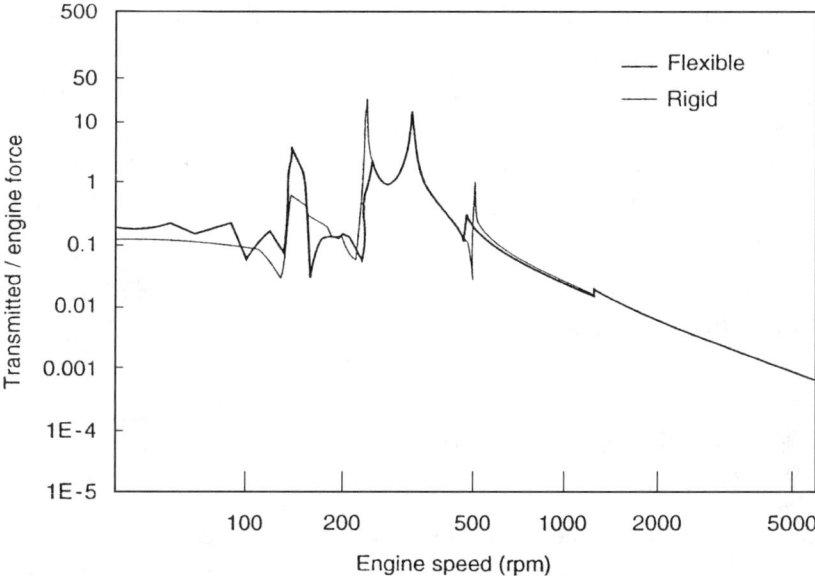

Fig. 2.5.1. Transmissibility for Rigid and Flexible Foundation Models for Engine Mounted in Vehicle (at Transmission Mount Location).

for recommendations on necessary parameter changes for desired system modifications. Some general considerations can be formulated as follows:

1. *Installation of vibration-sensitive objects.* The vibration levels on the upper floors are usually 1.5–3 times higher than those of the ground floor and, accordingly, criterion Φ_p from Eq. (2.2.15) should be 1.4–1.7 times less than it would be for the same device on the ground floor. Since a corresponding reduction in natural frequency may create excessive rocking motions, the proper way to comply with a reduced Φ_p for installation of a vibration-sensitive device on non-rigid structures is to increase damping while not significantly reducing stiffness of the isolators.
2. *Installation of a vibration producing or a general purpose machine.* When a vibration producing device is installed on the upper floor, response of the latter to the excitation is amplified because of small damping of the supporting structure. Compensating it by using compliant isolators may lead to excessive rocking motions. But, since the effective mass of the floor "attached" to the installed machine is much less than the mass of the machine, the dynamic

characteristics, especially damping, of the floor structure in a shop where several machines are installed, *can be controlled* by the proper choice of isolators. By using highly damped isolators it is possible to increase effective damping of the floor (or other non-rigid supporting structure, such as vehicle structure) 2.5–4 fold and, correspondingly, reduce its sensitivity to external excitations, see also [2.33].

Thus, the main feature of an isolator to be used for installation on upper floors or other non-rigid supporting structures is high damping.

2.6 ENGINE AND MACHINERY MOUNTING IN VEHICLES

Objects mounted in vehicles may experience large forces caused by shocks, wave motions for floating vehicles, road unevenness, load variations in drivetrains, etc. Vibration isolation is used to protect the objects. Excessive loads on or displacements of the isolated objects may also occur, therefore motion-limiting snubbers (see Section 1.9) or nonlinear hardening isolators (e.g., see Section 4.5) should be considered when designing the isolators. The loading of isolators on ships in rough conditions at sea occurs at a frequency of about 0.1 Hz and an amplitude of about one half of the object weight. Road excitations may result in loading isolators up to twice the object weight.

A rational engine mounting system design in on-road as well as off-road vehicles is one of the major techniques for improving the "Noise, Vibration, and Harshness" (NVH) environment for the occupants. A rational flexible engine mounting also reduces transmission of road shocks not only to the engine itself, but to the vehicle structure. Suspension parameters for the engine should provide for necessary natural frequencies and vibratory modes of the engine. To improve vibration isolation (both protection of the vehicle structure from the drivetrain excitation and of the engine itself from road-induced excitations, often amplified by a compliant vehicle frame), natural frequencies of the engine suspension and other engine attachments to the structure should be reduced. On the other hand, the attachment of the engine to the vehicle structure should be rather rigid, since it has to accommodate the engine torque without excessive deflections and misalignments. This contradiction (the suspension should be soft, but stiff—a typical contradiction defining many engineering problems—e.g., see [2.36]) is even more acute for front-wheel-drive vehicles, wherein the total torque transmitted to the driving wheels is acting on the engine/transmission mounting system. This situation requires a higher torsional stiffness of the mounting system. However, the other natural frequencies of the system must be low to provide the required isolation in the toughest case—idling of the engine at low rpm. "The low idle" is beneficial from a fuel-efficiency standpoint but low excitation frequencies (the 2nd unbalanced harmonic of engine rpm for four-cylinder engines)—associated with the low idle may cause severe vibrations of the steering wheel [2.37]. Also, it is desirable that the natural frequencies of the drivetrain do not become lower than the fundamental natural frequency of torsional vibrations of the transmission system. While for the rear-wheel-drive vehicles this natural frequency is in the vicinity of ~5 Hz, for the front-wheel-drive vehicles the transmission is shorter and this frequency may reach ~10 Hz. Because of this, the lower boundary of the frequency range of the natural frequencies of the engine suspension system should be raised. This range used to be about 5–30 Hz, but for the front-wheel-drive vehicles it is close to 10–30 Hz. It is considered beneficial to tighten this range down to 10–18 Hz. The range 5–12 Hz was recommended for four-wheel-drive vehicles [2.38].

Typical engine suspension systems use three or four isolators. Some or all of these isolators may be mounted directly onto the body of the vehicle or onto an isolated "antivibration subframe." To resolve the above contradiction, the majority of the developmental efforts related to engine mounting systems are directed to manipulations with locations of the mounts and changes in their arrangements (one approach—using the so-called focal arrangement—is described in Section 1.6). For the majority of production and measuring machinery and equipment, their principal axes of inertia are approximately parallel to the "natural" axes (vertical, longitudinal, transverse). For the vehicle engines, the longitudinal (x-x) principal axis of inertia (the minimum inertia axis) is usually, for a typical multicylinder engine, at an angle $\alpha \approx 10-15°$ to the horizontal crankshaft centerline, Fig. 2.6.1 (smaller values for vehicles with automatic transmissions, which are more massive than manual transmissions). There are many techniques, both experimental and computational, for determining radii of inertia of the engine-transmission unit. If the angle α

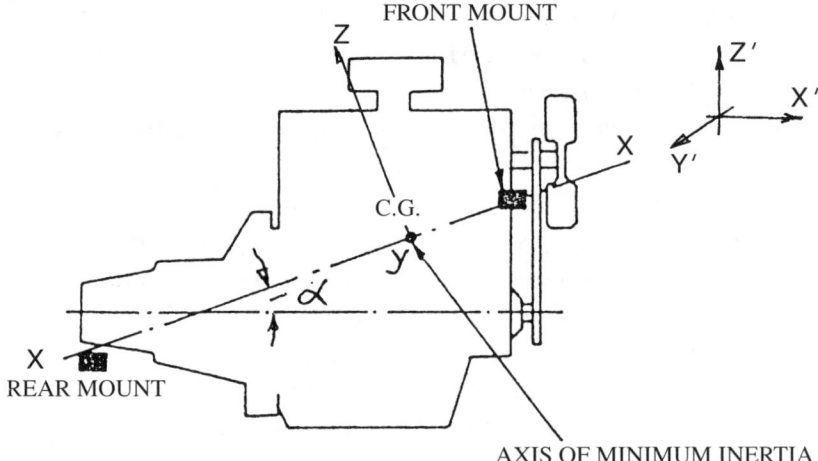

Fig. 2.6.1. Schematic of a Typical Automotive Engine.

is known, principal axes of inertia can be approximated as lines perpendicular to x-x, with the "vertical" axis z-z passing through the C.G. in the plane of the sketch in Fig. 2.6.1, and the transverse axis y-y perpendicular to this plane. After positions of the axes are established, the radii of inertia can be estimated using Eq. (1.1.3), where all the dimensions H are defined in the coordinate frame x, y, z.

Unless CNF isolators are used, three-isolator mounting seems preferable since it is a statically determined system; a side benefit is the lower cost of isolators. Six modes of vibration of the system "engine on its mounts" can be relatively easily decoupled in the coordinate system x, y, z using the approaches described in Section 1.3. However, the major vibration sources in the engine are gas explosions generating driving forces on the pistons, unbalanced inertia forces of the piston-crank mechanisms, and torsional vibrations of the crankshaft-flywheel-transmission system. The vectors of the translational excitations for an in-line engine coincide with the axes of the natural coordinate frame x', y', z', and for the torsional excitation with rotation about axis x'. Thus, if even ideal decoupling in the coordinate frame x, y, z is achieved, actual vibrations along and about axes x, y, z will still be coupled, since the right-hand sides (excitation functions) in the equations of motion contain components of the same excitation forces. It can be noted that the magnitude of the excitation term along the x-x axis, caused by the actual excitation along the axis x'-x', is proportional to $\sin \alpha$.

Principles and Criteria of Vibration Isolation • 195

For in-line engines this excitation-induced coupling is relatively small due to low magnitudes of the angle α. In V-block engines, the main excitation forces have additional inclinations in the y'-z' plane, thus creating additional interrelated excitation components in the equations of motion for different axes.

The same is true for torsional excitations about x-x axis whose components are present in the equations of motion about x' and z' axes.

Although these effects must be considered in the process of designing the engine vibration isolation system, in many cases they are alleviated due to: (a) smallness of angle α ($\sin 15° = \sim 0.26$, $\sin 10° = \sim 0.17$, $\cos 15° = 0.97$); and (b) high degree of balancing in V-block engines, as well as in six and more cylinder in-line engines, which reduces intensity of the excitation and the excitation-induced modal coupling.

Example. An example of the inter-coordinate coupling reduction achieved by narrowing the range of the natural frequencies is described in [2.38]. The changes were achieved by modifications in one (transmission) mount.

Fig. 2.6.2(a) shows natural frequencies and mode shapes for the original engine mounting system. The first mode is a combination of roll, lateral, and yaw motions; the second mode is a combination of pitch and longitudinal motions; the third mode is a relatively pure roll motion; the fourth is a combination of vertical and roll motions; the fifth mode is a relatively pure pitch motion; and the sixth is a predominantly yaw motion with some contribution of lateral motion.

Optimization can be, theoretically, achieved by repositioning of the power-train mounts together with modification of their parameters. However, repositioning of the mounts is usually impractical since the packaging of the engine compartment involves many contradictory design problems. In this case, it was found that just modification of the transmission mount (one-third reduction of its lateral stiffness and one-fifth reduction of its vertical stiffness) resulted in coupling reduction and in a better natural frequencies distribution [Fig. 2.6.2(b)].

The natural frequencies of the modified system are more uniformly distributed in the narrow and desirable range 5–11 Hz. The first mode is still mostly roll, but with less lateral motion

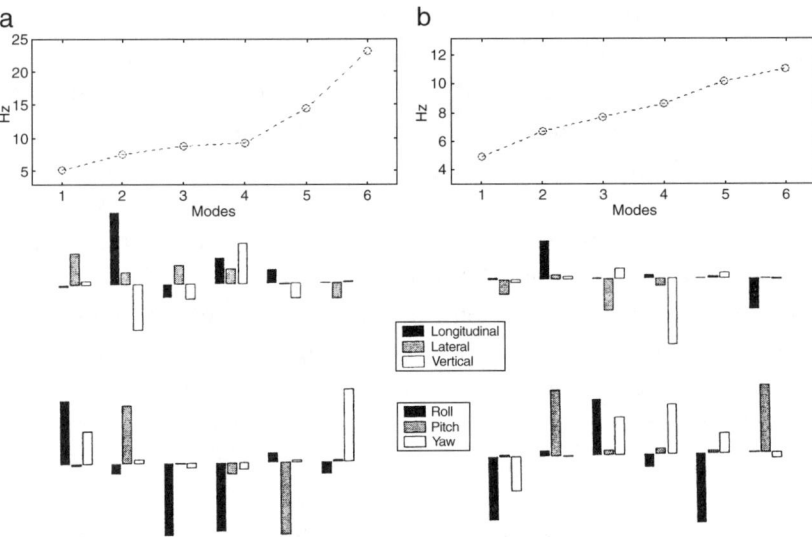

Fig. 2.6.2. Natural Frequencies and Mode Shapes of the Engine with: (a) Original; and (b) Modified Transmission Mount.

contribution; the second mode has an even larger pitch component and smaller longitudinal component; the third is more mixed, with a combination of intense roll, yaw, and lateral components; the fourth mode became a combination of vertical and yaw motions; the fifth became a predominantly roll mode; and the sixth mode became a combination of pitch motion with a significant contribution of longitudinal motion. Overall, the modal coupling in the natural coordinate frame for the engine is slightly reduced compared with that of the original system in Fig. 2.6.2(a).

The modification resulted in a slight reduction of vibration transmission to the cross-member supporting the transmission mount (7–17% reduction in the 850–4,000 rpm range, with 7% reduction at 850 rpm and 17% reduction at 2,000 rpm), and a slight (1.5 dBA) reduction of noise at the driver ear at low (850–1500) rpm.

Thus, uncoupling of the vibratory modes in the engine coordinate frame x, y, z together with a judicial modification of the natural frequencies are helpful for controlling the vibration isolation system of the engine.

It has to be noted that dynamics of the engine isolation system and influences of variation in engine and supporting frame designs are not yet clearly understood, e.g. see [2.39]. Thus, achieving decoupling in the engine isolation system in the natural coordinates for the engine (x, y, z) or natural coordinates for the vehicle (x', y', z') may not be the best approach. There have been claims of improved vibration conditions when the left and right front mounts were intentionally made with very different stiffness constants to *introduce coupling* into the system. Another similar opinion stated that coupling between the bounce (vertical) and pitch modes has a positive effect [2.1]. Partly, the disagreement may be caused by a more uniform damping distribution between the coupled modes. When the system is decoupled, the modes may have very different damping since the isolators may have widely differing damping in different directions. The isolators with elastomeric (rubber) elements usually have much higher damping in shear direction than in compression direction. While the efforts to develop a better dynamic model for the engine should continue, currently the most effective optimization technique is an experimental selection of the isolators' parameters as described in Section 2.7, using variable stiffness isolators (Section 4.5.1). This technique can be extended in the future by introducing "bang-bang" or continuous control of the passive variable stiffness mounts depending on the vehicle driving regime (idling, cruising, etc.).

Since the engines for surface vehicles such as cars, trucks, construction and agricultural machinery, etc., are mass-produced, their geometrical and inertia parameters can be economically determined with accuracy. Also, vibration isolators are not selected from a commercial line of isolators but usually are custom-designed for each engine model. This is quite different from the situation addressed in Section 1.3. Conservatively, there are three major uncertainty factors for the most widely used elastomeric mounts: variation of rubber hardness from the nominal values; variation of weight distribution between the mounts; and variation of stiffness ratios $\eta_{x,y}$ for mounts used at different mounting points of the engine. The first factor is the result of production tolerances (usually, ± 5 units of durometer or $\pm 17\%$ in mount stiffness). The second factor depends on the adopted mounting system, usually three-mount or four-mount installation. For the three-mount system the variation of weight distribution between the mounting points can be only due to the presence or absence of the optional accessories on the engine, such as an air-conditioner unit, super/turbo charger, etc. The uncertainty of this factor is rather small, $\sim \pm 5\%$. For the four-mount system the uncertainty is much wider due to static indeterminacy. There are potential deviations of levels of the mounting points on the chassis (less on the engine) caused by manufacturing tolerances. Variations of the levels can be assumed to be within $\pm 1.0-1.5$ mm. Since the typical natural frequency $f_z \approx 10$ Hz, it is equivalent to static deflection of the isolators under the weight load of the object $\Delta \approx 2.5$ mm, depending on K_{dyn} value e.g. see Eq. (1.4.33). Thus, the possible variations of the weight distribution between the mounting points can be conservatively assumed to be $\pm 1.0/2.5 = \pm 40\%$. Variation of the stiffness ratios $\eta_{x,y}$ between the engine mounts in one installation can be at least $\pm 20\%$. Accordingly, the ranges of effective scatter of stiffness between the engine mounts (see Section 1.3) are: for three-mount case $\pm 1.17 \times 1.05 \times 1.20 = \pm 1.47$; and for four-mount case $\pm 1.17 \times 1.40 \times 1.20 = \pm 1.97$.

There are numerous software modeling packages, some commercially available but mostly proprietary for vehicle manufacturers, that are used for optimization of the engine mounting system during the vehicle design process (e.g., [2.38], [2.39]). Often, these optimization algorithms do not consider dynamics of the supporting structure and also consider the engine-transmission as an independent unit, thus neglecting its connections with other structural units via drive shaft(s) and other connectors. Due to these factors as well as the above-noted isolator stiffness variation, use of engine mounts "as computed" often does not result in satisfactory performance. A lengthy trial-and-error "tuning" process using numerous engine mounts with slightly different characteristics is typical for many new vehicle models.

Use of CNF isolators would eliminate the second factor of the stiffness scatter (uncertainty of load distribution), reduce the first factor (uncertainty of rubber hardness) down to $\pm 2.5\%$ on the natural frequency or $\pm 5\%$ on stiffness (see Section 4.5.1), and reduce the third factor (uncertainty of the stiffness ratios) to less than $\pm 10\%$, provided that similar types of mounts are used. Thus the range of effective scatter of stiffness for CNF engine mounts would be about $\pm 1.05 \times 1.10 = \pm 1.16$. This would lead to a significant reduction of inter-coordinate coupling in the system (unless intentionally introduced) and shortening of the tuning process.

Decoupling in the engine mounting system can also be achieved by using the focalized system (use of inclined isolators, Section 1.6.3).

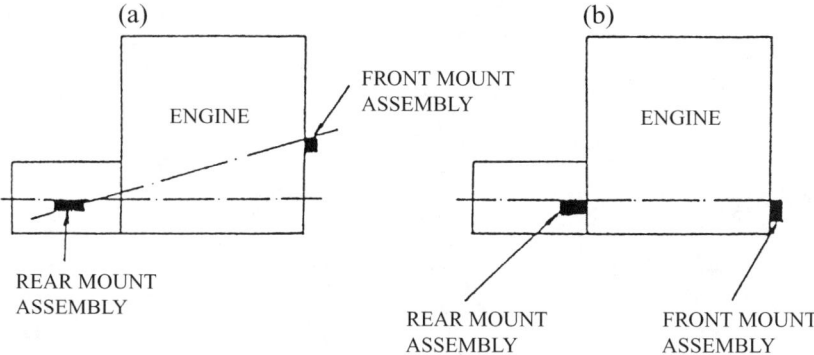

Fig. 2.6.3. Alternative Mount Locations for Automotive Engines.

Besides isolation of vibration generated by dynamic excitations within the engine, the engine mounting system has to comply with many constraints, including the following:

1. Control of engine movements caused by excitations from the road. On rough roads, dynamic loads on the vehicle structure may be up to three times greater than the static loads [2.26]. The minimum dynamic clearance between the engine and the adjacent chassis and body components is ~15 mm [2.38].
2. Control of excessive engine movements caused by large torque-reaction forces, especially in low gear, as well as control of the fore-and-aft engine movements caused by accelerating, braking, and cornering (accelerations up to 1 g) events. The mounting system should limit 3-D movements of the engine/transmission system to ±15 mm and roll to ±6°. To satisfy these constraints, shock-limiting snubbers or progressive stiffening of isolators may be required. The progressive stiffening load-deflection characteristic of CNF isolators makes them even more promising in engine mounting applications.
3. For safety reasons, the engine mounts must incorporate "captive" features so that a catastrophic failure of the flexible element does not result in separation of the engine from the subframe.

In the typical architecture of the engine mounting system, two mounts are located in the front on both sides of the x-x principal inertia axis, and one or two mounts are located under the transmission housing, Fig. 2.6.3(a) [2.26]. The spread between the front mounts (and, for the four-mount system, also between the rear mounts) results in a relatively

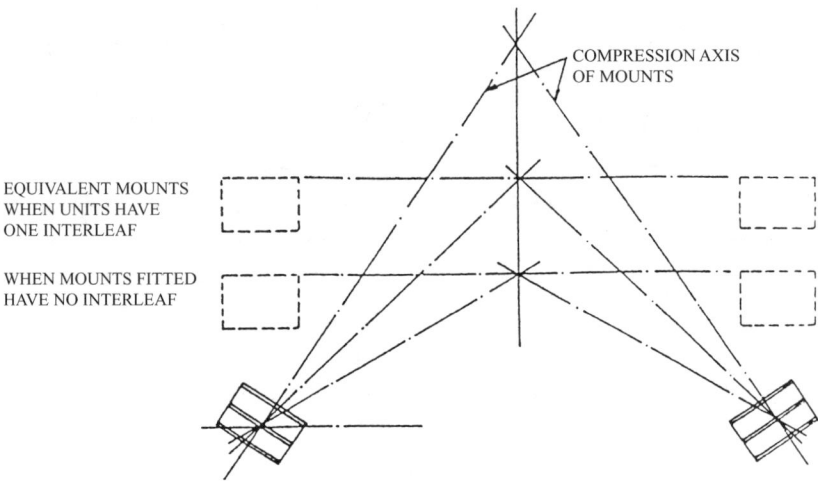

Fig. 2.6.4. Use of Inclined Mounts.

high torsional stiffness about x-x axis, which is required to accommodate torque pulses from the engine. While similar architectures are used for both gasoline and diesel engines, the intensity of torque oscillations from the diesel engine is much greater. Often, the front mounts are focalized and placed on the same level as the rear mounts, Fig. 2.6.3(b). The stiffness center of the focalized installation should be located on the x-x axis. It can be concluded from Eq. (1.6.18) that the greater the distance from the x-x axis, the greater must be the ratio of shear stiffness to compression stiffness of the mounts, Fig. 2.6.4 [2.26]. In order to increase $k_z/k_{x,y}$, the rubber block mounts with interleaves are used in the case of Fig. 2.6.4 (see Section 3.3.2C).

The engine mounting system must comply with many contradictory requirements. A low stiffness and high damping are required at low frequencies in order to isolate engine disturbances at the second order harmonic of the idle rpm (about 15–30 Hz range). A snubbing action and high damping are required to accommodate large amplitudes of engine bounce caused by shock excitation from the road. At high rpm the engine mounts should exhibit low damping in order to reduce high frequencies vibration transmission causing so-called "boom" noise in the vehicle (~200–500 Hz).

These contradictory characteristics can be achieved by utilization of dynamic nonlinearity (amplitude dependencies of stiffness and damping) of some materials used for vibration control applications, such as wire mesh materials, e.g. see [2.1] and Section 3.3.1A. Another alternative which became very popular, is the use of hydraulic mounts, see [2.39] and Section 3.6. In some luxury vehicles, semi-active mounts are used [2.39]. Hydraulic mounts, if properly designed, may improve performance by providing for a variation of mount parameters depending on the performance regime of the engine, but they are more expensive than elastomeric mounts. Hydromounts often are not adequately reliable in the tough environment of the engine compartment (especially high temperatures and aggressive environments such as salt spray [2.38]). New vehicles, especially popular trucks and SUVs, are subjected in the standardized tests to extreme loads simulating about 10 years of normal usage, e.g. [2.38]. While a hydromount can be precisely tuned for a given application, the tuning is very sensitive to tolerances on mounts manufacturing, see Section 3.6 and production variations between the vehicles, in addition to temperature variations influencing viscosity of the fluid inside the mount.

One example: when idle the torque requirements are minimal, and low stiffness constants of isolating mounts are required to provide for low natural frequencies, especially for four-cylinder engines that have unbalanced second harmonic of the engine RPM. On the other hand, during the cruising regimes the torque magnitude is high, but frequency of the excitation force is also high, so that stiffness of the mounts can be increased. Statically and/or dynamically nonlinear mounts may naturally resolve this situation.

High-damping mounts (mostly based on Butyl rubber compounds; see Chapter 3), were routinely used for engine mounting in the 1950 and '60s. Numerous studies (e.g., [2.40]) demonstrated that high damping of the engine mounts results in noticeable improvements in vehicle NVH (explanation of some positive effects can be found in Section 2.5). However, with increasing power density in the engine compartments and introduction of catalytic converters also contributing to temperature increases, the use of high-damping mounts have diminished. It is shown in Chapter 3 that recently developed high-damping/high-temperature-resistant rubber blends may radically improve the performance of engine mounting systems. This can be achieved faster and with better results by combining high damping with static and dynamic nonlinearities.

2.7 EXPERIMENTAL SELECTION OF ISOLATORS

Eqs. (2.2.15), (2.3.17), (2.4.11) and (2.4.13) do not take into consideration factors such as dynamic data of the floor; specific features of machine vibrations in different modes of the isolation system; dynamic data of the machine itself and isolators in a high-frequency range; real character of exciting forces; various regimes of machine performance, etc. Selection of isolators that consider these factors could result in substantial advantages, e.g., a reduction of noise levels and better vibration isolation, especially when the device operates in a broad range of regimes. However, an analytical approach for each individual object seems impractical. The best method of selection would be to compare several alternatives of the machine installation, with specific reference to the specified requirements, when each alternative is characterized by a single-rated parameter (selection criterion). Since each reinstallation of the isolated object, especially if it is a heavy machine, is a time- and labor-consuming procedure, *variable-stiffness isolators* (see Section 4.5.1) can be useful.

In cases where several factors should be considered, close-form expressions for the selection criteria can be used; in some cases subjective "common sense" judgment is adequate. When the object to be vibration-isolated is used in different modes of operation (e.g., a machine tool with many spindle and feed speeds, car engine operating in a broad rpm range), an obvious expression for the selection criterion is

$$B_\Sigma = \alpha_1 \beta_1 B_1 + \ldots + \alpha_i \beta_i B_i + \ldots,$$ Eq. (2.7.1)

where B_i is the selection criterion for the ith mode of operation; α_i is the weighting factor assigned to the ith mode; and β_i is a fraction of full working time during which the device is used in the ith mode of operation.

The "common sense" approach has been used in the selection of vibration isolators for a knife-type folding machine installed on the second floor of the printing shop. The office room was on the ground floor, directly beneath the machine. The goal of vibration isolation was to reduce noise transmission from the machine to the office room. Fig. 2.7.1 gives noise level readings in dBA both in the office and in the printing shop for installation of the machine on all-metal jack mounts (# 0) and on rubber mounts that have the same damping but different stiffness and, thus, provide for different f_v (26 Hz - #1; 20 Hz - #2; 16 Hz - #3; 10 Hz - #4).

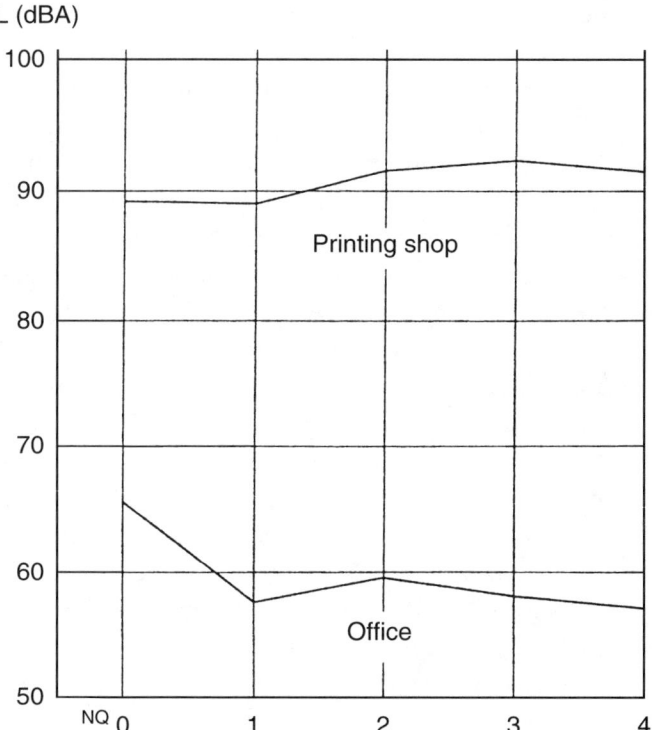

Fig. 2.7.1. Noise Level Readings vs. Mounting Conditions for Knife-Type Paper Folding Machine.

Unexpectedly, the #1 alternative (*the stiffest isolators*) was considered optimal since it resulted in the same (reduced) noise level in the office as that for the softest #4 isolators, while providing for less noise in the printing shop than #4 and causing less rocking in the machine.

2.8 ROLE OF DAMPING IN VIBRATION ISOLATION

There are very many textbooks on Vibration Theory and all of them have practically identical sections on Vibration Isolation. The considered system is a single-degree-of-freedom (SDOF) system with viscous friction damper as shown in Fig. 1.4.4a and its performance is characterized by transmissibility coefficients for cases of sinusoidal force and kinematic excitation, as presented by Eq. (1.4.6) with $m_f = \infty$, and Eq. (1.4.8), respectively. Such treatment of the vibration isolation system leads to a very basic conclusion that an isolation system is effective only at high frequencies $f > 1.41f_v$ of the sinusoidal excitation, where f_v is the natural frequency of the isolation system. While damping is reducing the resonance amplitudes in the considered system, this fact is of a low importance since the system normally functions only at the above-resonance frequencies. On the other hand, increasing viscous damping degrades the isolation effectiveness in the frequency range where the isolation effect is achieved. Unfortunately, these facts became a part of *"psychological inertia"* [2.36] of the overwhelming majority of engineers, including those who are involved in vibration control activities.

It was demonstrated in the preceding Sections of this Chapter that vibration isolation of real-life objects, both high precision/vibration sensitive, and vibration producing, and so-called "general purpose" machinery that the above notion is incorrect. The majority of the isolated objects in a majority of circumstances benefit from increasing damping.

There are several reasons justifying the above statement. One reason is, that the model analyzed in the textbooks (a mass sitting on a massless linear spring without damping and a linear viscous damper connected in parallel to the spring) is ideal for the close-form analytical treatment. However, this model represents only a minute fraction of the real-life vibration isolation systems (some systems having inertia blocks mounted on steel springs with oil dampers). The overwhelming majority of vibration isolation installations are using vibration isolators with elastomeric, polymer, wire mesh, etc. flexible elements having quasi-linear or strongly nonlinear damping characteristics of a hysteretic type. Such damping characteristics exhibit only a very mild deterioration of the high-frequency transmissibility with increasing degree of damping in the system, see Sections 1.4.2 and 3.3.

Another reason for the above statement is a complex spectral character of dynamic excitations in real-life vibration isolation systems. When the specter of the excitation has several spectral harmonics with commensurate amplitudes in a broad frequency range, it is often impractical to isolate (i.e., reduce intensity of) all the spectral components. The lowest frequency components are often the least objectionable and even their limited amplification is often acceptable. As a result, some of the low-frequency components may even resonate with natural frequencies of the vibration isolation system. The overall vibration intensity in such system can be reduced only by increasing damping, e.g. see Section 2.4.2.

Vibration isolation is of critical importance for precision vibration-sensitive objects. It is somewhat unexpected that vibration isolation systems even for these objects have resonance as their normal working regime. The explanation for this "paradox" is based on the fact that *vibration isolation systems for vibration-sensitive objects cannot be represented by SDOF systems*. A precision device has a stringent limitation for amplitudes of *relative vibration in its working area*, so the vibration transmissibility from the supporting surface (e.g., floor) to the object frame must be combined with transmissibility or transformation of absolute vibrations of the frame into relative vibration in the working area. The latter process is characterized by the very low transmissibilty (and thus very low amplitudes of relative vibrations) at low frequencies. Often a low frequency resonance in the vibration isolation system can be allowable, even for an extremely high precision object, if the limit for the amplitudes of relative vibration in the working area is not exceeded. And how can we reduce the resonance amplitudes and make this resonance even more tolerable? By increasing damping in the isolators, see Sections 2.2.4A and 2.2.4B.

Although damping does not influence transmissibility of pulse (shock) vibrations as much as it influences resonances of steady vibrations, damping enhancement in isolators is beneficial for isolation of pulse vibrations for both precision objects and for vibration-producing objects, see Sections 1.8, 2.2.4C, 2.3.4, 2.3.3.

When the isolated objects are installed on non-rigid supporting structures, increasing damping in isolators not only improves vibration but can increase damping of the supporting structure itself, see Section 2.5.

Damping of the isolators can also influence structural dynamics of the installed object as shown in Section 2.4.1.

It is important to note, that in the majority of cases discussed above in this Section, influence of damping change in isolators is intertwined with influence of stiffness of the isolators (or the natural frequencies of the vibration isolation system). In some cases reduction of damping in the isolators can be compensated by reduction of the isolators' stiffness (e.g., for isolation of precision objects). In other cases, reduction of damping in the isolators can be compensated by increase of the isolators' stiffness (e.g., for maintaining dynamic stability of machine tools, Section 2.4.1). Often, this interrelation can be expressed by formulating criteria combining stiffness (natural frequency) and damping for a given application. These criteria are presented in the following Eqs. in this Chapter: (2.2.15); (2.3.17); (2.4.7); (2.4.11); (2.4.13).

These criteria are typical for dynamic behavior of mechanical structures, e.g. see [2.16] and [2.41]. Although these criteria are based on analyses involving many assumptions and approximations, they are very useful in selecting isolator characteristics for a given application as well as for selecting materials for flexible elements of the isolators, e.g. see Section 4.2.

In some cases, high damping materials are not used for flexible elements of vibration isolators because of a concern that high damping would lead to a greater energy dissipation in the flexible element and to its overheating. While such a concern is justified for situations where a flexible element is subjected to a specified periodic strain, it is not justified for vibration isolators as shown in Section 2.1.1.

REFERENCES

[2.1] Rivin, E.I., 1985, "Passive Engine Mounts - Directions for Future Development," *SAE Transactions*, pp. 3.582–3.592.

[2.2] Rivin, E.I., 1965, "Review of Vibration Insulation Mountings for Machine Tools," *Machines and Tooling*, No. 8, pp. 37–46.

[2.3] DeBra, D.B., 1992, "Vibration Isolation of Precision Machine Tools and Instruments," *Annals of the CIRP*, Vol. 41/2, pp. 711–718.

[2.4] Amick, H. and Bayat, A., 2002, "Meeting the Vibration Challenges of New Photolithography Tools," *Sound and Vibration*, Vol. 36, No. 3, pp. 22–24.

[2.5] Amick, H. and Bui, S.K., 1992, "Review of Several Methods for Processing Vibration Data," *Vibration Control in Microelectronics, Optics, and Metrology*, SPIE, Vol. 1619, pp. 253–264.

[2.6] Rivin, E.I., 1995, "Vibration Isolation of Precision Equipment," *Precision Engineering*, Vol. 17, pp. 41–56.

[2.7] Pederson, D.L., 1992, "Vibration Control and Isolation Design for the Electrical Engineering/Computer Science Building, University of Minnesota-Minneapolis, Minnesota," *Vibration Control in Microelectronics, Optics, and Metrology*, SPIE, Vol. 1619, pp. 203–213.

[2.8] Gordon, C.G., 1991, "Generic Criteria for Vibration-Sensitive Equipment," *Vibration Control in Microelectronics, Optics and Metrology*, *SPIE Proceedings*, Nov., Vol. 1610.

[2.9] Ungar, E.E., Sturz, D.H., and Amick, C.H., 1990, "Vibration Control Design of High Technology Facilities", *Sound and Vibration*, No. 8, pp. 20–27.

[2.10] Gordon, C.G., 1987, "The Design of Low-Vibration Buildings for Microelectronics and Other Occupancies," *Vibration Control in Optics, and Metrology*, SPIE, Vol. 732, pp. 2–10.

[2.11] MIL-STD-810E, 1988, Military Standard, "Environmental Test Methods," U.S. Department of Defense, Washington, D.C.

[2.12] Myler, J.K., 1987, "The Influence of Machine Tool Vibration on the Surface Texture of Diamond Turned Components," *Vibration Control in Optics, and Metrology*, SPIE, Vol. 732, pp. 196–209.

[2.13] House, M.E. and Randell, R., 1987, "Some Measurements of Acceptable Levels of Vibration in Scientific, Medical, and Ophtalmic Microscopes," *Vibration Control in Optics, and Metrology*, SPIE, Vol. 732, pp. 74–80.

[2.14] Veprik, A.M., Babitsky, V.I., Pundak, N., and Riabzev, S.V., 2000, "Vibration Protection for Linear Stirling Cryogenic Cooler of Airborne Infrared Application," *Shock and Vibration*, Vol. 7, No. 6, pp. 363–381.

[2.15] Jendrzejczyk, J.A. and Wambsganss, M.W., 1992, "Vibration Considerations in the Design of the Advanced Photon Source at Argonne National Laboratory," *Vibration Control in Microelectronics, Optics, and Metrology*, SPIE, Vol. 1619, pp. 114–126.

[2.16] Rivin, E.I., 1999, "Stiffness and Damping in Mechanical Design," *Marcel Dekker Inc*.

[2.17] Dente, J.V., McCutcheon, J.L., and Jain, A.K., 1995, "Vibration Qualification of Commercial Computers for Use in Military Tactical Environments," *Presented at the 41th Annual Technical Meeting of the In-te of Environmental Sciences*, Anaheim, California, May 1995.

[2.18] Owen, N. and Hale, R., 1992, "Factors in the Design and Selection of Vibration-Sensitive Equipment," *Vibration Control in Microelectronics, Optics, and Metrology*, SPIE, Vol. 1619, pp. 56–70.

[2.19] Kwan, Y.B.P. and Loopstra, E.L., 2000, "Nullifying Acceleration Forces in Nano-Positioning Stages for Sub-0.1μm Lithography Tool for 300 mm Wafers," *Proceedings of the 15th Annual Meeting of ASPE*, pp. 525–528.

[2.20] Soong, T.T. and Costantinou, M.C. (Editors), 1995, "Passive and Active Structural Vibration Control in Civil Engineering," *Springer Verlag*.

[2.21] Walter, R.A., 1975, "Building on Springs," *Sound and Vibration*, No. 3, pp. 22–25.

[2.22] Lang, G.F. and Snyder, D., 2001, "Understanding the Physics of Electrodynamic Shaker Performance," *Sound and Vibration*, Vol. 35, No. 10, pp. 24–33.

[2.23] Eisele, F. and Schwaighofer, R., 1965, "Der Elektromotor als Störquelle an der Werkzeugmaschine [Electric Motor as Vibration Exciter for a Machine Tool]," *Maschinenmarkt*, Vol. 71, No. 79 [in German].

[2.24] Rivin, E.I., 1971, "Principles of Vibration Isolation of Mechanical Presses," *Kuznetchno-shtampovochnoe proizvodstvo [Metal stamping production]*, No. 10, pp. 20–22 [in Russian].

[2.25] Rivin, E.I., 1987, "Evaluation of Vibration Isolation Systems for Forging Hammers," *Meeting Handbook on "Vibration Isolation of Heavy Structures"*, Institute of Acoustics, London, pp. 91–97.

[2.26] Freakley, P.K. and Payne, A.R., 1978, "Theory and Practice of Engineering with Rubber," Applied Science Publishers, London.

[2.27] Belyaev, Y.V., 1970, "Peak Loads in Co-Impacting Parts of Forging Hammer," *Kuznetchno-shtampovochnoe proizvodstvo [Metal stamping production]*, No. 8 [in Russian].

[2.28] Jarausch, R., 1965, "Hammer, Fundament und Umgebung als Schwingungsystem," *Maschinenmarkt*, Vol. 71, No. 11 [in German].

[2.29] Rivin, E.I., 1974, "How Methods of Installation Affect the Vibration Resistance of Machine Tools," *Machines and Tooling*, Vol. 45, No. 11, pp. 25–27.

[2.30] Tobias, S.A., 1965, "Machine Tool Vibration," Blackie, London.

[2.31] Golec, Z. and Cempel, C., 1990, "Machine Vibroisolation and Dynamic Loads of Bearings," *Mechanical Systems and Signal Processing*, Vol. 4, No. 5, pp. 367–375.

[2.32] Rivin, E.I., 1983, "Cost-Effective Noise Abatement in Manufacturing Plants," *Noise Control Engineering Journal*, No. 6, pp. 103–117.

[2.33] Snowdon, J.C., 1979, "Handbook of Vibration and Noise Control," Report TM, U.S. Dept. of the Navy, 79-75.

[2.34] Kim, J.-H. and Lee, J.-M., 2000, "Elastic Foundation Effects on the Dynamic Response of Engine Mount Systems," *Proceedings of Institution for Mechanical Engineers*, Vol. 214, Part D, pp. 45–53.

[2.35] Ashrafiuon, H. and Nataraj, C., 1991, "Dynamic Analysis of

Engine-Mount System," *Structural Vibration and Acoustics*, ASME DE-Vol. 34, pp. 191–196.

[2.36] Fey, V.R. and Rivin, E.I., 1997, "The Science of Innovation," The TRIZ Group, W. Bloomfield, MI.

[2.37] Rivin, E.I. and Fey, V.R., 2000, "Case Study: Solving Rough Idle Problem for a Compact Car," *J. of the Altshuller In-te for TRIZ Studies*, Vol. 2, pp. 20–22.

[2.38] Zavala, P.A.G., Pinto, M.G., Pavanello, R., and Vaqueiro, J., 2001, "Powertrain Mounting Development Based on Computational Simulation and Experimental Verification Method," *Proceedings of the 2001 SAE Noise and Vibration Conference (Noise 2001 CD)*, SAE Paper 2001-01-1509.

[2.39] Yu, Y., Naganathan, N.G., and Dukkipati, R.V., 2001, "A Literature Review of Automotive Vehicle Engine Mounting Systems," *Mechanism and Machine Theory*, Vol. 36, pp. 123–142.

[2.40] Kruse, D.F. and Knable, J.J., 1984, "Mounting the Subframe—A Correlation of Cradle Mounting Dynamics and Vehicle Ride Performance," *SAE Technical Paper 840409*.

[2.41] Rivin, E.I., 2001, "Interrelation of Stiffness and Damping in Machine Tool Dynamics", in *Transactions of the North American Manufacturing Research Institution of SME*, pp. 137–143.

CHAPTER 3

Realization of Elasticity and Damping in Vibration Isolators

The techniques described in Chapter 2 and based on principles discussed in Chapter 1 allow specifying requirements for vibration isolators needed for a given object or for a group of objects. However, realization and implementation of these specifications are impossible without a clear understanding of properties of materials which are or can be used for constructing the vibration isolators. Since no concise description of the relevant properties is available, this Chapter attempts to fill this void.

Both static and dynamic characteristics of the principal materials are described, metal as well as polymeric. The greatest attention is given to elastomeric (rubber-like) materials due to their unique combination of an infinite variety of characteristics achievable by blending of the numerous component materials, and a great design flexibility.

However, there is a significant potential for metal flexible elements using both the appropriate designs and the new alloys possessing high damping. These materials, as well as pneumatic and hydraulic elasto-damping elements and fibrous and plastic materials are described.

3.0 GENERAL COMMENTS

Any vibration isolator has at least one elastic and/or damping element. In some cases the elastic and damping elements are separated and there is a spring (usually metal) and a damper. In the overwhelming majority of isolator designs, the required elastic and damping characteristics are combined in one flexible element. This can be achieved by using various elasto-damping materials, or pneumatic, hydraulic, electromagnetic, electrodynamic, etc., elements. It is relatively easy to obtain the required elastic and damping properties from each of these elements. This chapter discusses characteristics of basic elastic elements and dampers as well as important static and dynamic characteristics of elasto-damping materials.

3.1 METAL ELASTIC ELEMENTS (SPRINGS)

Metal springs can be used for large static deformations, which result in low natural frequencies. They practically exhibit no creep, their dynamic stiffness is equal to their static stiffness, and they are not sensitive to high or low temperatures (unless high damping shape memory alloys are used). The shortcomings of many metal spring designs include low damping—unless wire-mesh materials or cables (see Section 3.3.1) or special alloys, such as shape memory alloys [3.1], [3.2], are used—and relatively high costs. In cases where the maximum isolation efficiency at high frequencies is required, e.g., [3.3], low damping may become an advantage. However, this advantage can be negated by wave resonances (see Section 1.10). Coil springs, both coiled from steel wire or bars and machined from tubular blanks, are the most widely used metal elastic elements in vibration isolation systems. Other types, such as flat springs, Belleville springs, slotted springs, torsion bar springs, or specially designed flexible elements, e.g. see [3.3] are also used or can be beneficially used in some applications. Since vibration amplitudes in vibration isolation systems for stationary objects are usually small when compared with static deflections of the isolators, strength calculations of springs can be performed for static stresses resulting from the weight load of the supported object. Basic parameters of the springs for vibration isolation applications are stiffness, maximum stresses, and stability. Springs are usually made from an appropriately heat-treated spring steel having high yield stress and very low damping. Damping of Belleville springs is usually significant due to slippages between the spring and the supporting surfaces. Even more slippages (between the constitutive wires) are characteristic for wire-mesh/cable flexible elements. The damping of other types of metal springs can be increased if they are made from high damping metals (see Section 3.4).

3.1.1 Coil Springs

Coil springs are often characterized by *spring index* $C = D/d$. The axial (compression/tension) stiffness k_a of a cylindrical coil spring, with the constant pitch of coils and constant round cross-section of the wire along its length, is

$$k_a = \frac{Gd}{8C^3 n} = \frac{Gd^4}{8D^3 n}; \qquad \text{Eq. (3.1.1)}$$

Here D is median diameter of a coil, measured between center lines of wire; d—wire diameter; G—shear modulus of spring material; n—number of coils. For springs with ground face surfaces, the number n of active coils can be derived from the total count of coils minus 1.75.

Shear stress in the coils is determined by torsion of the wire caused by the axial force (asymmetrical due to the coil curvature), shear stress due to direct action of the axial force, and by additional stresses due to the

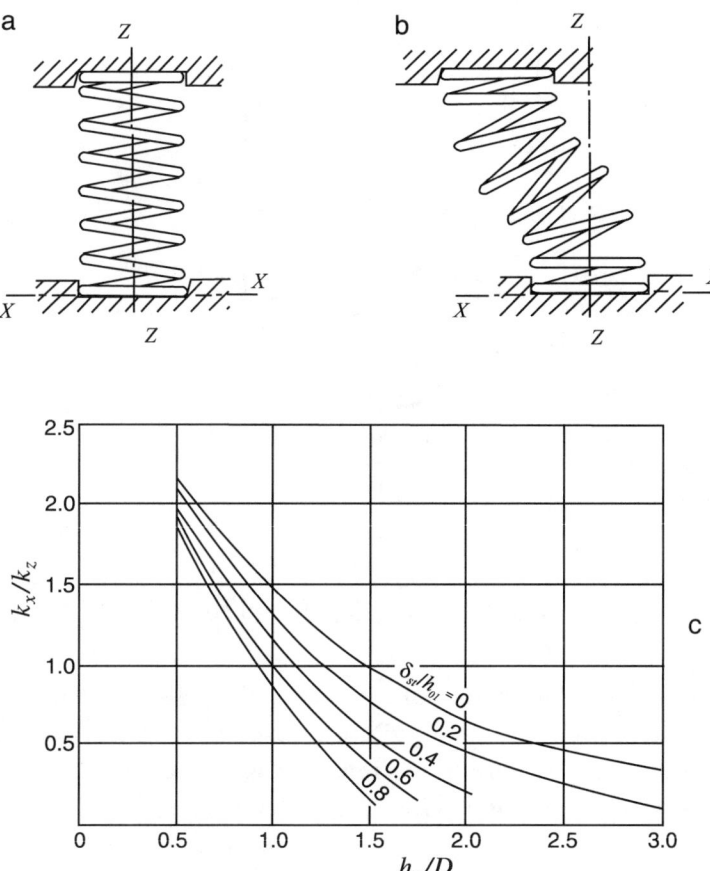

Fig. 3.1.1. (a) Mounting of Helical Coil Spring; (b) Lateral Deflection of Helical Coil Spring; (c) Ratio of Lateral to Axial Stiffness for Helical Compression Coil Spring.

inevitable eccentricity of the axial load decreasing with increasing n. The maximum shear stress considering these factors is

$$\tau = K\tau_0 = K\frac{8PD}{\pi d^3} = K\frac{8PC}{\pi d^2}, \qquad \text{Eq. (3.1.2)}$$

where τ_o is the nominal stress and K is the the *correction (Wahl) factor*

$$K = \frac{4C-1}{4C-4} + \frac{0.615}{C} \qquad \text{Eq. (3.1.3)}$$

Variations of the coiling operation as well as of installation techniques may result in a significant scatter of spring parameters. To improve

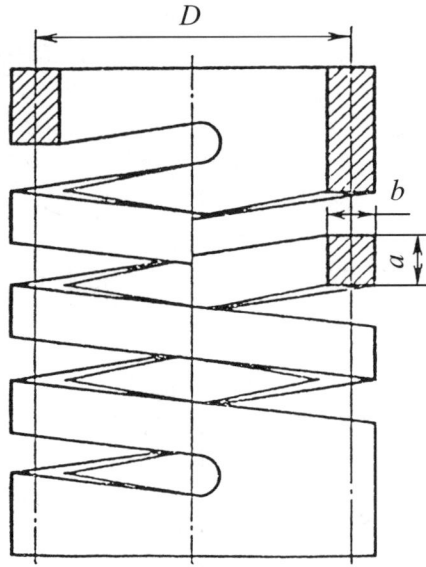

Fig. 3.1.2. Coil Spring Machined from Solid Bar.

performance and consistency of load-deflection characteristics, the end coils of compression springs in vibration isolation systems are flattened and restrained in the lateral directions, e.g. as shown in Fig. 3.1.1(a).

Springs with rectangular wire cross-section $a \times b$ can be fabricated by coiling them from rectangular cross-section wire or by machining (turning or milling) from tubular blanks (Fig. 3.1.2). The latter method ensures better accuracy of dimensions and stiffness, as well as better accuracy in assembly, since it can be directly fastened, e.g., by bolts, to the object and the base. In addition, due to more uniform stress distribution these springs can be made smaller although stress concentrations in the corners must be considered. Stiffness of and stress in springs with rectangular wire cross-section are

$$k_a = \frac{G\delta^4}{\nu D^3 n} \ ; \quad \tau_o = \frac{PD}{2\chi a^2 b}. \qquad \text{Eq. (3.1.4)}$$

Here P is axial force; δ is the smallest value from a, b; χ and ν are determined from Table 3.1.

While the axial deflection of a coil spring is mainly due to torsion of the wire, lateral deflection is due to both torsion and bending of the wire and of the spring itself, Fig. 3.1.1(b) [3.4]. Thus, lateral deflection depends on

Table 3.1. Coefficients for Calculations of Coil Springs with Rectangular Cross Section of Wire

b/a	χ	v	b/a	χ	v
1.0	2.40	5.57	3.0	0.625	1.0
1.5	1.44	2.67	4.0	0.44	0.7
1.75	1.2	2.09	6.0	0.28	0.44
2.0	1.02	1.71	10	0.25	0.16
2.5	0.78	1.26			

axial deflection of the spring. Fig. 3.1.1(c) gives the ratio k_x/k_z for the springs coiled from a steel round cross-section wire for different levels of axial loading (static deflection δ_{st} under the axial weight load), depending on the initial working height h_o of the spring. These plots are obtained for the conditions in which the ends of the spring are constrained to remain parallel in the lateral loading process, e.g., as in Figs. 3.1.1(a), (b).

Long (initial length $h_o > 2.5\,D$) coil springs can buckle at a large static deformation δ_{st}, unless one of the following stability conditions is satisfied:

$$\frac{\delta_{st}}{h_o} \leq 0.81\left[1 - \sqrt{1 - 6.9(D/h_o)^2}\right] \qquad \text{Eq. (3.1.5)}$$

or

$$\frac{k_x}{k_z} \geq 1.2\left(\frac{\delta_{st}}{h_s}\right). \qquad \text{Eq. (3.1.6)}$$

Since damping of metal coil springs is very low (unless they are made from high damping alloys; see Section 3.4), high-frequency resonance vibrations (*surge*) at natural frequencies of the spring, considered as a distributed parameters body (see Section 1.10), can be very intense. These natural frequencies can be calculated as

$$f_n = i\frac{2d}{\pi D^2 n}\sqrt{\frac{G}{32\rho}}, \qquad \text{Eq. (3.1.7)}$$

where ρ is density of the spring material, $i = 1, 2, 3, \ldots$.

A coil spring can also be used in vibration isolators while the inertia or weight load of the supported object is applied transversely to its axis, Fig. 3.1.3 [3.5]. This mode of loading (similar to radial loading of a stack of rings) enables accommodation of much higher loads (with much smaller deflections and higher stiffness values) and realization of a noticeable damping due to friction between the OD of the spring and the

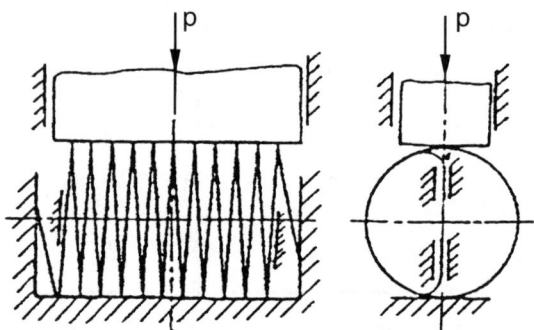

Fig. 3.1.3. Transversely Loaded Coil Spring.

supporting surfaces during the deformation process of the spring. Both extension springs (tightly coiled) and other types of springs can be used in this mode. Since elastic stability of the coils under the tranverse loading limits the maximum loads, high load capacity of such elements is better realized if the axial movements of the coils are prevented, e.g., by mechanical constraining elements (like walls shown in Fig. 3.1.3). Deflection δ_{st} (and stiffness $dP/d\delta_{st}$) and equivalent normal stress σ_{eq} can be calculated as

$$\delta_{st} = \frac{3.74 PD^3}{\pi^2 End^4} \frac{1 + 8.86 \sin^2 \alpha}{\cos^3 \alpha}, \qquad \text{Eq. (3.1.8)}$$

$$\sigma_{eq} = \frac{4PD}{\pi^2 nd^3} \sqrt{\pi^4 \tan^2 \alpha + (4 + \tan^2 \alpha)^2}. \qquad \text{Eq. (3.1.9)}$$

Here α is helix angle of the coils ($\alpha \cong 0$ can be assumed for tightly coiled springs); E is Young's modulus of the spring material.

The load-carrying capacity of this type of element can be further increased (at the price of increased stiffness) by modifying the contact surfaces (increasing their curvature, making rectangular or gothic arch contact grooves, etc.).

3.1.1A Statically Nonlinear Coil Springs

The term "statically nonlinear metal springs" is used here in order to distinguish these flexible elements from the metal mesh and cable elements described in Section 3.3.1. The latter are characterized by both static and, especially, dynamic nonlinearities (amplitude dependencies),

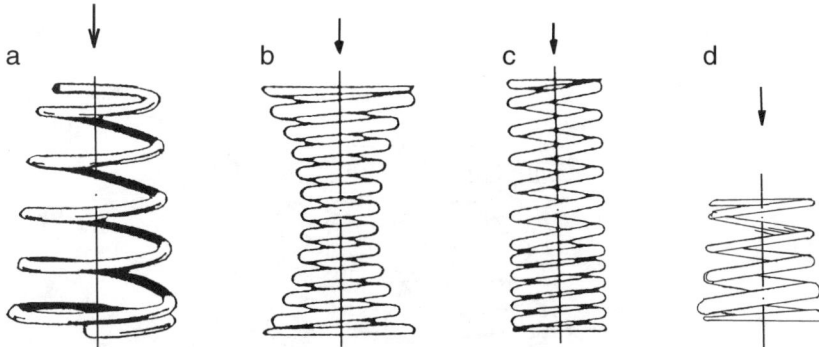

Fig. 3.1.4. Nonlinear Coil Spring Designs: (a), (b) Variable Diameter Springs Coiled from Constant Diameter Wire (a—Conical, b—Corset-Shaped); (c) Cylindrical Spring with Variable Pitch, Using Constant Diameter Wire; (d) Cylindrical Spring Using Tapered Wire and Constant Pitch.

in addition to high and nonlinear damping. The coil springs are the most frequently used statically nonlinear metal springs for vibration isolation (as well as for a special case of vibration isolation—vehicle suspensions). There are several modifications of nonlinear (compression) coil springs:

1. Springs that are coiled from constant cross-section wire but have non-cylindrical shapes, so that the coil diameter is not constant along the length, Fig. 3.1.4(a), (b). Usually, such springs have a conical shape as shown in Fig. 3.1.4(a) or a "corset" shape, as shown in Fig. 3.1.4(b) or "barrel" (double conical) shape. The largest diameter coil is the softest, see Eq. (3.1.1), and at the predetermined load its deformation is so large that it touches one of the bases or another (adjacent) coil, thus reducing the effective n and increasing stiffness, as shown in Fig. 3.1.5 for the conical spring in Fig. 3.1.4(a). If the generatrix of the nonlinear cone is described by an exponential function, the spring has a constant natural frequency (CNF) characteristic (see Section 1.3), whereas its axial stiffness is proportional to the axial load in a specified load interval.

2. Cylindrical springs that are coiled from constant cross-section wire but have variable pitch along their length, Fig. 3.1.4(c). At a predetermined compressive force the coils in the minimum pitch area touch one another, thus reducing the effective number of coils n and increasing stiffness of the spring.

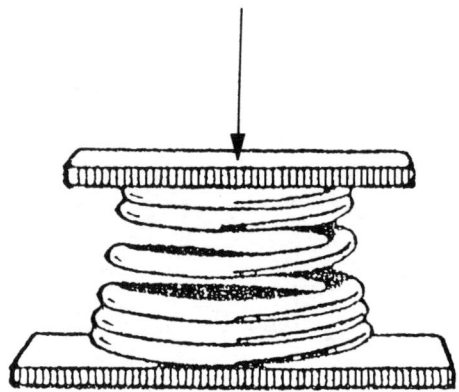

Fig. 3.1.5. Deformation Pattern of Variable Diameter (Conical) Coil Spring.

3. Cylindrical springs with constant pitch that are coiled from variable cross-section ("tapered") wire, Fig. 3.1.4(d). The coil made from the smallest cross-section wire segment is the softest, and at the predetermined load it touches the adjacent coil, thus increasing stiffness of the spring.
4. A combination of the above designs.

Cylindrical springs with variable pitch have nonuniform tangential stresses between the coils, thus resulting in larger size and weight [3.6], [3.7]. Coil springs made from tapered wire are also more bulky but may be made lighter due to more uniform stress distribution between the coils [3.6], [3.7]. Generally, variable stiffness cylindrical coil springs may develop noise and wear when the coils come in contact with one another during intense vibration. Non-cylindrical coil springs can be designed in such a shape that the low stiffness (large diameter) coils are deformed until they are in contact not with the adjacent coils, but with the supporting surface(s). This pattern results in the lower height of the spring, reduction of its weight, reduction of contact pressures, and elimination of noise and wear [3.6].

3.1.2 Slotted Springs

The cylindrical slotted springs shown in Fig. 3.1.6 combine high stiffness and load-carrying capacity with high accuracy in a small outline. The

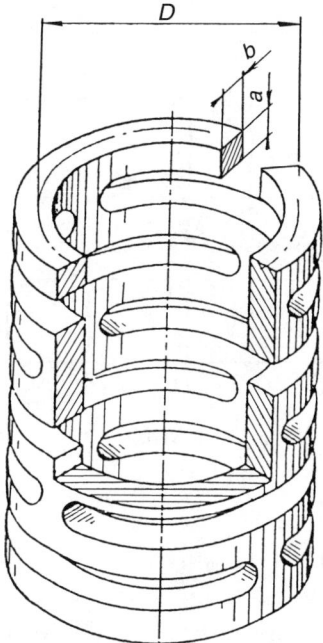

Fig. 3.1.6. Cylindrical Slotted Spring.

spring is very stable and sturdy since it has very high torsional stiffness around its axis and a high degree of symmetry. Usually, there are two slots in each transverse section of the springs, thus the spring is, essentially, a series connection of beams loaded at mid-span. Axial stiffness k_a and effective stress σ_{ef} are

$$k_a = \frac{Eab^3}{D^3 n}\frac{1}{\alpha_n}; \qquad \sigma_{ef} = k_\sigma \frac{PD}{ab^2}\beta \qquad \text{Eq. (3.1.10)}$$

Here E is Young's modulus of the spring material; α_n and β are determined from Table 3.2. The stress concentration factor k_σ for fillets between the beams is plotted in Fig. 3.1.7.

Table 3.2. Coefficients for Calculations of Slotted Springs

b/a	β	α_n	b/a	β	α_n
0.1	0.44	0.10	1.0	0.61	0.12
0.25	0.44	0.10	1.5	0.66	0.13
0.5	0.5	0.11	2.0	0.68	0.14
0.66	0.53	0.11	10	1.36	0.45

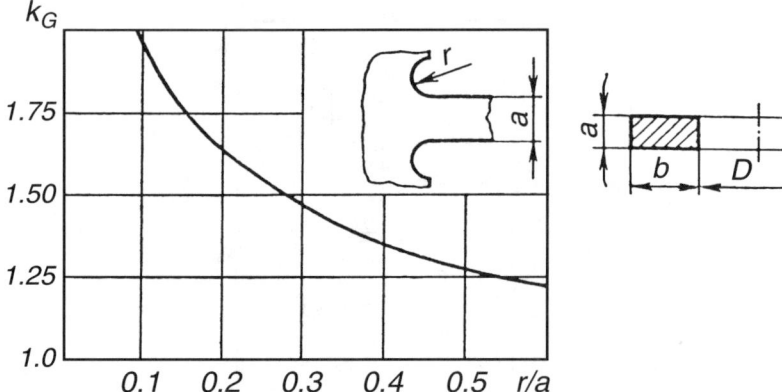

Fig. 3.1.7. Stress Concentration in Slotted Springs.

3.1.3 Belleville Springs

Belleville spring (Fig. 3.1.8) usually has $\alpha = 2$ to $6°$, $D/d = 2$ to 3. These springs provide high load-carrying capacity in a small outline. Their load capacity, stiffness, and deformations in the axial direction can be easily adjusted by connecting two or several springs (the lateral stiffness $k_{x,y}$ and tilting stiffness are very high as compared with coil springs). The series connection, Fig. 3.1.9(a) provides reduced stiffness/increased deformation in proportion to the number of springs but without changing the load rating. The parallel connection, Fig. 3.1.9(b), results in an increase of both rated load and stiffness in proportion to the number of stacked springs, but without changing the maximum deflection. A wide variety of stiffness/deflection/load characteristics can be obtained by combining the above

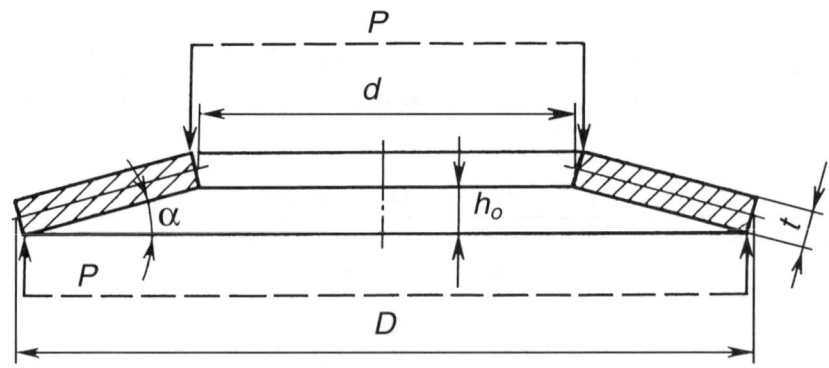

Fig. 3.1.8. Belleville (Disc) Spring.

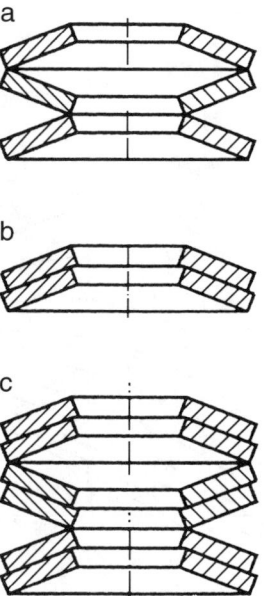

Fig. 3.1.9. Arrangements of Belleville Springs: (a) Series Arrangement; (b) Parallel Arrangement; (c) Combination Arrangement.

stacking patterns, Fig. 3.1.9(c). The Belleville springs provide significant damping since their deformation is accompanied by sliding between the spring and its support surface and between the springs in the stacks, as shown in Fig. 3.1.9, especially in configurations of Fig. 3.1.9(b), (c). In the configuration of Fig. 3.1.9(a), the amount of friction (and damping) is reduced to friction between the lowest spring and the supporting structure. For precision applications, influence of friction on load-deflection characteristics should be considered [3.8]. Belleville springs are produced in a huge variety of standardized dimensions and are easily available "off the shelf."

The load (P)—deflection (s) characteristic of Belleville springs is nonlinear,

$$P = \frac{\pi E t}{6(D-d)^2} s \left[\left(h_o - \frac{s}{2} \right)(h_o - s) + t^2 \right] \ln \frac{D}{d} \qquad \text{Eq. (3.1.11)}$$

As can be seen in Fig. 3.1.10 [3.9], where P_o is the load at which the spring is flattened, the character of nonlinearity changes in a wide range, depending on the spring's dimensional parameters. It may have linear parts, "hardening" nonlinear parts (stiffness increasing with increasing

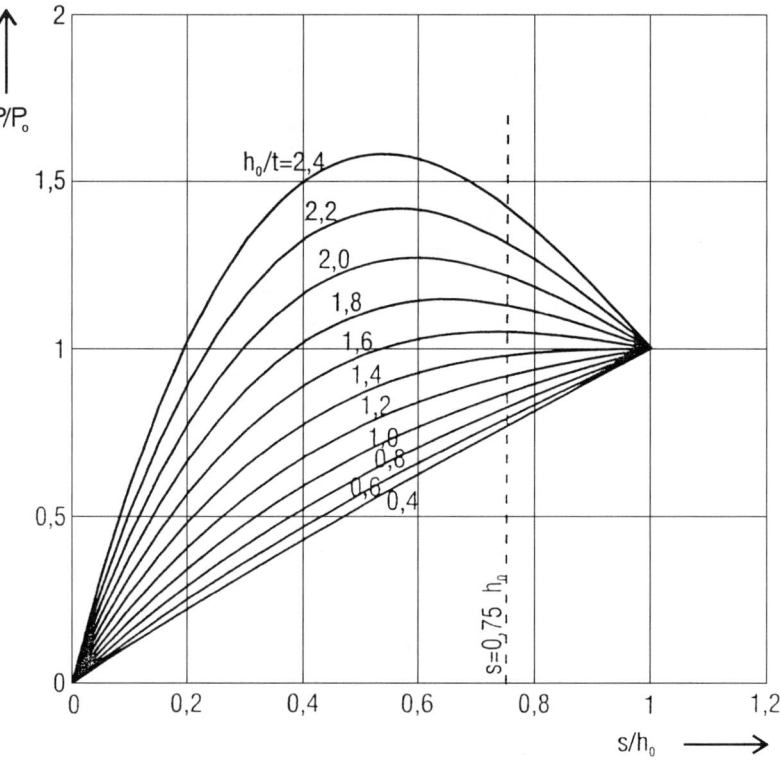

Fig. 3.1.10. Load-Deflection Plots for Belleville Springs with Various Parameters.

deformation), and "softening" nonlinear parts (stiffness decreasing while deformation is increasing). At $h_o/t > \sim 1.5$, there are even areas with quasi-zero stiffness.

The maximum normal stresses develop in the meridian cross-section, at the internal edge of the tapered shell,

$$\sigma_{\max} = \frac{4Es}{kD^2}(h_o k_o - sk_1 + t) \qquad \text{Eq. (3.1.12)}$$

where coefficients k, k_o, k_1 are taken from Fig. 3.1.11.

3.2 DAMPERS

Dampers are used as attachments to vibration isolators when damping in the elastic element of the isolators is not sufficient for optimization of the

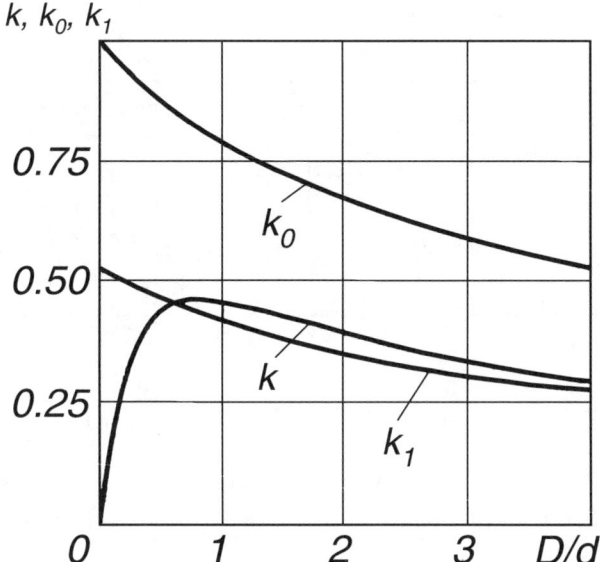

Fig. 3.1.11. Coefficients for Calculating Belleville Springs.

system. The most widely used are fluid (viscous) dampers, dry (Coulomb) friction dampers, and electromagnetic dampers.

3.2.1 Viscous Dampers

Viscous dampers utilize resistance to relative motion between two surfaces separated by a fluid or a gas (air), or resistance to fluid flow through thin channels/small holes, or resistance to movements of solid bodies in a viscous fluid.

In viscous dampers based on relative motion between two surfaces, as shown in Fig. 3.2.1, the resistance (damping) force P is proportional to the

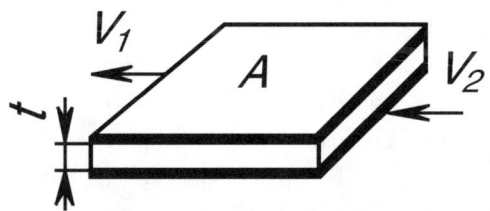

Fig. 3.2.1. Viscous Damper.

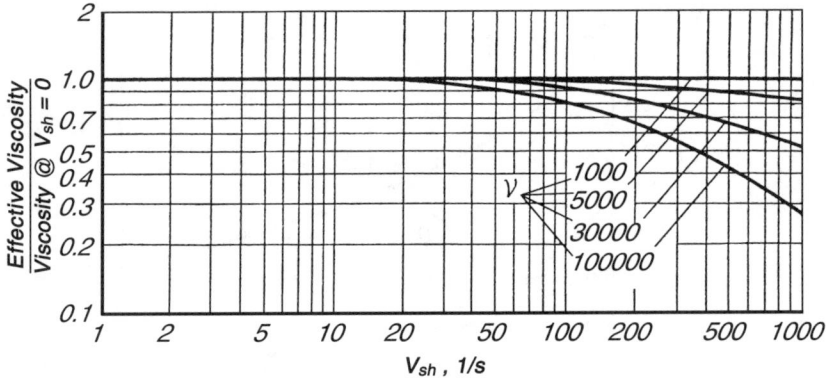

Fig. 3.2.2. Viscosity of Silicon Fluid vs. Shear Rate; in cm²/s.

velocity of relative sliding between the surfaces $\Delta V = V_1 - V_2$ if the flow is laminar, and the proportionality coefficient (damping coefficient) is

$$C = \frac{P}{V_1 - V_2} = \frac{A\mu}{t}, \qquad \text{Eq. (3.2.1)}$$

where A is surface area of the interacting surfaces; t—distance between the interacting surfaces; $\mu = \nu\rho$ is dynamic viscosity; ν—kinematic viscosity; ρ—density of the fluid (gas). The laminar flow develops if Reynolds number $Re = (Q/A)\, t(\rho/\mu) < 2{,}000$; however, if the laminarity is important for the right functioning of the damper, it is more reliable to design for $Re < {\sim}500$. Here Q is the flow rate between the surfaces. If the flow becomes turbulent, correlation between the velocity differential and the resistance force becomes nonlinear. Nonlinear viscous resistance force in self-leveling isolators, such as described in Sections 4.9.2 and 4.10.1, may lead to instability of the isolator [3.10].

Silicone fluids are often used in dampers since their viscosity does not strongly depend on temperature. A feature to consider is dependence of viscosity of silicone fluids on the sliding (shear) velocity, especially for high viscosity ν, (see Fig. 3.2.2). In some damper designs, e.g., for pneumatic isolators (Section 3.5) and hydraulic isolators ("hydromounts," see Section 3.6), flow of fluid/gas through a capillary or through an open-cell porous body (e.g., rigid foam) is used instead of relative sliding between the surfaces. Viscosity of specially formulated fluids can be controlled in a broad range by application of electrostatic field ("electrorheological" fluids) or magnetic field ("magnetorheological" fluids), e.g., see [3.11].

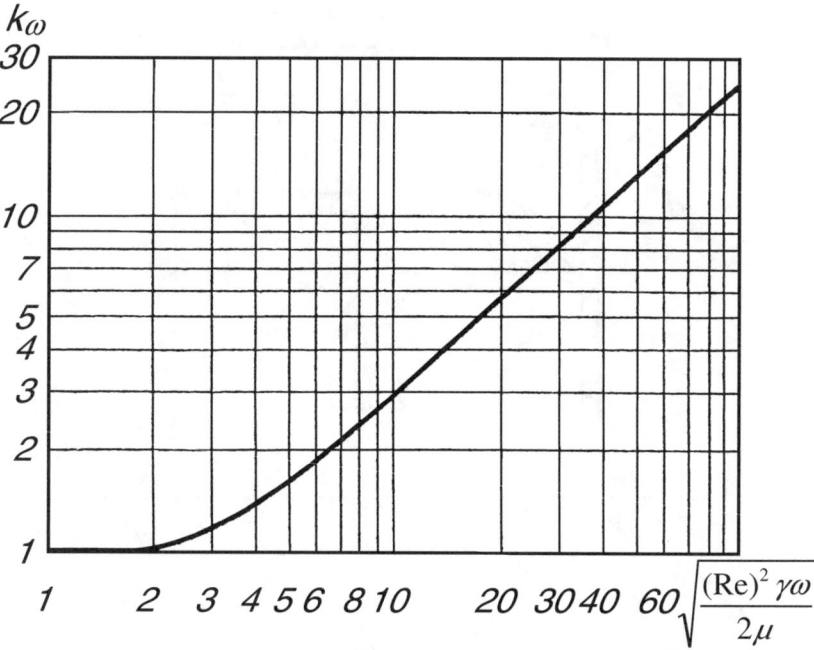

Fig. 3.2.3. Frequency-Dependent Coefficient k_ω for Capillary Resistance.

When a capillary through which the fluid is flowing (with velocity V) is used as a damping element, its damping coefficient at $Re < 2{,}000$

$$C = \frac{P}{V} = \frac{8\pi k_\omega l}{A}, \qquad \text{Eq. (3.2.2)}$$

where l, A are length and cross-sectional area of the capillary, respectively; and k_ω is the frequency-dependent coefficient plotted in Fig. 3.2.3 for round capillaries [3.12]. Some additional data is given in Section 3.5.

In cases when the damping effect should be generated in several directions simultaneously, the viscous dampers may use resistance to motion of a solid body inside the viscous fluid. Usually the solid body has protrusions, e.g., a brush-like structure (as shown in Section 4.5.2, Fig. 4.5.5).

3.2.2 Dry (Coulomb) Friction Dampers

These dampers are based on sliding between solid non-lubricated surfaces. The most consistent friction coefficients between the non-lubricated

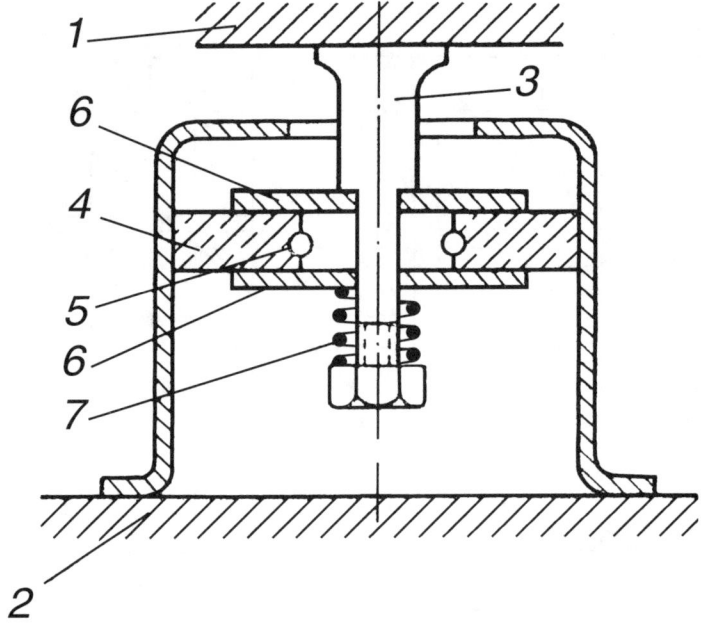

Fig. 3.2.4. Three-Dimensional Coulomb Friction Damper
(1—Isolated Object; 2—Supporting Structure; 3—Pushrod;
4—Ceramic Friction Sectors; 5—Lateral Preloading Spring;
6—Friction Discs; 7—Axial Spring).

surfaces can be obtained between surfaces of materials sintered from metal or ceramic powder, or surfaces impregnated by Teflon or meshed with Teflon fibers, e.g., [3.13]. Since the friction force is proportional to normal pressure between the surfaces, such dampers require a preloading device with a controllable preload force. To reduce the influence of inevitable wear on the performance of such dampers, the preloading devices have to use low stiffness elastic elements, even for generating large forces. Fig. 3.2.4 shows a Coulomb friction damper providing damping effects in three directions.

Such dampers can be used only in high-amplitude vibratory environments. At low amplitudes, typical for isolation of precision objects, the sliding process necessary for generating the damping force may not develop if the driving force does not exceed the static friction force magnitude between the damper surfaces. In such cases, the resistance forces are not velocity-dependent (damping) but are of the elastic nature due to inevitable deformations of the damper mechanism.

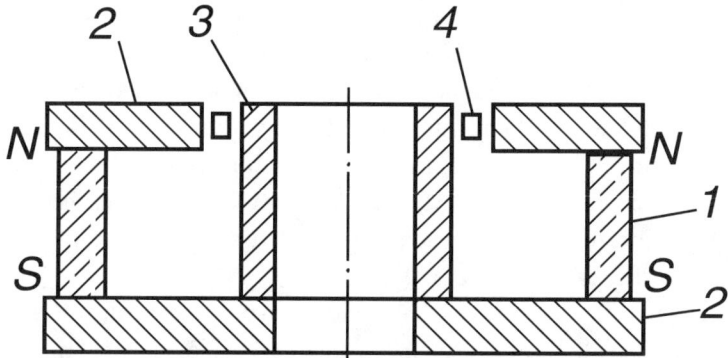

Fig. 3.2.5. Eddy-Current Electromagnetic Damper (1—Permanent Magnet; 2, 3—Magnetic Field Conductors, Mild Steel; 4—Electro-Conductive Coil).

3.2.3 Electromagnetic Dampers

Electromagnetic dampers utilize interaction between magnetic field (usually from strong permanent magnets) and eddy currents induced in a solid (short-circuited) conductor moving inside the magnetic field and attached to the vibrating object (see Fig. 3.2.5). The resistance (damping) force is proportional to velocity of the conductor, thus realizing linear, viscous-like damping force. The damping coefficient for the damper in Fig. 3.2.5 is

$$C = \frac{6.53 * 10^{-11} B^2 lS}{\rho}, \quad (N\text{-}s)/cm \qquad \text{Eq. (3.2.3)}$$

where B is magnetic induction in the clearance, Gauss; l—length of the conductor, cm; S—cross-sectional area of the conductor, cm^2; and ρ is specific resistance of the conductor, Ohm-cm.

3.3 ELASTO-DAMPING MATERIALS

"Elasto-damping materials" (EDM) are those that have relatively low elastic moduli, allowing large deformations and exhibiting relatively high-energy dissipation under vibratory conditions. Judicious selection of an EDM for an isolator simplifies its design since no damper is required. In many cases, static and/or dynamic characteristics of EDM are significantly nonlinear. This may make analysis and synthesis of an isolation system more complicated, but may result in substantial performance advantages. The EDM group includes high-damping metals, such as shape memory alloys; 3-D meshed or knitted fibrous materials (wire mesh, felt, etc.); elastomers (rubbers); some plastics. High-damping metals have not yet found a significant application for vibration isolators.

Elastic modulus of an EDM, and thus stiffness of elements made of the EDM, is often different for static and dynamic loading conditions. Dynamic stiffness coefficient $K_{dyn} = k_{dyn}/k_{st}$, where k_{st} is stiffness measured at static conditions (very slow, usually less than 0.05–0.1 Hz, load application), and k_{dyn} is effective stiffness at the vibratory conditions. K_{dyn} may be amplitude- and/or frequency-dependent and its magnitude may reach $K_{dyn} = 5-10$.

Energy dissipation in EDM is of a hysteretic nature and can be conveniently characterized by loss factor $\eta = \tan\beta$ or logarithmic decrement $\delta = \pi \tan\beta$ (see Section 1.4). The damping parameters also may be amplitude- and/or frequency-dependent; magnitude of δ may reach 1.5–3.0.

3.3.1 Meshed Fibrous Materials

3.3.1A Wire-Mesh Materials and Cables

Vibration isolators as well as other vibration control devices use wire-mesh materials in many ways. The flexible elements are usually made from stainless cold-drawn wire, in special cases from a nonmagnetic alloy. One technique is to knit hoses or bands from 0.1–0.6 mm diameter wire. The flattened hose or band is wrapped into round "pucks," which are placed in a die and pressed to the desired shape under ~100 MPa (15,000 psi) pressure. In another technique, stainless 0.03–0.25 mm diameter wire is formed into a long tightly coiled spiral of 0.15–1.0 mm diameter. Then the spiral is stretched four-to-six fold, crudely shaped into a "pillow," and placed into a die to be pressed to the required shape. Variation of the die pressure results in variation of density of the final product, thus allowing its stiffness and damping characteristics to be tuned.

Wire-mesh elements are mainly loaded in compression, sometimes in shear or bending. Increasing compression load results in increasing number of contacts between the wires, as well as in increasing contact pressure. As a consequence, static stiffness of the element also increases.

Doubling of the compressive load results in ~1.5 times increase in stiffness. Thus, the wire mesh isolators have a quasi-CNF load-deflection characteristic. Allowable specific static compression load is $p_{max} = 3$–30 MPa (450–4,500 psi), depending on the wire diameter and type of mesh. High dynamic overloads (eight-to-ten times p_{max}) can usually be tolerated. Creep rate (see Section 3.3.2F) of the wire-mesh elements under compression is 1–1.5% per decade, although the creep rate is much more significant in the first minutes after load application, possibly due to closing of the initial gaps between the wires.

Angles between the contacting wires as well as the forces in the contact areas are independent random variables. Due to high contact pressures, any lubricant present in the element is squeezed out, and friction between the contacting wires has characteristics of dry (Coulomb) friction. Due to the static friction forces, the 3-D mesh may become, especially at low vibration amplitudes, a very rigid body whose deformation characteristics are analogous to the basic elasto-frictional connections that are considered as models of structural damping, e.g. see [3.14]. For these models both stiffness and relative energy dissipation do not depend on loading frequency but are strongly dependent on vibration amplitudes (energy dissipation increases, stiffness decreases with increasing amplitude).

These amplitude dependencies can be explained by the fact that under vibratory loading the wires are slipping relative to one another. The energy of motion in the areas of slippage is dissipated, and the wire spans between the contact points increase during the process of slippage. The latter effect increases effective lengths of the wires involved in bending and/or stretching, thus increasing compliance and reducing stiffness of the whole element. Each joint between the wires represents a basic elasto-frictional connection. At small relative vibration amplitudes, a (the ratio of vibration amplitude to the element's height) dynamic loads between the wires are low. Slippage develops only in the contacts where either contact pressures (set during pressurizing in the die as well as induced by the compression load on the element) are low, or where there is a combination of magnitude and direction of the angle between the contacting wires beneficial for the development of sliding. With increasing a, both the number of the slippage areas and amplitudes of the slippage increase, thus energy dissipation increases and K_{dyn} decreases. With continuous increase of a, both the stiffness and the absolute energy

232 • PASSIVE VIBRATION ISOLATION

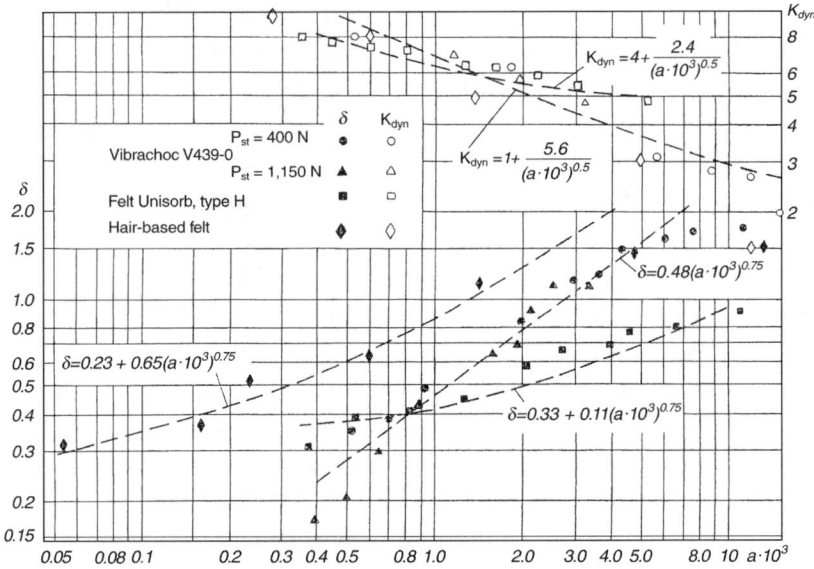

Fig. 3.3.1. Dynamic Characteristics of Wire-Mesh and Felt Materials as Functions of Vibration Amplitude.

dissipation asymptotically approach their limits at which the resources of additional contact areas coming to action are exhausted. Since the total energy (maximum potential energy or maximum kinetic energy) of a vibratory cycle is proportional to a^2, the relative energy dissipation $\psi \cong 2\delta$ during the cycle should have a maximum.

Fig. 3.3.1 shows typical plots $\delta(a)$ and $K_{dyn}(a)$ for wire-mesh elements Vibrachok at $a = (0.4-15)10^{-3}$ obtained from oscillograms of decaying free vibrations. Small values of a correspond to relatively small log decrement $\delta = 0.15$–0.2, and to very large $K_{dyn} = 8$–10. Increasing a results in δ reaching a plateau at $\delta = 2.0$–2.2. Other data indicates that this plateau indicates a peak of δ which tend to decrease at further increase of a. K_{dyn} is monotonously decreasing with increasing a, approaching $K_{dyn} = 1.0$. At small amplitudes a,

$$K_{dyn} \approx 1 + \frac{A}{a^{0.5}}; \quad \delta \approx Ba^{0.75}, \qquad \text{Eq. (3.3.1)}$$

where constants A and B depend on parameters of wire-mesh element.

Influence of static loading P_{st} on the element (within the working range) on δ and K_{dyn} is not very pronounced (see Fig. 3.3.1, plots for two values of P_{st}).

Effective reduction of dynamic stiffness (and thus natural frequency) with increasing vibration amplitude is typical for elements with softening nonlinear load-deflection characteristics, e.g., see [3.4]. Thus, wire-mesh elements show a paradoxical combination of strongly hardening nonlinear characteristic for static loading and strongly softening nonlinear characteristic for dynamic loading.

The advantages of using wire-mesh elements for vibration isolators include their large rated loads; quasi-CNF characteristic of their static stiffness; insensitivity to aggressive environments; and insensitivity to both high and low temperatures. Neither performance characteristics, nor reliability depend on the environment. Heat generated due to high intensity of energy dissipation at large amplitudes is easily evacuated. Some densification of the mesh under continuous loading at the upper end of the rated load range has been observed. It can be said that wire mesh elements are superior for high vibration intensity environments (also see Section 4.2, Table 4.1). Some designs of vibration isolators using wire-mesh elements are presented in Section 4.6.3.

Cable isolators [3.15] have multistrand cables, usually made from stainless-steel wire, as their flexible elements. The dynamic characteristics of cables are similar to those of wire-mesh elements. However, due to relatively long spans of the cables constituting the flexible elements of typical cable isolators, (see Section. 4.6.3), the static load-deflection characteristic at low loads is hardening nonlinear, similar to wire mesh elements, but at high loads (usually associated with shock loading), the cable elements may buckle. The buckling results in a softening nonlinear characteristic, often with quasi-zero stiffness segments, which is beneficial for shock isolation.

3.3.1B Felt

Felt is a fabric produced by meshing natural or synthetic fibers by a combination of mechanical movements, chemical actions, and application of moisture and heat, but without using looms or knitting operations. Typical felt is made from one or several brands of wool, possibly with addition of plant or synthetic fibers. Felt mats can be 1 to 75 mm thick and up to 2,000 mm wide. The best brands of felt are not sensitive to mineral oils, greases, organic solvents, low temperatures, dryness, ozone, and sunlight. Felt pads are usually attached to the support surfaces by adhesives.

In its basic structure (chaotic mesh of thin fibers), felt is analogous to wire-mesh elements. However, the materials of the constitutive fibers of felt mats have much lower stiffness and much higher damping properties than those of steel wires, thus the properties of the fiber material modify dynamic characteristics of felt. Also, a typical felt pad comprises a variety of fiber diameters, although the range of fiber diameters in one brand is relatively narrow. This structural similarity results in similarities of dynamic characteristics between felt and wire mesh.

Load-deflection characteristics of felt pads are quasi-linear up to relative compression $\varepsilon = {\sim}0.25$, then the stiffness increases fast; stiffness at $\varepsilon = 0.5$ is about ten times greater than at $\varepsilon = 0.25$. Elastic modulus in compression is $E = 5-28$ MPa (750–4,200 psi), with higher numbers for felt mats that are made with a large content of animal hair fibers. After a prolonged loading the value of E increases. Allowable specific compressive loads for thick pads (thicker than 13 mm or 0.5 in.) are $p_{max} = 0.05-0.35$ MPa (7–50 psi), for thin pads they are up to 0.8–2MPa (120–300 psi). The highest load rating is for the animal hair felt.

Just as with the wire-mesh materials, K_{dyn} and δ for felt are strongly dependent on vibration amplitude. However, this dependence is less pronounced, especially for felt made from fine fibers, due to the influence of K_{dyn} and δ of the fiber material (natural and/or synthetic polymeric materials). A felt pad can be considered as a parallel connection of two damped springs: one representing dynamic characteristics of the fiber material; another, the dynamic characteristics of the fiber conglomerate. Stiffness of the first spring does not depend noticeably on vibration amplitude (as it is typical for polymeric material without fillers). As a result, K_{dyn} and δ for felt are characterized by "diluted" expressions similar to Eq. (3.3.1),

$$K_{dyn} = \frac{1}{N}\left(M + \frac{A_1}{a^{0.5}}\right), \frac{M}{N} > 1 \; ; \quad \delta = C + B_1 a^{0.75}, \qquad \text{Eq. (3.3.2)}$$

where A_1, B_1, C, M, N are experimentally determined constants. Test data for the fine fiber felt isolation pads (Unisorb Co.) shown in Fig. 3.3.1 can be closely approximated by $K_{dyn} = 4 + [24/(a \times 10^3)^{0.5}]$; $\delta = 0.33 + 0.11(a \times 10^3)^{0.75}$. Similar expressions for K_{dyn} and δ for animal hair felt are close to Eq. (3.3.1) since the fibers are very stiff.

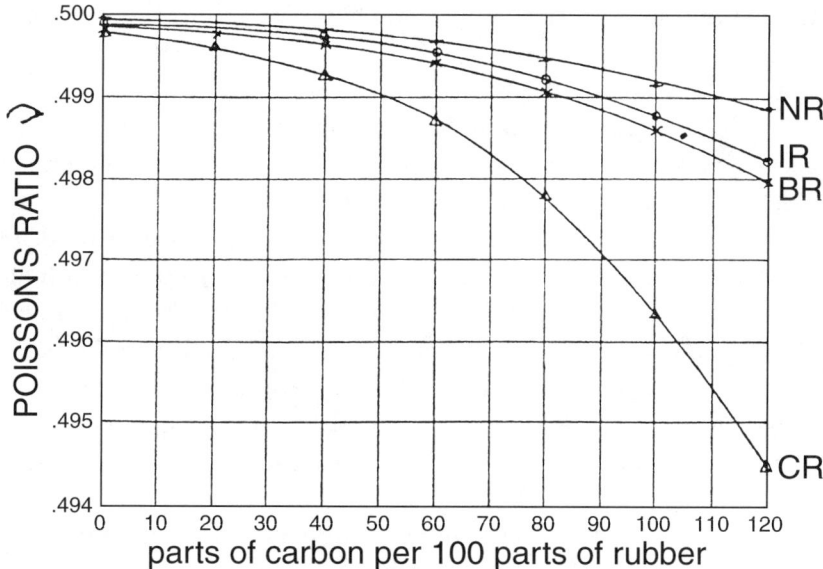

Fig. 3.3.2. Poisson's Ratio for Various Rubbers as a Function of Carbon Black Content (NR—Natural Rubber; IR—Cis-polyisoprene, "Synthetic Natural Rubber"; BR—Butyl Rubber; CR—Chloroprene Rubber).

3.3.2 Elastomeric (Rubberlike) Materials

Rubber makes up a unique family of materials. Elastic modulus values of rubber are low, and rubber endures without failure static stretch up to 1,000% (for soft blends of rubber). After such deformation the rubber part quickly restores its initial dimensions. While rubber parts are easily deformable, rubber is a practically volumetric incompressible material (Poisson's ratio $\nu = 0.49$–0.4999, depending on the type of rubber and on carbon black content; e.g. see Fig. 3.3.2, [3.16]). As a result of the closeness of the Poisson's ratio to 0.5, deformation characteristics of rubber parts in different directions can be independently controlled by shaping the element (see below). Rubber elements can be easily produced in various shapes and can be easily bonded to external and internal rigid (usually metal or hard plastic) structural covers or inserts. Damping in rubber elements is of the hysteretic type, thus it can be characterized by magnitude of logarithmic decrement δ.

A suitable base elastomer or blend is selected based upon the required dynamic and time stability performance and upon expected environmental conditions. Rubber compounds based on *natural* rubber (NR) or analogous synthetic *isoprene* (IR) have high tensile strength, relatively low damping capacity ($\delta \cong 0.05-0.7$, depending on reinforcing fillers content), relatively low dynamic-to-static stiffness ratio K_{dyn}, and good bonding properties to metals. However, this family of rubber is characterized by low resistance to mineral oils and gasoline. Always present in the air micro quantities of ozone cause cracking in the stressed areas of parts made from NR and IR rubbers. The temperature range of use for natural rubber parts is low, usually below 90–100 °C (190–212 °F). Recently developed, so-called *epoxidized natural* rubber (ENR) family, e.g., see [3.17], has much higher damping (2–4 times higher than for NR) and higher resistance to temperature. Highly polar elastomers, such as *acrylonitrile butadiene* (*nitrile* rubbers, NBR) are resistant to oils, have higher damping ($\delta \cong 0.4-1.2$), and have higher resistance to temperature. *Chloroprene* (*chlorobutadiene*, *neoprene*) rubbers (CR) have good mechanical properties similar to those of natural rubber but also have a good resistance to ozone and mineral oil products. Their temperature resistance is rather poor (significant stiffening at temperatures below −10 °C, which can be somewhat reduced by proper blending). *Ethylene-propylene-diene* (EPDM) rubbers are resistant to oxidation and ozone attack, have relatively low creep and compression set, and may be blended for relatively high damping. *Butyl* elastomers (BR or IIR) have very high damping ($\delta \cong 0.3-3.0$), high aging resistance, are very temperature-resistant, are practically impermeable for gases, but may have poor creep and compression set characteristics, are not oil-resistant, and have poor bonding properties. Recently developed BROMO modification of BR [3.18] has even better temperature resistance and significantly lower set (creep), which is the characteristic especially important for vibration isolators. *Silicone* rubbers are extremely heat- and cold-resistant, but their mechanical characteristics and oil resistance are relatively low. *Fluoroelastomers* are very resistant to environmental conditions and can be blended for high damping. The above comments are very general since many parameters of rubber can be significantly upgraded by proper blending. For example, usually high damping is associated with high creep rates. However, an excellent combination of both can be achieved by judicious blending (e.g., [3.19]). It should be noted that rubber flexible elements for vibration isolators should be selected primarily based on their dynamic and creep characteristics and in special cases, such as engine mounts, by temperature tolerance. Protection from the aggressive environments can be provided by embedding them in a close-cell foam matrix resistant to the environmental factors [3.20] or by other types of protective coatings.

Rubber properties can be changed in a very broad range by blending various additives with the base rubber compound. However, rubber properties may be sensitive to relatively minor deviations from quality and quantity of the ingredients and from the curing regime (temperature and duration of curing). The main goal of the curing process is to bind separate molecules into a 3-D structure, thus transforming plastic raw rubber into an elastic material that is much more stable to aggressive fluids dissolving the raw rubber. The curing process is performed by adding *vulcanizing agents* to the raw rubber, usually 1–5% of sulfur or other compounds, and heating to 120–150 °C. Some *accelerators* are sometimes added to reduce the curing temperature or time. An increase of the vulcanizing agent content results in stronger intra-molecular connections and in higher stiffness (hardness). However, usually hardness is adjusted by adding *active* and *inert fillers* and *plasticizers*. The most common active filler is *carbon black* with particle sizes that are 0.02–0.1 μm and special *clays*. The active fillers enhance tensile strength, wear resistance, and damping. Inert fillers do not have a strengthening effect but reduce the cost of rubber parts. Plasticizers (*softeners*) are usually liquids added to the rubber blend for improving processing characteristics and cold temperature properties. *Anti-aging agents* are added to the rubber blend to reduce oxidation and ozone cracking of the rubber parts, thus lengthening their useful life.

The volumetric coefficient of thermal expansion for vulcanized rubber is of the order of magnitude of 0.05% per 1 °C, corresponding to reduction of volume by ~6% during cooling of the part from the curing temperature to the room temperature. Due to presence of other, opposite effects resulting in some increase of the part volume, the overall volumetric shrinking is about 5%, corresponding to the linear shrinkage of 1.6–1.7%. If the rubber is bonded along a significant surface area to metal (or some other rigid material) components having linear shrinkage 0.1–0.25% in the same temperature interval, significant tensile stresses may develop in the rubber part, which is sometimes detrimental to its durability.

Some highly damped materials for vibration control devices use different polymers. EAR Corporation makes *Isodamp*™ and *Isoloss*™ materials,

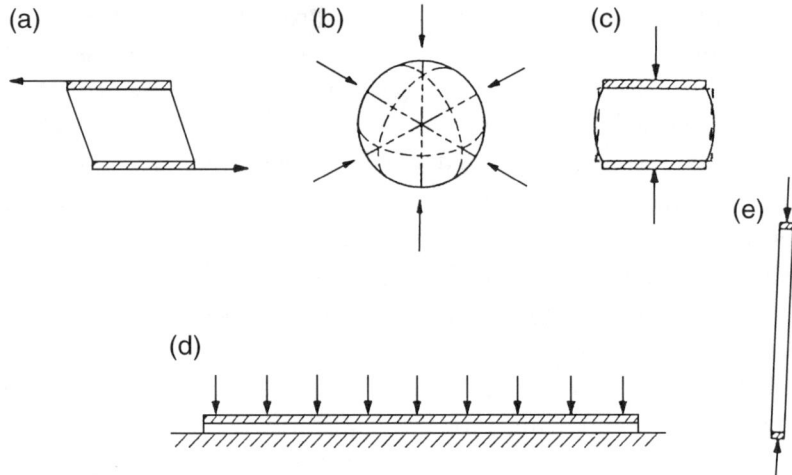

Fig. 3.3.3. Typical Modes of Deformation of Elastomeric Elements: (a) Pure Shear; (b) Omnidirectional (Hydrostatic) Compression; (c) Compression of Element Bonded to Rigid Surfaces (Dotted Lines—Compression of Lubricated Unbonded Element); (d) Highly Restrained Compression; (e) Unrestrained Compression.

which are polyvinyl chloride-based materials [3.21]. Stiffness of these materials is highly dependent on temperature. One brand, EAR C-1002 has stiffness (moduli) increasing 8.6 times from room temperature (70 °F) to 45 °F, and decreasing to 0.15 of the room temperature value at 140 °F. Sorbothane™ is a pure polyurethane [3.21] characterized by extremely high damping and very low values of elastic modulus.

3.3.2A Static Deformation Characteristics of Rubberlike Materials

A rubberlike material may experience two basic types of deformation, described by two independent elastic moduli. The shear modulus G describes a shear deformation, not associated with a change in volume, Fig. 3.3.3(a) [3.22], and the bulk modulus K describes a volume deformation, not associated with a change in shape and characterizing an omnidirectional hydrostatic compression, Fig.3.3.3(b). Rubbers without reinforcing fillers, such as carbon black, have G and K of the order of 0.2–1 and 1,000–5,000 MPa (30–150 and 150,000–750,000 psi), respectively.

A rubber specimen sandwiched and bonded between parallel flat rigid surfaces like those in Fig. 3.3.3(c) is not in a homogeneous compression governed by bulk modulus K, unless the lateral dimensions of the specimen are very large relative to its thickness as in Fig. 3.3.3(d), so that the free surface for outward expansion (bulging) of the material during compression is relatively insignificant (also see 3.3.2C). In such a case, both volume and shape are changing and the stress/strain ratio is governed by a modulus [3.22]

$$M = K + 4G/3 \cong K. \qquad \text{Eq. (3.3.3)}$$

Properties of such thin-layered rubber-metal laminates are described in Section 3.3.2C; they are very different from the resilience characteristics normally associated with rubber. In the other geometric extreme presented in Fig. 3.3.3(e), the lateral dimensions are small relative to the specimen thickness so that the bulging is not restrained (provided that buckling stability of such bar or rod is assured). The similar effect of non-restrained bulging can be achieved in a more realistic case of the configuration as shown in Fig. 3.3.3(c), if there is no bonding and no friction between the rubber element and the rigid surfaces, so that ideal sliding conditions exist, as represented by broken lines in Fig. 3.3.3(c). In such cases the stress-to-strain ratio is governed by the Young's modulus and the ratio of the lateral to axial strain is described by Poisson's ratio, ideally

$$E_o = \frac{9KG}{3K+G} \approx 3G\ ; \qquad \nu = \frac{E_o}{2G} - 1 \approx 0.5\ . \qquad \text{Eq. (3.3.4)}$$

The effective Young's modulus depends on the shape and composition of the rubber element [see Eq. (3.3.4)]. Measuring of E_o or G on the finished rubber part without its destruction is often impossible, and mechanical characteristics of rubber elements strongly depend on their shape and size. Accordingly, the main characteristic of rubber stiffness or modulus of the element is its hardness measured in Shore durometer units H or in similar British Standard (BS 903) hardness degrees. The shear modulus G depends on H somewhat differently for different compositions of the rubber blend. Fig. 3.3.4 [3.23] shows the correlation between BS hardness degrees and values of G and E_o as stipulated in the British Standard BS 903. Fabrication of rubber parts involves many uncertainties, such as uniformity of properties and concentrations of various ingredients, uniformity of mixing these ingredients, deviations of nominal temperature

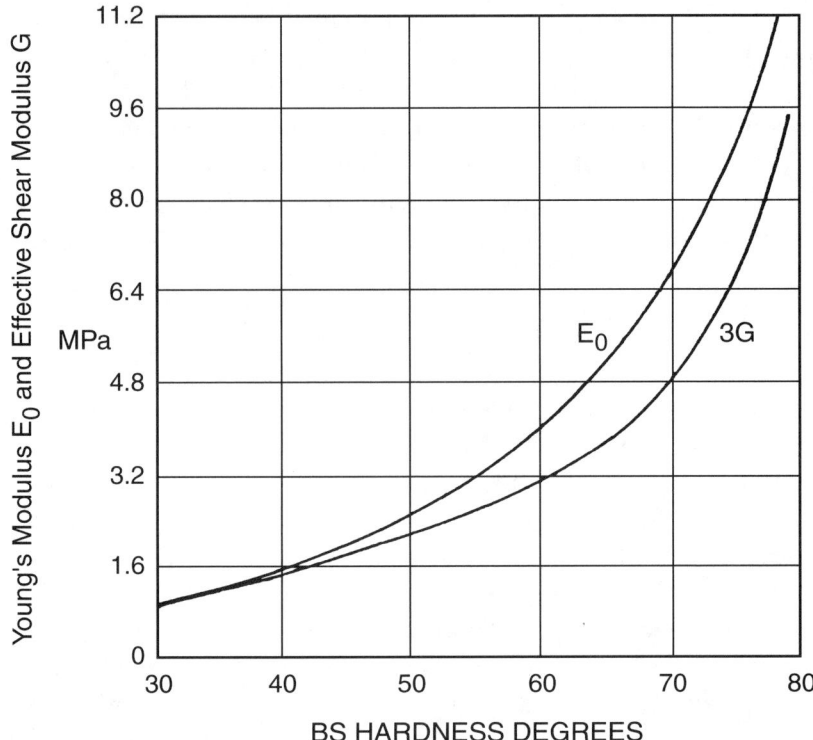

Fig. 3.3.4. Correlation Between Compression E and Shear G Moduli of Rubber Element and Its Hardness.

and pressure in the mold, etc. Accordingly, some variation of rubber parameters, especially hardness, is inevitable. The "normal" variation is within ±5 durometer units ($\sim$±17% in stiffness/modulus variation). More stringent tolerances can be maintained, such as ±2 durometer units, but at a significantly higher price. The effective parameter variation can be reduced without tightening the tolerances in constant natural frequency isolators, as shown in Section 4.5.

The load-deflection characteristic of rubber in shear is softening nonlinear at relative shear $\gamma \geq 0.05$. For vibration isolator design, it is important to know deformation characteristics at low relative shear $\gamma \leq 0.01$–0.02, which are significantly different than the characteristics at larger γ. Fig. 3.3.5 [3.24] shows values of G in two ranges of the relative shear: $\gamma = 0$–0.014 (plain marks) and $\gamma = 0.025$–0.35 (encircled marks). The measurements were performed on rubber cylinders (30 mm diameter,

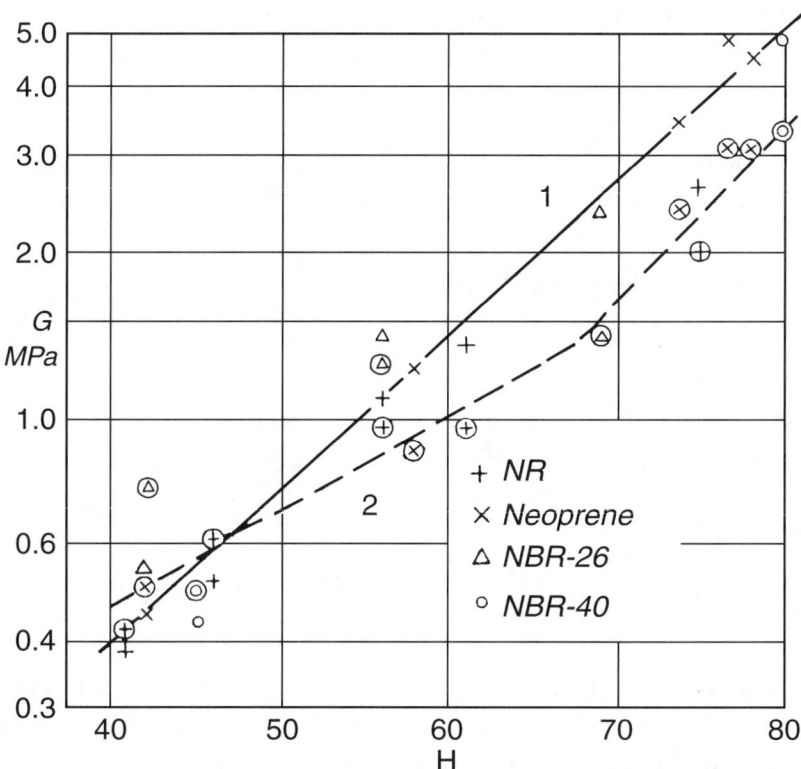

Fig. 3.3.5. Shear Modulus G of Rubber as Function of Relative Shear γ (Plain Marks—Test Data for y = 0–0.014 Fitted by Line 1; Encircled Marks—y = 0.025–0.035 Fitted by Line 2).

14 mm high) made from various types (NR, Neoprene CR, NBR with 26% and 40% of acrylonitrile contents) and hardness of rubber, and precompressed to axial strain $\varepsilon = 0.07$. At these ranges of γ, there is no significant shear nonlinearity at $H = 45$–50; there is a weak hardening nonlinearity at $H \leq 45$; and there is a significant softening nonlinearity at $H > {\sim}50$ (G is greater for $\gamma = 0.014$ than for $\gamma = 0.025$–0.035). For hard rubber blends the nonlinearity is very substantial, e.g., for $H = 80$, $G_{\gamma \approx 0.007} \approx 3 G_{\gamma \approx 0.03}$. The shear modulus G depends on the relative compression for ratios of the smaller lateral dimension a to thickness h $a/h > 3$ and $h = 2$–5 mm. It was noticed (and confirmed analytically, e.g., [3.25]) that shear stiffness decreases with increasing compression load/deformation. This effect is clearly pronounced for rubber-metal laminates (Section 3.3.2C), precompressed rubber bushings (Section 3.3.2D), etc.

Dependence of G and K on compression for thin rubber layers is addressed in Section 3.3.2C.

If a rubber element of a significant thickness h and bonded to rigid (metal) plates on both faces is subjected to shear deformation, some bending deformation also occurs. It can be considered by using a bending correction term in calculating shear deformation x caused by shear force P [3.23],

$$x = \frac{Ph}{GA}\left(1 + \frac{h^2}{36i^2}\right), \quad \text{Eq. (3.3.5)}$$

where A is cross-sectional area of the loaded specimen; and i is bending radius of inertia of the cross-sectional area about its neutral axis.

A rubber element configured as in Fig. 3.3.3(c) is characterized by an effective modulus of elasticity E_{eff}, intermediate between magnitudes of K and M [3.22],

$$E_{eff} = \frac{E(1+\beta S^2)}{1 + E/K(1+\beta S^2)}. \quad \text{Eq. (3.3.6)}$$

Here S is the *shape factor*, the ratio of one loaded area of the element to the total of load-free surface areas; and β is numerical constant dependent on rubber hardness H, $\beta \cong 2.68 - 0.025H$ @ $H = 30-55$; $\beta \cong 1.49 - 0.006H$ @ $H = 60-75$. The shape factor of a rubber cylinder (diameter D, height h) is $S_{cyl} = D/4h$; for a prismatic rubber block (sides a, b, height h), $S_{rec} = ab/h(a+b)$. For rubber elements with reasonably uniform dimensions (D/h, a/h, $b/h < \sim 10$), Eq. (3.3.6) can be written as

$$E_{eff} \cong E(1+\beta S^2) = 3G\left(1+\beta S^2\right). \quad \text{Eq. (3.3.7)}$$

The rubber elements with the uniform dimensions have linear load-deflection characteristics up to $\sim 10\%$ compression deformation, and quasi-linear characteristics up to $\sim 15\%$ deformation. If $b >> a$ for a rectangular shape element, Eq. (3.3.7) should be replaced by [3.22]

$$E_{eff} = (2/3)E(2+\beta S^2) = 2G(2+\beta S^2), \quad \text{Eq. (3.3.8)}$$

where $S = a/2h$.

While designers of metal parts are usually limited by allowable maximum stresses under specified loading conditions, it is customary to design rubber components using allowable strains, or relative deformations. For rubber components bonded to metal inserts or covers, a conservative strain limit for shear loading is relative shear $\gamma = 0.5-1.0$ for the load applied and held for a long time. Some blends can tolerate higher values of relative shear, up to $\gamma = 3.0$. A compression loading of a bonded rubber specimen is usually associated with the allowable relative compression $\varepsilon = 0.1-0.15$, with a somewhat larger limit for softer rubber blends.

3.3.2B Streamlined ("Ideal Shape") Rubber Elements

Elastomeric elements of vibration isolators are required to satisfy many constraints, often contradictory. They have to accommodate high loads and large deflections while being small in size and exhibiting low permanent deformations (creep). Typical elastomeric elements of vibration isolators have a rectangular shape and are bonded to rigid (metal) plates or, sometimes, have a cylindrical shape with end faces of the cylinder bonded to rigid plates. The rigid plates transmit loads to the element, thus loading it in compression or shear mode. If the loading is in compression, such loading pattern creates high nonuniformity of stress distribution and intense stress concentrations in the corners, e.g., see Figs. 3.3.6(a), (b), which present stress distribution in one quarter of a rubber cylinder loaded by the bonded face plate in compression (compression deformation 15% and 30%, respectively). These stress concentrations are responsible for high rate of creep, for high heat generation under dynamic loading, and eventually for starting disintegration of the element. As a result, in the design practice for vibration isolators only relatively small deformations (10–15% of the component thickness) are allowed in compression-loaded rubber elements bonded to metal plates. Thus, to obtain large deflections required for low natural frequency vibration isolators, the thickness of the flexible elastomeric elements loaded in compression must be 7–10 times larger than the required deformation. Such excessive height of isolators designed for low natural frequencies (requiring large deformations) causes packaging problems and may lead to elastic instability (buckling) of the isolator. While buckling can be avoided by using flexible elements loaded in tension (e.g., rubber cords or long rubber bars bonded to holding metal end plates), such elements loaded by weight of the object are prone to failure due to microcracks or bonding imperfections.

244 • PASSIVE VIBRATION ISOLATION

It was established in Chapter 2 that high damping is beneficial for many applications of vibration isolators. However, increase of material damping in elastomeric components may have undesirable consequences. Usually (although not universally) high damping elastomeric compounds have higher creep rates than lower damping compounds. While a damping increase in the elastic element of a resonating oscillatory system results in a *reduction* of heat generation due to reduction of the resonance amplitude (see Section 2.1.1, [3.26]), heat generation increases with increasing damping in cases of forced motion, such as in elastic/isolation elements of vibratory technological machines and in misalignment compensating

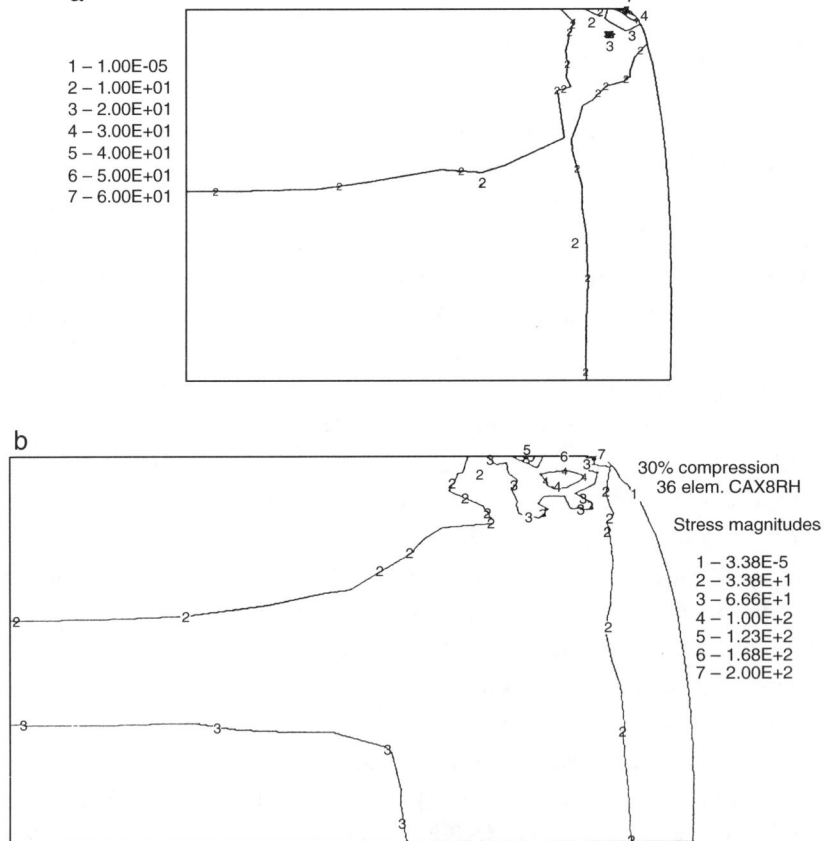

Fig. 3.3.6. Stress Distribution in Cylindrical Rubber Element Loaded in Compression at: (a), (c) $\varepsilon = 15\%$; and (b), (d), 30% Along Its Axis (a, b) and Radially (c, d).

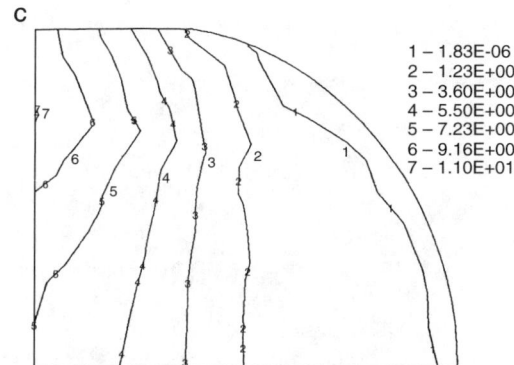

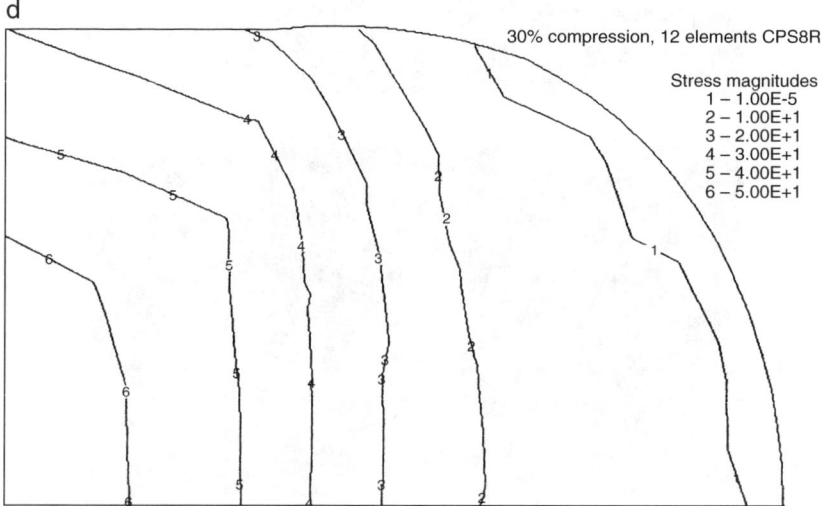

Fig. 3.3.6. (*continued*)

couplings. If there are stress concentrations in such loaded elastomeric elements made of high damping elastomers, they may overheat.

These contradictions of bonded elastomeric elements loaded in compression can be resolved by using streamlined or so-called *ideal shape* elastomeric elements, such as *radially loaded* cylinders, spheres, toruses (O-rings), and ellipsoids. It was demonstrated experimentally [3.27] that creep rates of two identical rubber cylinders—one loaded axially, the other radially—are different, with the axial loading associated with creep rates 25–40% higher than the radial loading. It was shown analytically [3.28] that peak tensile stresses in an elastomeric cylinder axially loaded in

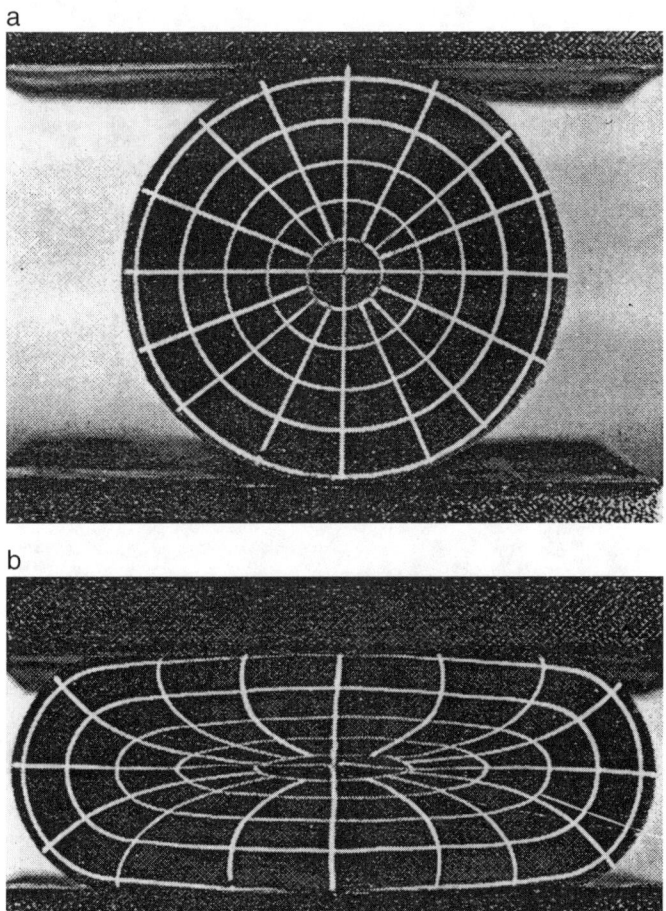

Fig. 3.3.7. Rubber Roller ($D = 60$ mm) with Grid Lines on the Face Surface: (a) Undeformed and (b) Loaded to 50% Compression.

compression between bonded end plates, Figs. 3.3.6(a), (b) are about five times higher than stresses in the same element without bonded plates and loaded in the direction radial to 15% compression [Fig. 3.3.6(c)], and four times higher for 30% compression [Fig. 3.3.6(d)]. It can be seen from Fig. 3.3.6 that in addition to reduced maximum stresses, the stress distribution in the case of radial compression is much more uniform than in axial compression of the same element bonded to the end plates. Due to the combination of the lower maximum stresses and the more uniform stress distribution, much higher relative compression values, up to and

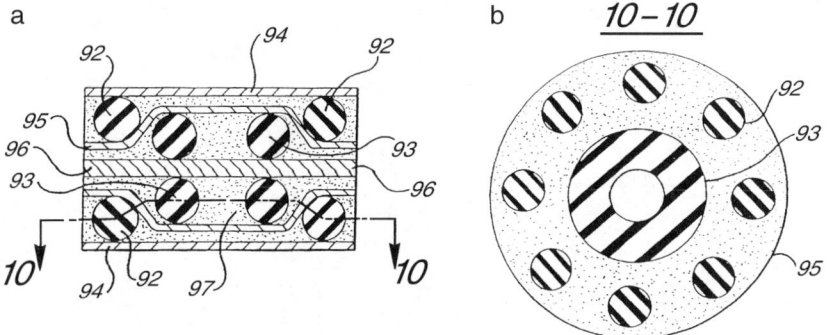

Fig. 3.3.8. Packaging of Streamlined Rubber Elements 92, 93 and Rigid Inserts 95, 96 by Foam Matrix 97 without Bonding.

exceeding 40% (see Section 4.5), can be tolerated for the radially loaded elements. Fig. 3.3.7 [3.29] shows strains developing under the radial deformation of a rubber cylinder at 50% relative compression. It shows, as was also illustrated in Figs. 3.3.6(c), (d), that the surface layer of the cylinder circumference suffers practically no stress while the internal circle undergoes high tension. The part subjected to the highest stress in tension is embraced by layers, like bandages, which are less and less stressed the closer they are to the periphery. Thus, thanks to the practically non-existent tension on the surface, the element is insensitive to dirt, less sensitive to oil and ozone, etc., thus explaining its high robustness as studied also in [3.30]. Introduction of surface interruptions in a streamlined rubber element, or its bonding to rigid components, disturbs the "natural" deformation process. Still, a quasi-streamlined vibration isolator Lastosphere (see Section 4.5), produced for many years by Lord Corp., has cutouts and bonded metal plates for attachment to the isolated object and to the supporting surface, but still tolerates compression deformations at rated static loads 40–45% and can be exposed to dynamic loads having magnitudes four times greater than those of the rated static load.

The ultimate advantages of using ideal shape elements for vibration isolators and other vibration control devices materialize when these elements are not bonded to rigid supporting or constraining components, since the bonding is associated with creating stress concentrations. However, bonding is a prevalent technique of packaging of the rubber flexible elements. The packaging problem was solved in [3.20] by fixating the desired positions of ideal shape elements and constraining rigid

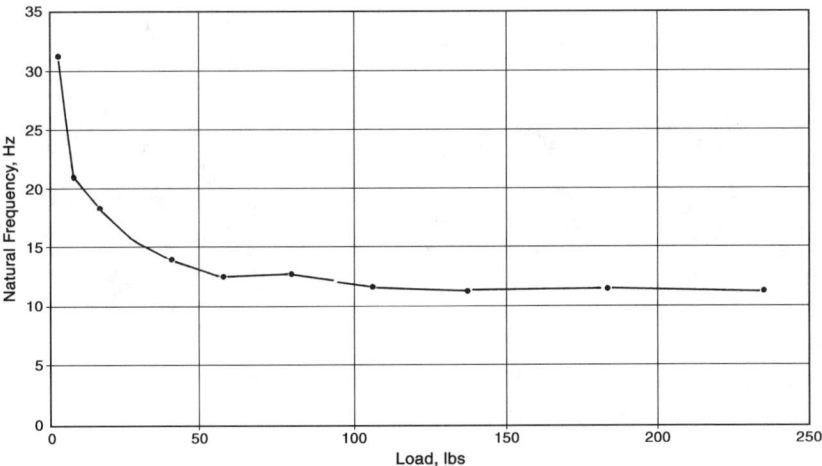

Fig. 3.3.9. Natural Frequency vs. Axial Compressive Load for Rubber Torus ($H = 30$, $D = 50.8$ mm, $d = 12.7$ mm).

(metal) components by means of embedding all the components in a foam matrix, e.g., as shown in Fig 3.3.8. The foam maintains the relative positions of all components but does not change noticeably their load-deflection characteristics. It also makes interaction between the ideal shape rubber elements and the rigid components more consistent, since it eliminates friction between the rubber elements and the rigid components and replaces it by elastic deformation of the foam particles. The external shape of the foam matrix can be easily tailored to conform with the surrounding structural systems, such as a suspended/supported object, a foundation, etc. The foam can also protect the rubber elements from aggressive environments, e.g., from oil in case of natural rubber components installed in an oil-mist environment.

"Ideal shape" elastomeric elements also exhibit substantial nonlinearity of their load-deflection characteristics due to continuous increase of the shape factor with increasing compression load (the loaded surface area increases while the free area decreases). This process results in the "constant natural frequency" characteristic, beneficial for vibration isolators as shown in Section 1.3.

Fig. 3.3.9 [3.30] shows the load-natural frequency characteristic for a rubber torus with cross-sectional diameter 12.7 mm (0.5 in.) and $K_{dyn} \cong 1$. The natural frequency is 12 Hz ± 6% in the broad range of weight load, $P_{max}/P_{min} \cong 4.4$, with compression deformation $\Delta_{max} = 30\%$ at P_{max}.

While this load range is significantly narrower than the load ranges realized for specially designed CNF isolators (Section 4.5), it would increase to at least $P_{max}/P_{min} \cong 6$–8 if used to 40–45% deformation. Due to larger allowable compression deformations (40–45% vs. 10–15% for conventional bonded elements loaded in compression), use of ideal shape unbonded elastomeric elements allows for much more compact designs. This is especially important for low natural frequency isolators (see Section 4.5). The natural frequency in the CNF range can be changed by changing cross-sectional diameter of the cord/cylinder or of the O-ring. The square of the natural frequency for a given load (i.e., stiffness) is proportional to the cross-sectional diameter. Spherical and ellipsoidal rubber elements have similar load-deflection characteristics and can be used for lightly loaded vibration isolators.

Deformation characteristic of a radially loaded cylinder or a torus is described reasonably well by the following analytically derived expression [3.31]

$$\frac{P}{2LdG} = \frac{\lambda^{-3}}{16}(\Psi\lambda^{3/2} + 1.5\Psi\lambda^{1/2} + 1.5\sin^{-1}\Psi)$$

$$+ \frac{\lambda^{-2}-1}{16}\left(\frac{\pi}{2} - \Psi\lambda^{1/2} - \sin^{-1}\Psi\right) - \frac{\lambda}{6}\sin^{-1}\Psi - \Psi\frac{\lambda^{1/2}}{16} \quad \text{Eq. (3.3.9)}$$

or an empirical expression [3.32]

$$\frac{P}{Ld} = E_o\left[1.25\left(\frac{x}{d}\right)^{1.5} + 50\left(\frac{x}{d}\right)^6\right] \quad \text{Eq. (3.3.10)}$$

where P is compression force; x is compression deformation; L is length of the element (πD_o for torus, where D_o is its mean diameter); d is cross-sectional diameter; x is compression deformation; $\Psi^2 = x/d$, $\lambda = 1 - \Psi^2$; G is shear modulus; E_o is Young's modulus.

Stiffness of the radially-loaded cylinder can be obtained by differentiating the load-deflection expressions Eq. (3.3.9) or Eq. (3.3.10). Differentiation of Eq. (3.3.10) results in

$$k = \frac{dP}{dx} = LE_o\left[1.88\left(\frac{x}{d}\right)^{0.5} + 300\left(\frac{x}{d}\right)^5\right] \quad \text{Eq. (3.3.11)}$$

This expression leads to an interesting and somewhat unexpected conclusion that the radial stiffness of a rubber cylinder of length L made from a certain blend of rubber (E_o) does not depend on its diameter d, only on

Table 3.3. Coefficients for Calculating Deformations of Rubber Sphere

x/D	α_s	β_s
0.225	0.0138	0.0007
0.293	0.024	0.002
0.367	0.039	0.004
0.452	0.063	0.010
0.553	0.101	0.025
0.684	0.175	0.083

the relative radial compression x/d. Of course, one has to remember that Eq. (3.3.10) is an approximate one.

Deformation characteristic of a compressed rubber sphere is described for small deformations (up to ~20%) by [3.32]

$$\frac{P}{D^2} = 0.44 E_o \left(\frac{x}{D}\right)^{1.5}; \quad k = \frac{dP}{dx} = 0.66 D \left(\frac{x}{D}\right)^{0.5} \qquad \text{Eq. (3.3.12)}$$

and for large deformations (20–65%) by [3.32]

$$P = E_o \pi D^2 (\alpha_s + k\beta_s), \quad \text{where} \quad \alpha_s = (\gamma - \ln\gamma - 1)/8;$$

$$\beta_s = (\gamma^{-1} + 2\ln\gamma - \gamma)/64. \qquad \text{Eq. (3.3.13)}$$

Here $1 - \gamma^{1/2} = x/D$ and α_s and β_s are given in Table 3.3.

Nonlinear load-deflection characteristics of the ideal shape rubber elements results in their low sensitivity to inevitable variations of durometer within the batch and in their suitability for variable stiffness isolators (see Chapter 4). It was demonstrated in [3.30] that the radial loading of cylinders results in about 15% higher damping than for conventionally loaded elements made from the same rubber blend. This effect can be explained by a larger role of shear deformations/stresses in deforming unbonded cylinder under the radial compression than in compression of a bonded element, since the bulk (hydrostatic) deformation, characterized by modulus K and playing a large role in compression of bonded elements, is not associated with a significant energy dissipation.

3.3.2C Thin-Layered Rubber-Metal Laminates

If a rubber element in compression has a large value of the shape factor S, as in Fig. 3.3.3(d), its free surface is very small, thus the bulging process is

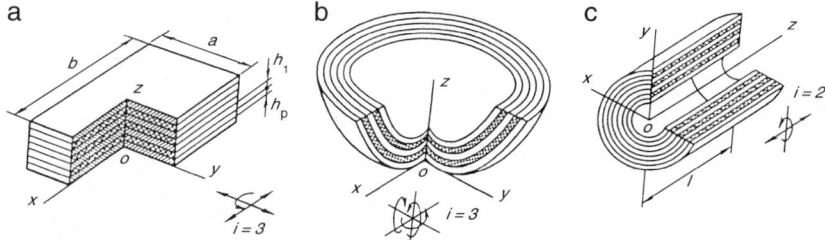

Fig. 3.3.10. (a) Prismatic. (b) Spherical. (c) Cylindrical Rubber-Metal Laminates (I—Number of Degrees of Mobility).

very restrained and deformation of the element in compression is governed by the bulk modulus K, while deformation in the perpendicular shear direction is governed by the much smaller shear modulus G. As a result, this type of element has highly anisotropic stiffness characteristics [3.33] that may be useful for special types of vibration isolators (Section 4.7). Usually, thin rubber layers alternate with thin rigid layers made of strong metal or fabric woven from high-strength fibers, with rubber and rigid layers connected by bonding. The multilayer structure is needed to allow for larger shear displacements. Although the laminates for various applications may have different shapes—flat, Fig. 3.3.10(a), spherical, Fig. 3.3.10(b), cylindrical, Fig. 3.3.10(c)—their application for vibration isolators is mostly limited to flat elements. The laminate usually consists of n layers of metal or other rigid material and n-1 layers of rubber; for packaging purposes, the end metal layers are often made much thicker than the intermediate layers.

The bulk modulus K for different blends of rubber varies significantly (see above). This variation is correlated to deviation of Poisson's ratio from the theoretical value for an incompressible material $\nu = 0.5$. The main requirement for the laminated elements for vibration isolator applications is high compression stiffness, which is determined by four major factors: deviation of ν from $\nu = 0.5$, see Fig. 3.3.2 (the more the deviation, the lower compression stiffness); shape factor S (the larger the value, the greater compression stiffness); deformations in the bonding (adhesive) layer between the rubber and the rigid layers, which effectively reduce compression stiffness; and deformations within the rigid layers under the hydrostatic pressure in the rubber layers (determined by Young's modulus of the rigid layers and/or their thickness). Usually, ν is closer to 0.5 for softer (lower durometer) rubber blends that have fewer fillers, as illustrated in Fig. 3.3.2. For example, a neoprene rubber with hardness $H = 40$ had $\nu = 0.4997$ vs. $\nu = 0.4990$ for $H = 75$ [3.32]. This phenomenon is also reflected in values of β in Eq. (3.3.6). As a result, using a higher durometer rubber for thin-layered laminates results in much less of an increase in compression stiffness than increase in shear stiffness. It also can be concluded from Fig. 3.3.2 that the highest compression stiffness and the greatest ratio between compression and shear stiffness values for a laminate can be achieved by using soft (low-filled) natural rubber blends. The low oil and ozone resistances of the natural rubber blends are not as critical for thin-layered laminates as they are for rubber elements with more uniform dimensions, since penetration of these aggressive substances into the element is limited by the small thickness of the rubber layer. In addition, the exposed edges of the rubber layers can be protected by appropriate coatings or by foam strips.

Effective compression modulus E_{eff} of the laminate vs. specific compression load $p_z = P_z/A$, where A is the surface area of the laminate, is plotted in Fig. 3.3.11 [3.33]. E_{eff} is related to the total thickness of rubber h_r and was calculated from experimentally obtained load-deflection plots as

$$E_{eff} = \frac{\Delta p_z h_r}{\Delta z}, \qquad \text{Eq. (3.3.14)}$$

where Δz is compression deformation caused by increment Δp_z of the specific compression force (pressure). While compression of rubber elements of more uniform dimensions, like those in Fig. 3.3.3(c), is characterized by a linear load-deflection characteristic up to relative deformations $\varepsilon = 0.1$–0.15, compression of the thin-layered laminates can be highly nonlinear (progressive increase of E_{eff} in Fig. 3.3.11, starting from the smallest deformations down to $\varepsilon = \Delta z/h_r = 0.001$). In the highly constrained environment of a thin rubber layer bonded to the rigid surfaces, the major factor is hydrostatic pressure. Both shear modulus G and bulk modulus K increase with the increasing hydrostatic pressure in thin rubber layers [3.34]. Increase of K is much more pronounced than increase of G. Consequently, increase of modulus E_{eff} is much more significant than increase of G, about one decimal order of magnitude per 1.5 decimal orders of magnitude increase of compression load. The thin-layered laminates can accommodate very high static compression loads,

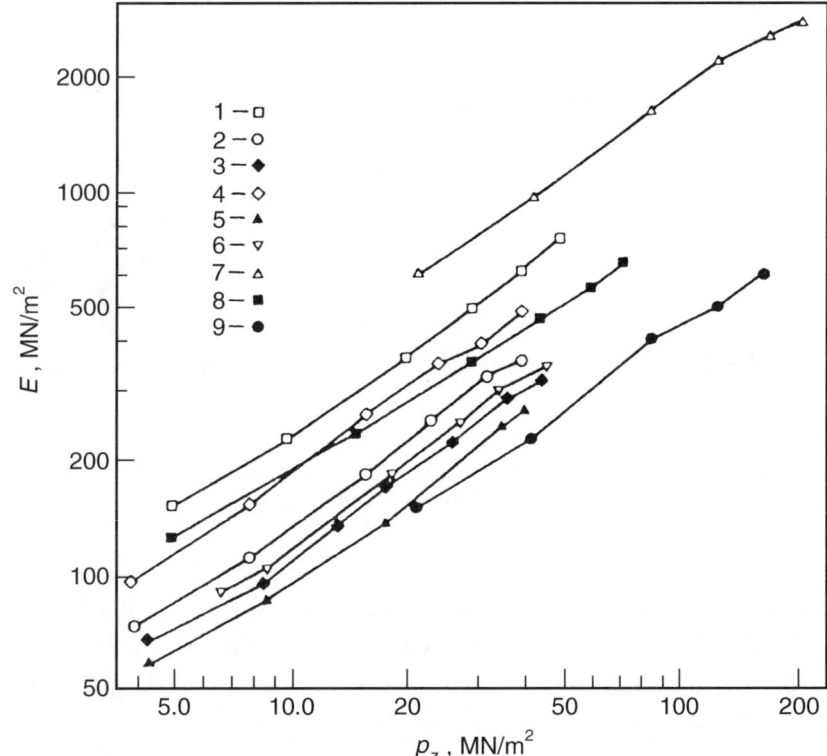

Fig. 3.3.11. Effective Compression Modulus for Thin-Layered Rubber-Metal Laminates as Function of Compressive Load (pressure p_z) (1–4, 7–9, $H = 42$; 5, $H = 58$; 6, $H = 75$; 1–6, 0.05 mm Brass Metal Layers; 7–9, 0.1 mm Steel Metal Layers; 1, $A = 21.3$ cm^2, $t_r = 0.16$ mm, $n = 33$; 2, $A = 26.4$ cm^2, $t_r = 0.33$ mm, $n = 33$; 3, $A = 23.7$ cm^2, $t_r = 0.39$ mm, $n = 17$; 4, $A = 25.9$ cm^2, $t_r = 0.25$ mm, $n = 9$; 5 $A = 23.5$ cm^2, $t_r = 0.53$ mm, $n = 17$; 6, $A = 23.1$ cm^2, $t_r = 0.58$ mm, $n = 11$; 7, $A = 12.3$ cm^2, $t_r = 0.106$ mm, $n = 15$; 8, $A = 36$ cm^2, $t_r = 0.28$ mm, $n = 15$; 9, $A = 12.3$ cm^2, $t_r = 0.44$ mm, $n = 14$. Here H is Rubber Hardness, A is Loaded Surface Area, t_r is Thickness of One Rubber Layer, n is Number of Rubber Layers).

at least up to 500–600 MPa (75–100,000 psi). The degree of nonlinearity weaken and the maximum compression loads decrease for thicker ($\geq \sim 2$ mm) rubber layers. The ultimate static load is limited by strength of the rigid (metal) layers, since the mode of failure at static loading was observed to be disintegration of the rigid (metal) layers. Thus, strength enhancement of a laminate for compression loading can be achieved by using a stronger steel or thicker steel layers. Absolute values of

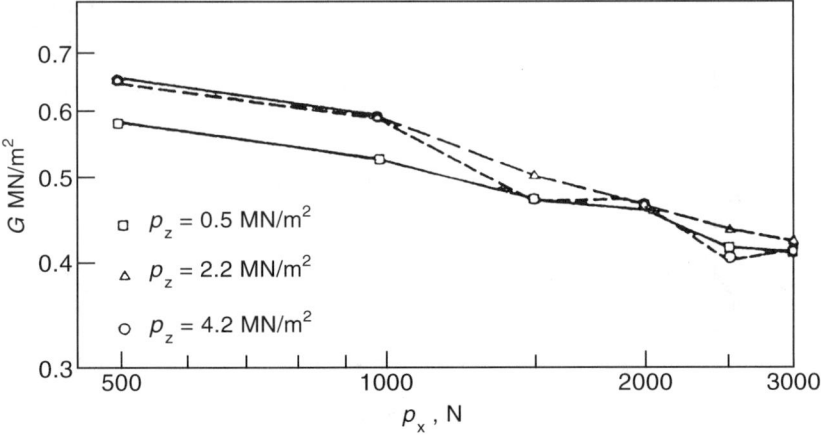

Fig. 3.3.12. Effective Shear Modulus of Compressed Thin-Layered Rubber-Metal Laminates.

compression stiffness are quite high. Thus, for sample 7 in Fig. 3.3.11, E_{eff} = 1,800 MPa at p_z = 100 MPa, which is equivalent to 1 μm deflection per 1 MPa increase in compression pressure, about the same as contact deformation between two pressed-together ground flat steel surfaces [3.14].

Effective shear modulus $G_{eff} = \Delta p_x h_r/\Delta x$ is plotted in Fig. 3.3.12 vs. shear force $P_x = p_x A$, where Δx is shear deformation caused by increment $\Delta p_x = \Delta P_x/A$. The precompressed laminates loaded in shear demonstrate a slight softening nonlinearity. For elements with thickness of one layer $h < \sim 1.5$ mm, it was observed that there was only a relatively weak correlation between the magnitude of compression pressure on the specimen (in the range of $p_z = 0.5–150$ MPa) and its effective shear modulus (Fig. 3.3.12 shows G_{eff} in the range of $p_z = 0.5–4.2$ MPa). For $h = 2–4$ mm, G_{eff} increases as $\sim p_z^{0.5}$.

It was shown in [3.34] that stiffness of cylindrical (diameter d) or prismatic ($a \times b, a \leq b$) thin-layered rubber-metal laminates with $0.01 \leq \alpha \leq 0.1$ ($\alpha = h_r/d$ or h_r/a) can be determined as

$$k = \frac{A}{h} \frac{(1+\Psi)G}{(\beta + c\alpha^2 + \gamma)}, \qquad \text{Eq. (3.3.15)}$$

where $\gamma = (\chi G H_r/E_m h_m)$, $\chi = 1 + (h_m/h_r)[\nu_m/(1-\nu_m)]$, $\Psi = [(h_m + 2h_a)/h_r]$. Here h_r, h_m, h_a is thickness of rubber, metal, adhesive layer, respectively;

$h = nh_r(1 + \Psi)$ is total thickness of the element; A is surface area of the element; $\beta = G/K$ for the rubber layers; E_m, ν_m are Young's modulus, Poisson's ratio for the metal layers; c is a numerical constant ($c = 2.57$ for cylindrical elements, $c = 1.95$ for prismatic elements). This expression gives stiffness at small deformations when the moduli G and K are not yet distorted by high hydrostatic pressures. Some nonlinearity is reflected by the term $c\alpha^2$ in the denominator.

Since compression deformation of the thin-layered laminates is due to so-called hydrostatic compressibility with, practically, no shear deformation involved, compression damping of the laminates is very low, even when the laminates are made from high-damping rubber blends. Damping associated with shear deformation is the damping characteristic for the used rubber blend.

3.3.2D Elastic Stability of Laminated Rubber Parts

When a rubber element is subjected to compression forces, there may be a possibility of its elastic instability or buckling. While buckling of a "squat," low profile isolator cannot occur under the practically feasible load/deformation conditions, the danger of buckling is increases with increasing ratio between the height h of the element and its smallest dimension in the lateral direction a. Such elements are often employed in vibration isolators for low natural frequency isolation systems. A typical configuration of the buckled element is shown in Fig. 3.3.13 [3.32] for the typical case wherein metal face plates bonded to both ends of the element could not shift. While a laminated rubber element is shown in Fig. 3.3.13, a similar buckling pattern is observed also for solid elements. It can be seen that the rubber part of the element is "shearing out" at the buckling event. Treatment of buckling in most Strength of Materials textbooks implies that the buckling process is a "bang-bang" process, whereas the element has its initial characteristics at compression forces $P < P_{cr}$ and abruptly changes them at $P = P_{cr}$, where P_{cr} is *critical* or *Euler* force. However, the shear stiffness is actually decreasing even at $P << P_{cr}$. This effect is identical to the effect of compression force application to a metal column, e.g., see [3.14]. The schematic in Fig. 3.3.13 illustrates a general case where, in addition to compressive force P_c, shear force P_s is also applied to the element (at its center). The critical force P_{cr}, of course, decreases with increasing P_s.

256 • PASSIVE VIBRATION ISOLATION

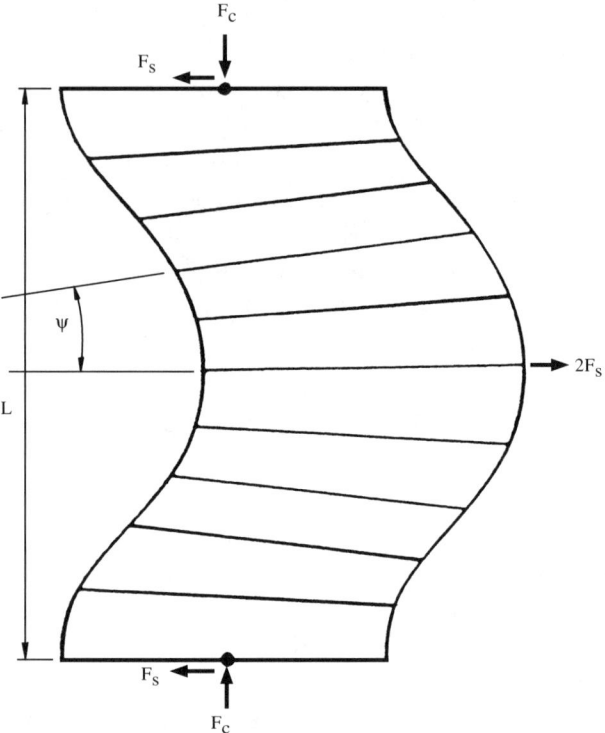

Fig. 3.3.13. Column of Bonded Rubber Blocks Between Flat Surfaces Subjected to Compression and Shear.

The expression from which the critical compressive force P_{cr} can be determined is given in [3.32],

$$\frac{P_{cr}}{T'}\left(1 + \frac{P_{cr}}{K'}\right) = \frac{4\pi^2}{L^2} \qquad \text{Eq. (3.3.16)}$$

where K is shear stiffness of a unit column of total thickness t_T (one rubber layer of thickness t and cross-sectional area A together with bonded to its metal plate); $K = AG/t$ is shear stiffness and $K' = Kt_T$ is "reduced" shear stiffness; T is experimentally determined bending stiffness of the unit column; and $T' = Tt_T$ is "reduced" bending stiffness.

The overall shear stiffness $K_o = K'\varepsilon$ (ε is the compression strain induced by force P_c) in the center of the column, comprising n unit columns and compressed by force P_c, is presented in Fig. 3.3.14 for columns with various number n of the unit columns (shown at the respective lines), each

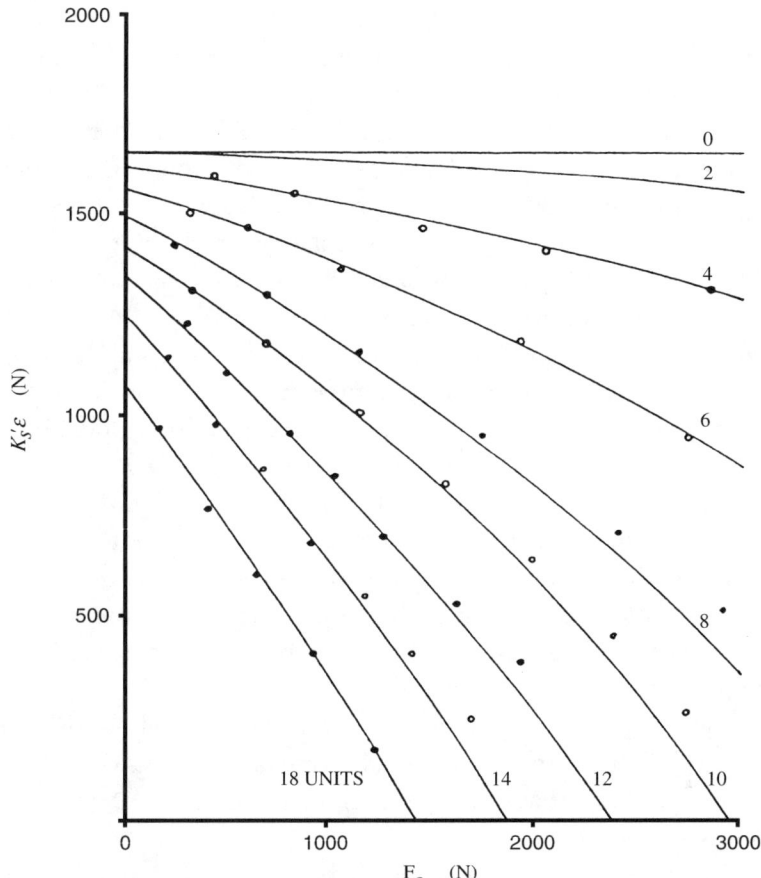

Fig. 3.3.14. Apparent Shear Stiffness $K_s'\varepsilon$ of Column of Bonded Rubber Blocks as Function of Compressive Force.

of which has the following parameters: cross-section $A = 76.5 \times 25.6$ mm^2; rubber thickness $t = 4.4$ mm; total thickness of each unit $t_T = 7.5$ mm; compression stiffness of one unit $k_c = 7$ MN/m; shear stiffness of one unit $K_s = 0.22$ MN/m; and "reduced" shear stiffness $K_s' = K_s t_T = 1,670$ N bending stiffness of one unit $T = 11.4$ N-m/rad. The ordinate is expressed as "apparent" shear stiffness $K_s'\varepsilon = K_s n t_T \varepsilon$. It is obvious that the shear stiffness can be controlled in a broad range by varying the compressive force P_c.

For a column that has a rectangular or elliptical cross-section, the instability occurs in the direction of the shorter dimension of the cross-section. A specimen with a "tubular" cross-section is generally more stable than a specimen with a solid cross-section of a similar shape and similar dimensions. If the compressive load is below the critical load but close to it, the creep rate of the rubber is increasing, and after some time the specimen buckles. The critical compression deformation at which the buckling event occurs does not depend significantly on hardness (elastic modulus) of the rubber. Instability is not observed if the dimension of the rubber body in the direction of compression is less than ~0.6 of the diameter (for a cylindrical specimen) or of the smaller side of the rectangular cross-section. For the taller elements, the instability occurs at relative compression values decreasing with increasing height of the rubber body. For the column of given dimensions, the critical compression deformation (but not the compressive force!) decreases approximately proportionally to the number of rigid plates dividing the rubber column into the separate units. For the rubber columns not bonded to the face plates, and lubricated so that they may slide or otherwise expand in the direction transverse to that of the compressive force, the shape of the rubber body can change so much with the increasing compressive force that the instability does not develop.

Stability of rubber components is often determined by the "critical strain," rather than by the critical load. For example, an axially loaded in compression hollow rubber cylinder (height $h = 75$ mm, OD and ID 75 and 50 mm, respectively) buckles at $\varepsilon = 0.25$ regardless of rubber blend.

While the unexpected collapse of a vibration isolator due to excessive compressive forces is unacceptable, the effects accompanying compressive loading of rubber columns must be considered in the process of isolator design; in some cases, a controlled use of these effects can be very beneficial. Two applications of these effects can be noted: adjustment/tuning of horizontal stiffness of vibration isolators and designing flexible elements of isolators to perform in the buckled condition to attain extremely low horizontal stiffness. The effect of decreasing shear stiffness in the compressed rubber columns should be considered in the isolators whose rubber elements are simultaneously under compression and under shear (e.g., ones shown in Figs. 4.4.9 and 4.4.10). A judicious preloading in compression of isolators containing the rubber-metal laminates described in Section 4.7 may reduce their shear (horizontal) stiffness while further increasing their compression (vertical) stiffness due to the nonlinearity addressed in Section 3.3.2C.

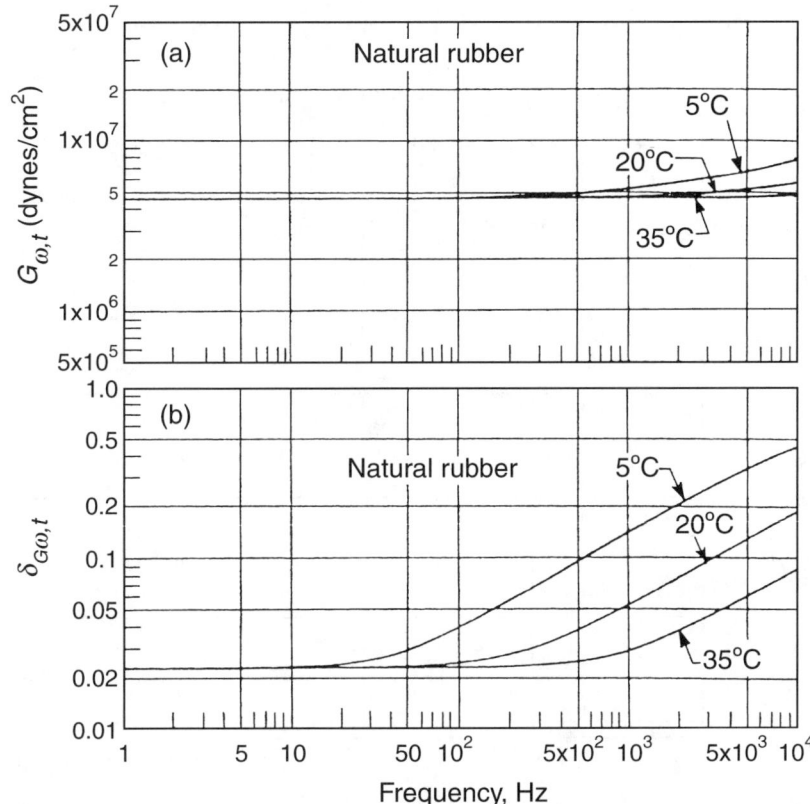

Fig. 3.3.15. (a) Shear Modulus G; and (b) log Decrement δ for Unfilled (gum) Natural Rubber.

Buckling as it relates to low stiffness vibration isolators is addressed in Section 4.8.1.

3.3.2E Dynamic Characteristics of Rubberlike Materials

Since energy dissipation in rubberlike materials has a hysteretic character, their dynamic characteristics are often expressed as *complex dynamic shear modulus* and *complex Young's modulus*,

$$G^*_{\omega,t,a} = G_{\omega,t,a}(1 + j\eta_{G_{\omega,t,a}}), \quad E^*_{\omega,t,a} = E_{\omega,t,a}(1 + j\eta_{E_{\omega,t,a}}). \qquad \text{Eq. (3.3.17)}$$

Here *dynamic moduli* $G_{\omega,t,a}$ and $E_{\omega,t,a}$ are the real parts of the complex moduli $G^*_{\omega,t,a}$ and $E^*_{\omega,t,a}$; and $\eta_{G_{\omega,t,a}}$ and $\eta_{E_{\omega,t,a}}$ are energy *"loss factors"* associated with shear deformations and with the Young's modulus of the

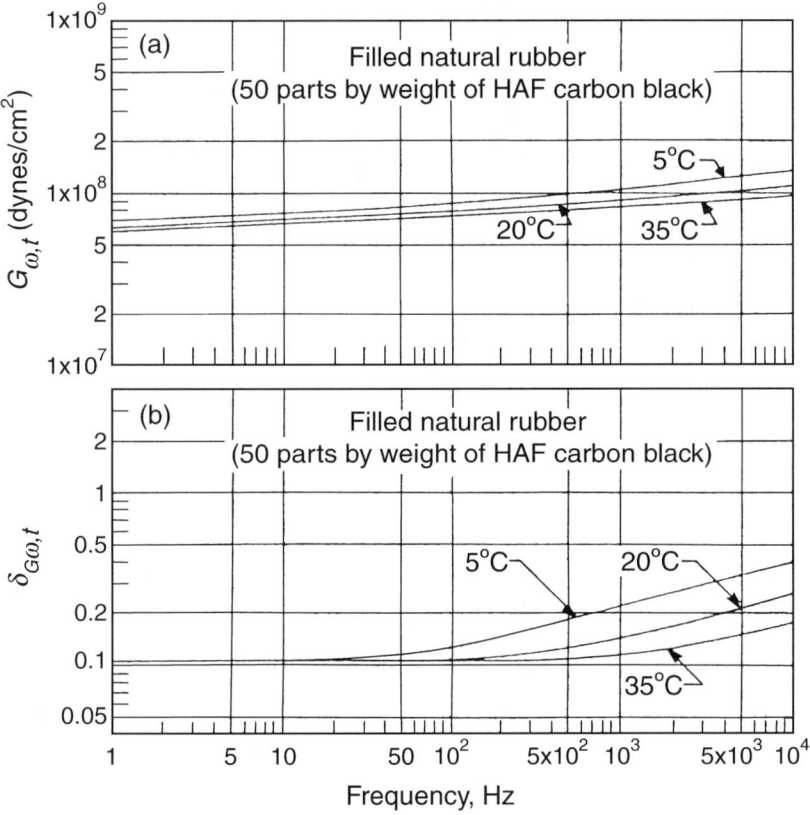

Fig. 3.3.16. (a) Shear Modulus G; and (b) log Decrement δ for Natural Rubber with 50% Carbon Black.

element (log decrement $\delta = \pi\eta$). The subscripts ω, t, a indicate that both dynamic moduli and loss factors are functions of vibration frequency, temperature, and vibration amplitude. In most cases, the dynamic moduli in Eq. (3.3.17) are experimentally found to increase with increasing frequency and decreasing temperature, and to decrease with increasing vibration amplitude. Many rubberlike materials have their dynamic moduli and loss factors practically independent of frequency for frequencies below ~100 Hz (unless these materials are close to the glasslike state; see below). This frequency range is especially important for most of the vibration isolation applications. Dependencies of the loss factors/log decrements are usually more complex. Figs. 3.3.15–3.3.17 [3.22] show typical dependencies of G and δ on frequency ω and temperature t. The

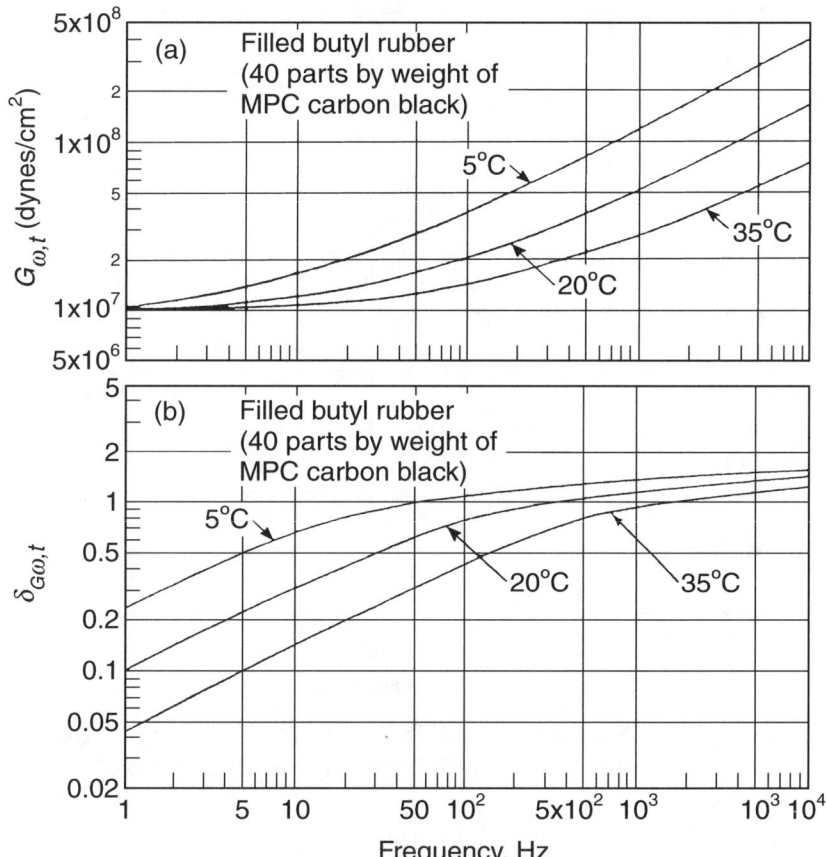

Fig. 3.3.17. (a) Shear Modulus G; and (b) log Decrement δ for Butyl Rubber with 40% Carbon Black.

"transition" frequency ω_t and temperature t_t refer to "glass" transition of rubberlike materials (at high frequencies or low temperatures) to a glasslike state, wherein they are loosing their elastomeric properties. The dynamic shear modulus and the corresponding log decrement are plotted in Fig. 3.3.15 [3.22] for unfilled natural rubber ("gum"), in Fig. 3.3.16 for NR filled with 50% by weight of reinforcing carbon black, and in Fig. 3.3.17 for butyl rubber reinforced by 40% carbon black. One can see differences in behavior of damping δ plots for natural rubber and butyl, since for the latter the glass transition region is less abrupt. The data in Figs. 3.3.15–3.3.17 are shown from frequency 1 Hz. The transition from "static" to "dynamic" rate of loading starts at about 0.05–0.1 Hz

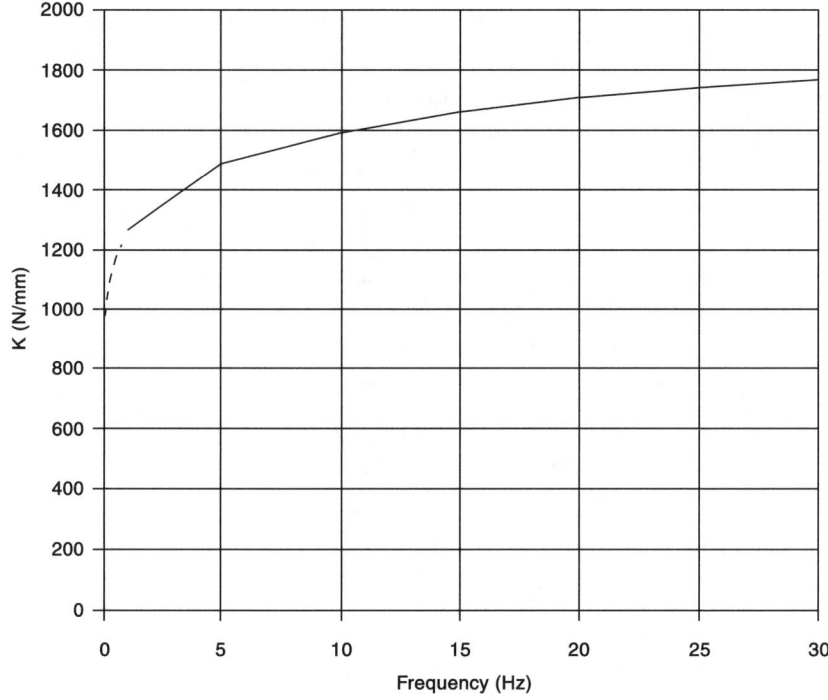

Fig. 3.3.18. Axial Stiffness *K* for NBR Rubber O-Ring Showing Transition from "Static" to "Dynamic" Stiffness.

(Fig. 3.3.18). As was mentioned, the ratio K_{dyn} of dynamic k_{dyn} to static k_{st} stiffness is a very important parameter for vibration isolation. Generally, rubber blends with higher damping have larger values of K_{dyn}. However, a judicious use of ingredients can result in increasing damping without increasing K_{dyn}, e.g., see Fig. 3.3.19 where the correlation between K_{dyn} and loss factor η is shown for carbon-black-filled NR specimens and specimens made from "epoxidized" natural rubber ENR [3.17]. Lower values of K_{dyn} result in smaller vibration isolators having higher static stiffness for the given natural frequencies, thus reducing cost and weight and increasing sturdiness of the vibration isolation installation.

The data of Figs. 3.3.15–3.3.17. were obtained for very small vibration amplitudes, for which the rubberlike materials exhibit linear behavior. Unfilled and lightly filled rubber blends remain linear up to relatively large strains, but moderately and heavy filled (hard) rubber blends show strong amplitude dependencies for K_{dyn} and δ. Rubberlike materials comprise a polymeric base, and both inert and active (usually, various brands of

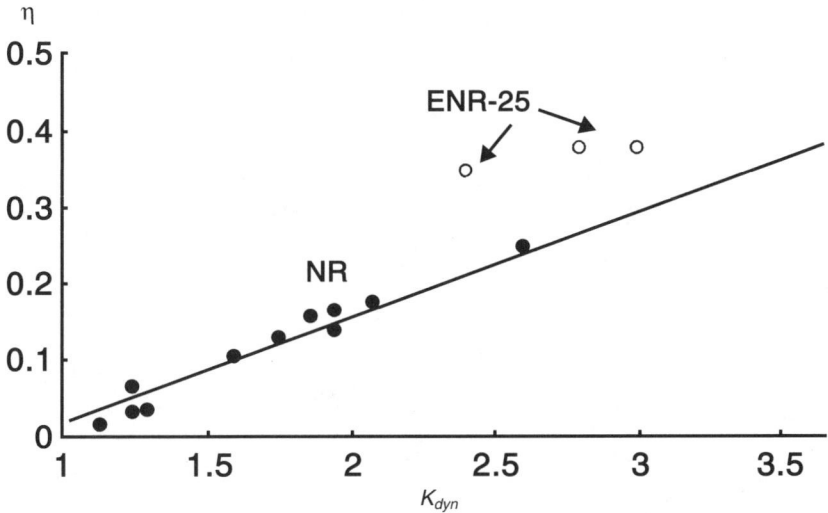

Fig. 3.3.19. Dynamic-to-Static Stiffness Ratio vs. Damping (Loss Factor η) for Natural Rubber (NR) and Epoxidized Natural Rubber (ENR-25) Blends.

carbon black) fillers. The active fillers combine with the base polymer and develop reinforcing structures inside the rubber. These structures break down during deformation, quickly restoring back into somewhat different configurations. The destruction and reconstruction of these structures have a discrete character similar to dry (Coulomb) friction, whereas the external force cannot move the body until it reaches the limit—the static friction force. When the body stops, the friction force quickly reaches the static friction force magnitude. Thus, the mechanism of deforming the carbon black structure is somewhat similar to the deformation mechanism of wire-mesh structures. Dynamic characteristics of the filled rubbers are determined, like in fibrous felt-like materials, by dynamic characteristics of active carbon black structures (K_{dyn} and δ depend on the amplitude but not on the frequency of vibration) and dynamic characteristics of the polymer base (K_{dyn} and δ do not depend on the vibration amplitude and do not strongly depend on frequency below $\sim$100 Hz). Relative importance of these two factors is mostly determined by the specific characteristic and concentration of the active fillers. Some rubber blends contain ingredients preventing development of the carbon black structures. In such blends amplitude dependencies of K_{dyn} and δ are weak even for large carbon black content values.

264 • PASSIVE VIBRATION ISOLATION

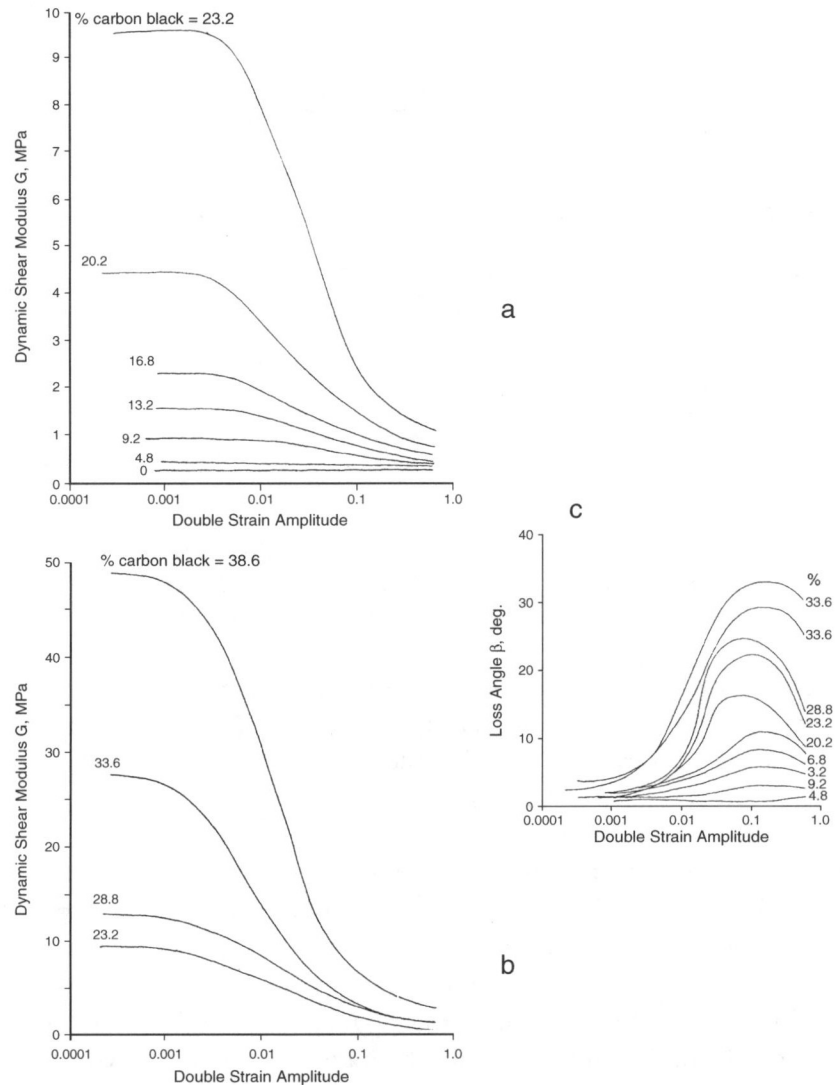

Fig. 3.3.20. (a), (b) Dependence of Shear Modulus G; and (c) Loss Angle β on Vibration Amplitude for Butyl Rubber Blends with Different Carbon Black Content (Shown in % by Weight).

Fig. 3.3.20 [3.23] shows amplitude dependencies of (a), (b) the dynamic shear modulus G and (c) loss angle $\beta = \tan^{-1}\eta$ for butyl rubber blends having the same composition and different active carbon black content. Fig. 3.3.21 shows amplitude dependencies of K_{dyn} and δ for various

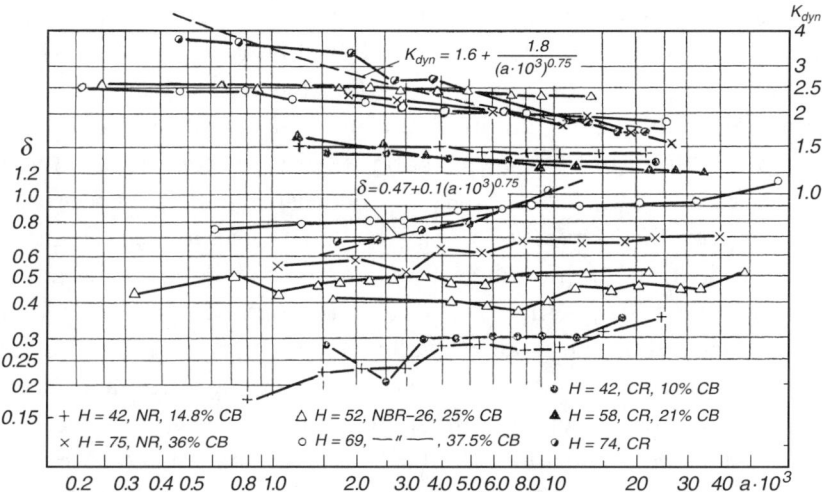

Fig. 3.3.21. Dependence of Shear Modulus K_{dyn} and log Decrement δ on Vibration Amplitude for Various Rubber Blends Used in Vibration Isolators.

rubber blends developed for vibration isolators, obtained by testing in free vibration conditions on specimens 30 mm diameter, 14 mm thick loaded in compression. The concentration of the active carbon black corresponding to development of the uninterrupted carbon black structure (about 20–40%) can be detected by a sharp increase in electroconductivity of the rubber; this condition is usually accompanied by sharp increases of both K_{dyn} and δ as well as by a steeper slope of their amplitude dependencies.

The dynamic characteristics of some rubber blends are not very sensitive to the static load P_{st} on the specimen. For other blends, such as ones based on NBR, K_{dyn} may significantly increase (and δ decrease) with increasing P_{st}, e.g., see [3.35]. In some cases, the "scale factor" may be of importance. Increase of the loaded surface A_l of the element was observed in some cases to correlate with reduction of δ (in one case, increase of A_l by a factor of 5 resulted in 1.3–1.6 times reduction of δ), possibly due to a greater role of hydrostatic component of deformation with larger bonded areas of the rubber specimen.

An important dynamic characteristic is *fatigue resistance* of isolators under vibratory loads. Usually, it is not as important as creep and/or chemical degradation caused by environmental factors for isolators used for the protection of vibration-sensitive equipment due to small vibration amplitudes. However, fatigue failures should be considered for isolators exposed to large vibration amplitudes (vibration-generating production equipment, automotive engine mounts and isolation bushings for steering/handling linkages, etc.). "Fatigue" of a rubber element means a gradual change in its performance characteristics caused by a prolonged action of vibratory stresses [3.32]. The most pronounced change is a gradual reduction of modulus (thus, increasing deflection), although sometimes the stiffness increases during the service life of the element. This may depend on the mode of loading (see *"Example"* for compression loaded engine mounts). In some cases the fatigue criteria relate to a designated amount of change in the dynamic stiffness, loss factor, etc. This is a form of relaxation similar to the gradual increase of strain/ deflection under a constant stress (creep). A plot of modulus vs. logarithm of the number of stress cycles is a straight line, similar to the plot of creep vs. logarithm of time (or vs. the number of stress cycles, e.g., see [3.27]). While a quantitative prediction of fatigue failure for rubber elements is a rather futile exercise due to major influences of the base rubber, additives, variations of production processes, etc., basic qualitative correlations related to fatigue of rubber elements are well-represented in Fig. 3.3.22 based on an extensive experimental study [3.36] as summarized in [3.32]. Reduction of stiffness constant or modulus by 10% is considered as failure of the rubber specimen. Fig. 3.3.22 presents results of the study wherein rubber specimens were exposed to torsional shear, and the deformation cycle is described by the ratio of double amplitude of the shear deformation (strain) $\Delta\theta$ to the maximum shear strain θ_m or $\gamma = \tan\theta_m$. Same maximum strain θ_m (abscissa in Fig. 3.3.22) would result in very different number of cycles before the failure, depending on $\Delta\theta/\theta_m$. Solid lines in Fig. 3.3.22 represent numbers of cycles before the failure (10% modulus reduction) occurs. The horizontal line $\Delta\theta/\theta_m$ represents the static situation (amplitude $\Delta\theta = 0$) and the point A represents the ultimate static strain ($\theta_m = 86°$ or shear strain $\gamma = \sim 14$).

The worst condition is when $\Delta\theta/\theta_m = 1$, i.e., the minimum strain during the loading cycle is zero (complete unloading). Many studies confirmed the critical importance of avoiding this type of a condition. While it may occur only in a very severe loading environment, it emphasizes usefulness of preloading of rubber elements exposed to dynamic loading. Preloading is frequently used for isolating bushings. A design described in [3.37] is a

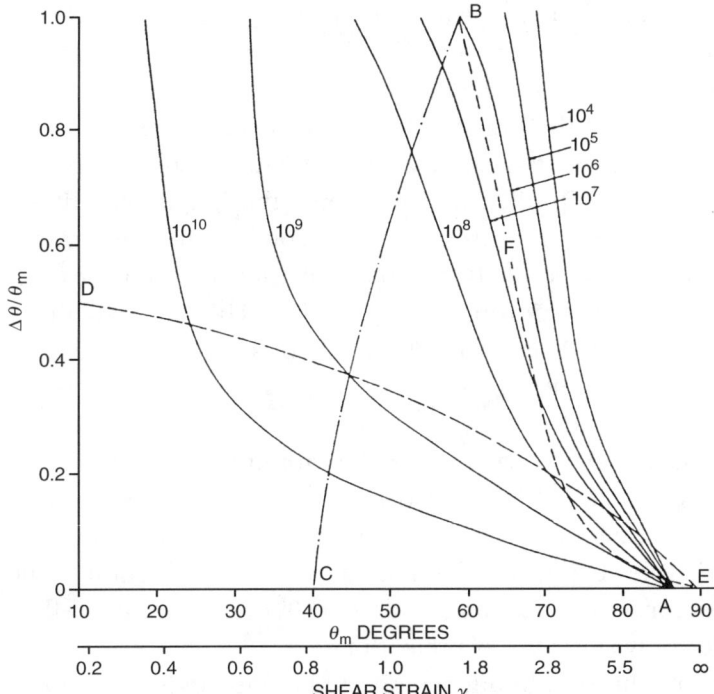

Fig. 3.3.22. Influence of Maximum Angular Strain θ_m and Double Strain Amplitude $\Delta\theta$ on Fatigue Life of Mountings Stressed in Torsion.

bushing-shaped engine mount in which the rubber element is molded between the inner and outer steel pipes. After molding, the bushing is processed by swaging, which reduces its outer diameter from 92 mm to 87 mm, thus creating compression preload of the rubber element. As was mentioned before, some reduction of shear stiffness as a result of the preload should be considered.

Plots in Fig. 3.3.22 also demonstrate other effects of design changes and of various operating conditions on fatigue life (assuming linearity of the rubber element, $\tau = G\gamma$, where τ is shear stress). If the mean load is constant but the dynamic load amplitude varies, the fatigue life can be presented by line BC. As the amplitude increases from 0 at C to the maximum at B (momentary unloading), the fatigue life falls from 10^{10} to $\sim 10^6$ cycles.

If the ratio of the dynamic load amplitude to its mean value is constant, the fatigue life varies as shown by line DE, which represents the case when the ratio "dynamic load/mean load" is 2/3 (or the ratio "dynamic load/maximum load" is 1/2). At the left side (D), the stresses are small and the fatigue life is $> 10^{10}$ cycles. Point E represents an infinite stress ($\theta_m = 90°$, tan $\theta_m = \infty$), and the fatigue life is zero. Even if the stress distribution inside the component is non-uniform, $\Delta\theta/\theta_m$ is constant throughout the volume, thus the DE-like line represents the fatigue conditions for the whole part. Thus, the stress (θ_m) in some point determines fatigue endurance at this point. It can be seen that doubling of θ_m results in about ten-fold reduction of the fatigue life. This fact explains the experimentally observed superior fatigue life of streamlined rubber elements having relatively uniform stress distribution and low maximum stresses [see Fig. 3.3.6(c), (d)].

If the mean load varies while the vibratory load intensity does not change, a line like BFE demonstrates the effects of such variation. It can be seen that the fatigue life increases with the increasing mean load (e.g., from 10^6 cycles for max deflection 59° and min deflection 0°, point B, to 10^7 cycles for max deflection 72°, and to 10^8 cycles for max deflection 80°). This also shows benefits of preloading.

Effect of vibration frequency on the fatigue life is not very significant although the higher frequency may result in somewhat faster reduction of the modulus. Of course, a higher frequency means a shorter fatigue life for a given number of cycles. Temperature variation in a not very broad range (like 40–90 °C) is also not a significant factor.

Example. While Fig. 3.2.22 gives a useful presentation of various factors influencing fatigue of vibration isolators with elastomeric flexible elements, it has to be admitted that the process of deterioration of dynamically loaded elastomeric components is not yet well understood. The following is an interesting practical study described in [3.38]. First, dynamic stiffness and damping of a rather large batch of new truck engine mounts were measured as functions of frequency (Fig. 3.3.23, lines 1). The mounts are identified by their stiffness $k = 1,500$ N/mm at low frequency ~ 1 Hz. The standard deviation of stiffness from the mean value was found to be 4.8%, well within the 17% tolerance range (see Section 3.3.2A). Then the mounts were installed on *"fleet"* trucks and on *"durability"* trucks. The fleet trucks are used on a daily basis for commercial business. The accumulated mileage may be from any part of the country,

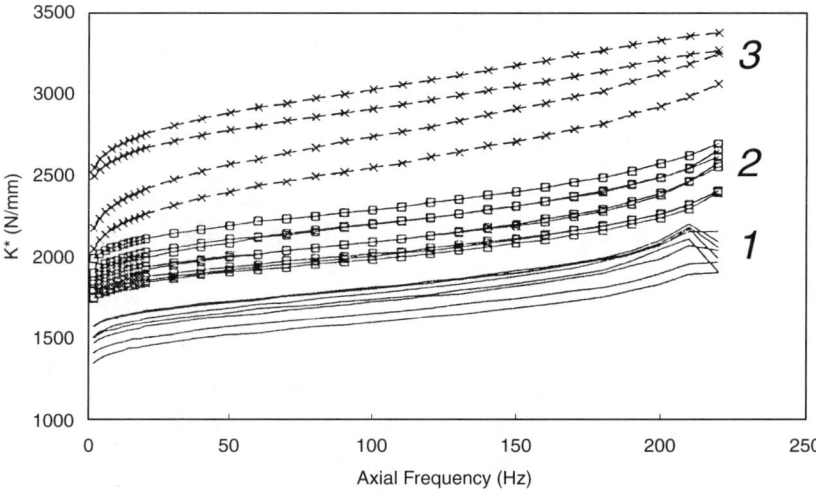

Fig. 3.3.23. Dynamic Axial Stiffness of New (1), Fleet-Tested (2) and Durability-Tested (3) Engine Mounts.

under a variety of load conditions, and can be any combination of on- and off-highway driving. Durability vehicles are driven on a standard durability test track, normally located in one area of the country. The accumulated miles are more "severe" and are designed to compress the durability test cycle. One durability test mile is equivalent to about four miles of customer usage. Both fleet and durability trucks accumulated between 30 and 40,000 miles.

It can be seen from Fig. 3.3.23, line 2, that the mounts removed from the fleet-tested trucks have significantly higher stiffness than the new mounts, line 1, (about 21% average increase). Even larger increase, about 31%, was exhibited by the mounts from the durability-tested trucks, line 3 in Fig. 3.2.23. In addition, after such use the scatter of stiffness values within each tested group has also widened—up to 5.3% for the fleet trucks and up to 7.8% for the durability trucks, as compared with 4.8% for the new mounts. The 31% increase in stiffness is equivalent to about 15% increase in natural frequencies, thus significantly degrading the performance of the mounts.

It can be expected that isolators employing streamlined elastomeric elements would exhibit slower aging due to lower stresses/strains.

3.3.2F Creep of Rubberlike Materials

A very important parameter of rubber blends for use in vibration isolators is *creep*, i.e., gradual increase of deformation while loaded by a constant magnitude load (e.g., weight load). *Creep rate* is an important performance parameter both for vibration isolators for precision machinery and equipment, and for tightly packaged installations (such as engine compartments in vehicles). The total deformation can be approximated as

$$\Delta' = \Delta + b \log t, \qquad \text{Eq. (3.3.18)}$$

where t is time, min; Δ is relative deformation of the element at 1 min after load application; b is constant characterizing creep rate (intensity), expressed as a percent increase of deformation in one *decade of time* (i.e., from 1 min to 10 min, or from 8 days to 80 days, etc.). The creep rate depends on the type of the rubber base, composition of fillers (especially, type of carbon black), type of deformation, presence of rigid (metal) inserts, geometry of the element, and does not strongly depend on the initial deformation, provided that it does not exceed 10–15% for compression and 50–75% for shear for elements bonded to rigid load-carrying components. It was demonstrated experimentally in [3.27] and analytically in [3.28] that the creep rate of radially loaded unbonded cylindrical rubber specimens is 20–30% lower than the creep rate of the same elements axially loaded via rigid (metal) plates bonded to their faces. This effect was explained by reduced and more uniform stresses for the former mode of loading as illustrated by Fig. 3.3.6. Typical values of b are for NR-based rubbers $\sim$2.5%; for CR (neoprene)-based rubbers $\sim$11%; NBR $\sim$7% [3.23]. However, these values can be radically smaller for properly formulated rubber blends. Thus, several NBR rubbers used in isolators were found to have $b = 0.5$–1.5%. While there is an opinion that the creep rate is correlated with damping of the rubber blend, in addition to the earlier quoted development of reasonably high damping/low creep NR-based rubber blends in [3.19], there are blends of very high-damping BR-based rubber having creep rate on the order of $b = 1\%$. The extensive creep tests of cylindrical specimens made from butyl rubber bonded to metal plates and loaded in compression with the specific load 129 psi (9 MPa) for more than 3.5 years are described in [3.39]. Five blends, all having $H = 50$, demonstrated average creep rate $\sim$0.76%. The lowest rate of creep was

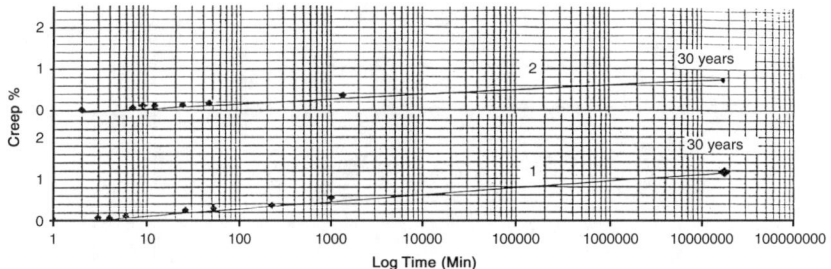

Fig. 3.3.24. (a) Creep History for Natural Rubber O-ring ($D = 63$ mm; $d = 13$ mm; $H = 45$; $P = 1,400$ N); and (b) Silicone Rubber Ball ($D = 38$ mm, $H = 9$, $P = 75$ N) Testing Performed for 1,000 min (Three Decades of Loading).

recorded for compression. If the creep rate for compression is 1.0, then it is usually ~1.2–1.25 for shear deformation, and ~1.5–1.6 for tension [3.23]. Fig. 3.3.24 [3.40] shows creep test results for a low-filled NR O-ring specimen (63 mm diameter, 13 mm cross-sectional diameter, $H = 45$, compression load 1,400 N), line 1, and for a very soft unfilled silicone rubber ball (38 mm diameter, $H = 9$, compression load 75 N), line 2. These curves indicate the creep rate to be only 0.1–0.2%, due to a combination of the elastomer properties and streamlined shapes of the test specimens. Also shown are creep magnitudes projected (by extrapolation) for 30 years.

3.3.3 Plastic Materials

Plastics (other than elastomers) are slowly gaining popularity as materials for vibration isolators. One reason for the plastics not being used more widely is their relatively low damping values (usually $\delta \leq 0.15$–0.3) and high elastic moduli of most of the structural plastics. The exceptions are poly vinyl chloride (PVC)–based plastics [3.21] having $\delta = 0.9$–1.0 (e.g., EAR Isodamp™ [3.21]) and some non-filled polyurethanes (like Sorbothane™ [3.21]), $\delta = \sim 3$. PVC materials are used both without fillers, such as EAR isolators for computer peripherals [3.41], and as a matrix for isolation pads reinforced by jute fibers (Air-Loc™), by glass fibers (Vibra-Check™), or by cork particles (Tico™). Fig. 3.3.25 shows amplitude dependencies of K_{dyn} and δ for these materials.

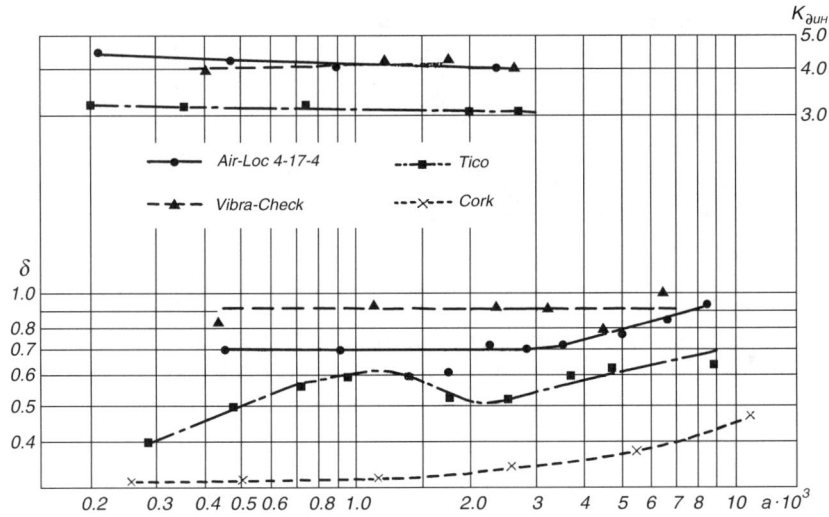

Fig. 3.3.25. K_{dyn} and log Decrement δ vs. Vibration Amplitude for Plastic Composite Vibration Isolation Pads and Cork.

Cork is a natural polymer used sometimes for vibration isolation purposes. Its amplitude dependencies of K_{dyn} and δ are also shown in Fig. 3.3.25. The best type of cork for vibration isolation purposes is natural cork (bark of the cork oak tree). It is cut into strips and assembled into plates for ease of handling. More than 50% of its volume is occupied by air. If natural cork is substituted for by so-called compressed cork comprising ground cork particles bound by adhesives into sheets, the sealed air cells are destroyed. Vibration tends to destroy the binding adhesive and thus to reduce the resilience of the material. The natural cork is available in 50–75 mm (2–3 in.) thick slabs. The recommended loading range is 10–20 MPa (1,500–3,000 psi).

3.4 HIGH-DAMPING METALS

High-damping metals can be only marginally included into the EDM group since their elastic moduli are rather high. However, they are (or can be) used in structures having relatively low stiffness by design, such as metal springs. Presently, use of high-damping metals for vibration isolators is very limited, to a large extent due to lack of information and also due to "psychological inertia" [3.42]. However, they have significant

Table 3.4. Rounded-Up Values of Damping and Elastic Characteristics of Metal and Alloys

Material	$\eta \times 10^3$	$E \times 10^{-5}$, MPa	$E\eta$
Magnesium (99.9% Mg, cast)	39	0.45	1,800
Magnesium alloy (Mg, 0.5% Zr)	50	0.45	2,250
Incramute (Cu, 44% Mn, 1.8% A1)	31	1.2	3,700
Sonoston (Mn, 36% Cu, 4.5% A1, 3% Fe, 2% Ni)	40		
Silentalloy (Fe, 12% Cr, 3% A1)	36	2.0	7,200
Nivco (72% Co, 23% Ni)	30	2.0	6,000
Nitinol (∼50% Ni, ∼50% Ti)	28	0.4	1,600
Same, optimally heat treated and prestressed	60–100	0.4	2,400–4,000
Nickel (Ni, pure)	18	2.0	3,600
Iron (Fe, pure)	16	2.0	3,200
High carbon gray iron	19	1.5	2,850
Gray cast iron	6	1.1	660
Cast iron, nodular (malleable)	2	1.3	260
Steel stainless, martensitic	8	2.0	1,600
Steel stainless, ferritic	3	2.0	600
Steel stainless, austenitic	1	2.0	200
Steel, low carbon	4	2.0	800
Steel, medium carbon	1	2.0	200
Aluminum, 1100	0.3	0.7	21
Aluminum, 2024-T4	<0.2	0.7	<14
Superalloys, Ni-based	<0.2	2.0	<40
Titanium alloys	<0.2	1.1	<22

potential for vibration isolation applications since they combine the advantages of metal springs (very low creep, relatively wide temperature range, high load-carrying capabilities in small size, insensitivity to many environmental factors) with a relatively high damping. A serious drawback of conventional metal springs—relatively low frequency and high intensity of the wave resonances (see Section 1.10) is also alleviated in high-damping metal springs since amplitudes of high-frequency resonances are reduced.

Damping of metals and metal alloys is due to various mechanisms, mostly associated with their crystalline structure. However, all these mechanisms result in "hysteretic" type damping, whereas the loss factor (and log decrement) can be assumed to be constant in a rather broad frequency range and often is amplitude-dependent. Some representative values of η for various metal alloys were compiled from several sources, e.g. [3.43] and are given in Table 3.4. While damping of typical structural metals, such as steel and aluminum alloys, is usually low (loss factor $\eta = \delta/\pi = 10^{-3}-10^{-4}$; Table 3.4), there are many metal alloys that combine reasonable strength with high damping. It was shown in Chapter 2 that performance of vibration isolation systems is sometimes determined by the criterion $k\delta$, where k is effective stiffness of vibration isolators. A similar criterion is important for use of elasto-damping materials for other vibration/noise control applications, e.g., see [3.44]. Accordingly, Table 3.4 includes, besides values of Young's modulus E and η also values of product $E\eta$. Table 3.4 gives the "ballpark" values of η and rounded-up values of E since usually η depends on vibratory strain amplitude, d.c. stress, heat and mechanical treatment, small variations in contents of alloying elements, etc. Modulus E also depends (albeit to a lesser degree) on alloying elements content and in some cases (e.g., in shape memory alloys like Ni-Ti) on the crystalline structure.

The largest $E\eta$ values are observed for a ferrous alloy "Silentalloy," and the largest values of η are for the optimally prestressed Nitinol alloys ($\sim$50% Ni + $\sim$50% Ti) [3.2]. Actual fractions of Ni and Ti vary within $\sim$5% each in order to tune performance of the alloy for the desired temperature range, and small amounts of other metals are usually added. These alloys are representative of so-called shape memory alloys whose design applications are growing, especially for biomedical devices. These alloys have transformations from one crystalline structure (martensitic) to another (austenitic) at certain transformation temperatures. Since such transformations are associated with significant changes in Young's modulus E, a spring made from this type of material changes its stiffness within the specified temperature change, thus providing a potential self-adaptation that might be desirable in some vibration isolation applications, in addition to high damping.

Study in [3.2] has shown that damping of NiTi materials is strongly dependent on their stress condition. Fig. 3.4.1 shows damping of a NiTi alloy specimen (a cylinder 0.5 in. diameter, 2.5 in. long) as the function of prestress caused by application of constant (d.c.) axial compression and of cyclic stress caused by sinusoidal vibratory axial compression at 20 Hz. It was found that the maximum loss factor $\eta = 0.06-0.1$ ($\delta = 0.2-0.3$) is observed at the d.c. stress $\sigma = 10.000$ psi (65 MPa). At this level of the d.c. stress, dependence of η on the cyclic stress amplitude is not very pronounced; it varies $\sim$50% for a six-fold change of the cyclic stress amplitude.

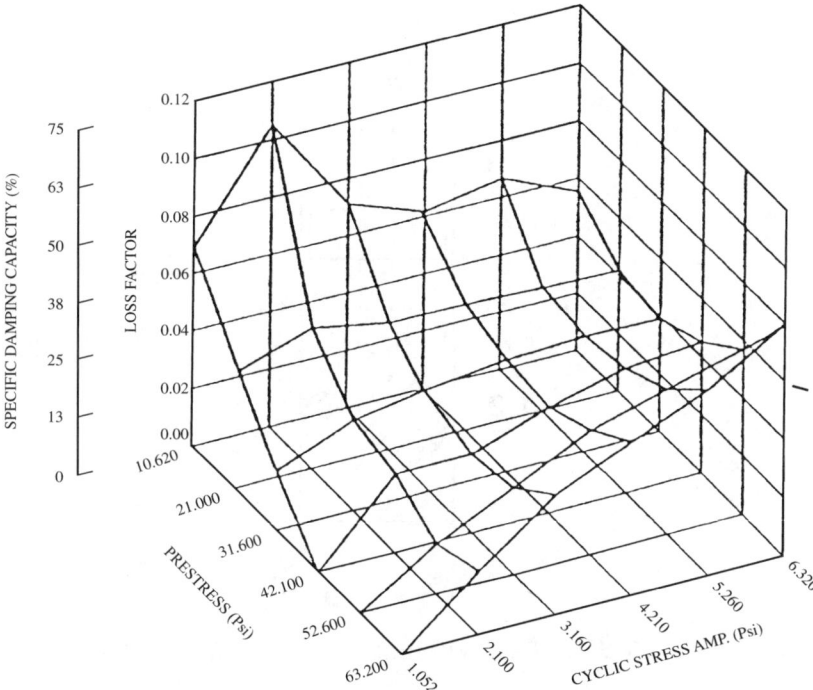

Fig. 3.4.1. Loss Factor vs. Prestress and Cyclic Stress for NiTi Specimen 0.5 in Diameter, 2.5 in Long Under Compression.

Due to their high damping, NiTi alloys are already used in some vibration-control devices, e.g. [3.45].

3.5 PNEUMATIC FLEXIBLE ELEMENTS

Vibration isolators using compressed air as flexible element are widely used both for vehicle suspensions and for installation of stationary machinery and equipment, especially in cases when relatively low natural frequencies are required. While often they are used "as is," they are also uniquely suitable for leveling installation systems comprising simple servo-control subsystems.

Fig. 3.5.1 shows a generic pneumatic flexible element—a linear cylinder-piston system. Piston 1 is moving inside cylinder 2 of any cross-sectional shape 3 (usually, round or square) with cross-sectional area A.

276 ● **PASSIVE VIBRATION ISOLATION**

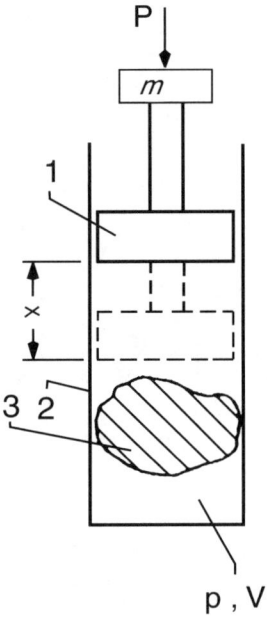

Fig. 3.5.1. Basic Pneumatic Flexible Element.

To reduce friction, the piston can be connected with the cylinder by a rolling diaphragm, e.g., see 2 in Fig. 3.5.2(a).

The instantaneous internal volume between piston 1 and cylinder 2 is V. The initial absolute pressure in the cavity V, accommodating weight of the isolated object $P_o = mg$, is

$$p_i = p_a + mg/A \qquad \text{Eq. (3.5.1)}$$

where p_a is environmental (atmospheric) pressure. The external force P is applied to piston 1. Assuming that the process of compression in the cavity V is adiabatic (no energy exchange between the cavity and the environment), then

$$p_i V_i^n = p_x V_x^n, \qquad \text{Eq. (3.5.2)}$$

where V_i is initial volume, p_x, V_x are absolute pressure and volume after the piston moved by x; and n is the ratio of specific heats ($n = 1.4$ for air at adiabatic condition). Since

$$p = P/A, \; V_x = V_i - Ax, \qquad \text{Eq. (3.5.3)}$$

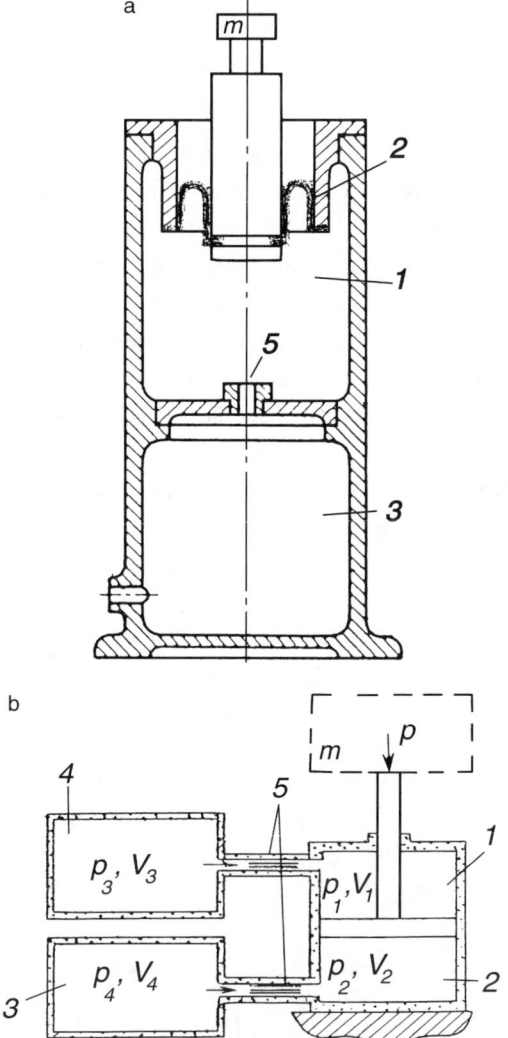

Fig. 3.5.2. (a) Single action; and (b) Double Action Pneumatic Flexible Elements with Damping Chambers.

with the assumption of constant piston surface area A, stiffness of the pneumatic flexible element is

$$k_o = \frac{dP}{dx} = \frac{np_i A^2}{V_i} \left[\frac{1}{1 - (\frac{A}{V_i})x} \right]^{n+1} .$$ Eq. (3.5.4)

If $Ax \ll V_i$, then

$$k_o \cong \frac{np_i A^2}{V_i}.$$ Eq. (3.5.5)

In pneumatic isolators, usually $p_o = P_o/A = mg/A \gg p_a$, thus stiffness k_o for a generic pneumatic flexible element is approximately proportionate to the weight of the supported object, or the pneumatic flexible element has a natural "constant natural frequency" characteristic.

The element in Fig. 3.5.1 does not have any significant damping. The damping in pneumatic isolators can be provided by the cylinder material, as in the elastomeric/pneumatic isolator shown in Fig. 4.9.1, or by adding a special damping or "surge" chamber. A self-damped pneumatic spring in which main cylinder (chamber) 1 is connected with damping chamber 3 through a capillary 5, Fig. 3.5.2(a), is analyzed in [3.46]. A double-action pneumatic cylinder with two main chambers 1, 2, each of them connected with the respective damping chambers 3, 4 by capillaries 5 [Fig. 3.5.2(b)], is analyzed in [3.47].

For the system in Fig. 3.5.2(a), the weight flow rate into chamber 1 is

$$W_{t1} = \frac{g}{RT_o}\left(\frac{V_{t1}}{n}\frac{dp_1}{dt} + p_1\frac{dV_{t1}}{dt}\right),$$ Eq. (3.5.6)

where R is the universal gas constant [$R = 286.4$ N-m/(kg K) for air]; T_o is absolute temperature of gas; V_{t1} is instantaneous cylinder volume; p_1 is cylinder pressure. The weight flow rate from the damping chamber 3 is

$$W_{t1} = -\frac{gV_3}{nRT_o}\frac{dp_3}{dt}.$$ Eq. (3.5.7)

Weight flow rate, rather than mass flow rate, is used in Eq. (3.5.6) and Eq. (3.5.7) so that the formulation in [3.46] is consistent with [3.47].

The flow through a capillary (a passage whose length l is much greater than its cross-sectional dimension, e.g. diameter d) is

$$W_{t1} = \frac{gC_c}{2RT_o}(p_1^2 - p_3^2).$$ Eq. (3.5.8)

The capillary resistance coefficient for a round capillary is

$$C_c = \frac{\pi d^4}{128\mu l}$$ Eq. (3.5.9)

and for a rectangular cross-section capillary

$$C_c = \frac{wh^3}{12\mu l} \qquad \text{Eq. (3.5.10)}$$

where μ is dynamic viscosity of gas; and w, h are width (larger dimension) and, height (smaller dimension) of the capillary. Since at small displacements $p_1 + p_3 \approx 2p_i$, where p_i is the initial pressure and the mean pressure around which small oscillations occur, Eq. (3.5.8) can be linearized as

$$W_{t1} = \frac{p_i g C_c}{RT_o}(p_1 - p_3). \qquad \text{Eq. (3.5.11)}$$

This linearized relationship is valid only if the flow through the capillary is laminar; if the flow is turbulent, the correlation becomes nonlinear, pneumatic damping becomes amplitude-dependent, its effectiveness diminishes, even instability can develop in the active leveling systems (see Section 4.9 [3.3], [3.49]). The laminar flow develops if the Reynolds number $Re < 2{,}000$ or, taking a prudent safety margin, $Re < 500$ [3.49]. The Reynolds number is $Re = vd\rho/\mu$ for a round passage, where v is flow velocity and ρ is gas density, or for a rectangular passage $Re = (Q/wh)h(\rho/\mu) = (Q/w)(\rho/\mu)$, where Q is the flow rate.

Assuming p_o is a constant, solving Eq. (3.5.11) for p_3 and substituting into Eq. (3.5.7) gives

$$W_{t1} = -\frac{gV_3}{nRT_o}\left(\frac{RT_o}{gp_iC_c}\frac{dW_{t1}}{dt} + \frac{dp_1}{dt}\right). \qquad \text{Eq. (3.5.12)}$$

Following [3.47], using operator notation $(d/dt) = D$, and solving Eq. (3.5.12) for W_{t1} gives

$$W_{t1} = -\frac{gV_3}{nRT_o}\frac{Dp_1}{1 + \frac{V_3}{nC_cp_i}D}. \qquad \text{Eq. (3.5.13)}$$

Upon letting $p_1 = p_i + \Delta p_1$ and $V_1 = V_i - Ax$ (where V is the initial volume of the cylinder chamber 1) and assuming $Ax \ll V_i$ and $\Delta p_1 \ll p_i$, Eq. (3.5.6) becomes

$$W_{t1} = \frac{g}{RT_o}\left(\frac{V_i}{n}Dp_1 - p_iADx\right). \qquad \text{Eq. (3.5.14)}$$

Equating Eq. (3.5.13) and Eq. (3.5.14) and solving for Dp_1 gives

$$Dp_1 = \frac{np_i A Dx}{V_i}\left(1 + \frac{V_3/V_i}{1 + \frac{V_3}{nC_c p_i}D}\right). \qquad \text{Eq. (3.5.15)}$$

Given that the applied force on the piston is

$$P = (p_i - p_a)A + A\Delta p_1, \qquad \text{Eq. (3.5.16)}$$

and the dynamic stiffness is

$$k = DP/Dx, \qquad \text{Eq. (3.5.17)}$$

then [3.46]

$$k = \frac{np_i A^2}{V_i}\left(1 + \frac{V_3/V_i}{1 + \frac{V_3}{nC_c p_i}D}\right). \qquad \text{Eq. (3.5.18)}$$

If the right-hand side of Eq. (3.5.18) is multiplied by 2, the formula would represent stiffness for the double-action pneumatic flexible element in Fig. 3.5.2(b), identical to the expression in [3.47]. If $C_c = \infty$ (the damping chamber 3 is separated from chamber 1), Eq. (3.5.18) becomes identical to k_o from Eq. (3.5.5), and if $C_c = 0$ (chambers 1 and 3 are combined), then Eq. (3.5.18) becomes identical to Eq. (3.5.5) in which V_i is replaced by $V_i + V_3$.

It is convenient to introduce "volume ratio" N and "base frequency" ω_o, where

$$N = \frac{V_3}{V_i} \quad \text{and} \quad \omega_o = \frac{nC_c p_i}{V_i} = \frac{C_c}{A^2}k_o. \qquad \text{Eq. (3.5.19)}$$

Then, for a sinusoidal forcing function ($D = j\omega$), Eq. (3.5.18) becomes

$$k = k_o \frac{1 + N\frac{j\omega}{\omega_o}}{1 + N + N\frac{j\omega}{\omega_o}}. \qquad \text{Eq. (3.5.20)}$$

From Eq. (3.5.20) the real and imaginary components of k can be determined as [3.46]

$$Re[k] = k_o \frac{1 + N + N^2\left(\frac{\omega}{\omega_o}\right)^2}{(1+N)^2 + N^2\left(\frac{\omega}{\omega_o}\right)^2}, \quad Im[k] = k_o \frac{N^2\left(\frac{\omega}{\omega_o}\right)}{(1+N)^2 + N^2\left(\frac{\omega}{\omega_o}\right)^2}.$$

$$\text{Eq. (3.5.21)}$$

To better understand effects of the design parameters, it is convenient to present the complex stiffness as

$$k = Re[k](1 + j\eta), \qquad \text{Eq. (3.5.22)}$$

where the loss factor η is defined by

$$\eta = \frac{Im[k]}{Re[k]} = \frac{N^2 \frac{\omega}{\omega_o}}{1 + N + N^2 \left(\frac{\omega}{\omega_o}\right)^2}. \qquad \text{Eq. (3.5.23)}$$

By solving $d\eta/d\omega = 0$, it can be found that the maximum value of η occurs when

$$\left(\frac{\omega}{\omega_o}\right)_{\eta_{max}} = \frac{\sqrt{N+1}}{N} \qquad \text{Eq. (3.5.24)}$$

and

$$\eta_{max} = \frac{N}{2\sqrt{N+1}}. \qquad \text{Eq. (3.5.25)}$$

Thus, the maximum loss factor depends only on the damping tank/cylinder volume ratio. Eq. (3.5.19) shows that the capillary resistance coefficient C_c determines the frequency at which the maximum damping (loss factor) occurs if k_o and A are kept constant. The magnitude of the complex stiffness from Eq. (3.5.20) is

$$|k| = \frac{k_o}{(1+N)^2 + \left(N\frac{\omega}{\omega_o}\right)^2} \sqrt{\left[1 + N + N^2 \left(\frac{\omega}{\omega_o}\right)^2\right]^2 + N^2 \left(\frac{\omega}{\omega_o}\right)^2}.$$

Eq. (3.5.26)

The stiffness at maximum damping can be found by substituting Eq. (3.5.24) into Eq. (3.5.26) to get

$$\left(\frac{|k|}{k_o}\right)_{\eta_{max}} = \frac{1}{\sqrt{N+1}} = \frac{1}{N\frac{\omega}{\omega_o}}. \qquad \text{Eq. (3.5.27)}$$

Eqs. (3.5.23) and (3.5.27) are plotted for $N = 1, 2,$ and 3 in Figs. 3.5.3(a), (b). Fig. 3.5.3(a) shows the locus of maximum damping given by

$$\eta_{max} = \frac{1}{2\frac{\omega}{\omega_o}}. \qquad \text{Eq. (3.5.28)}$$

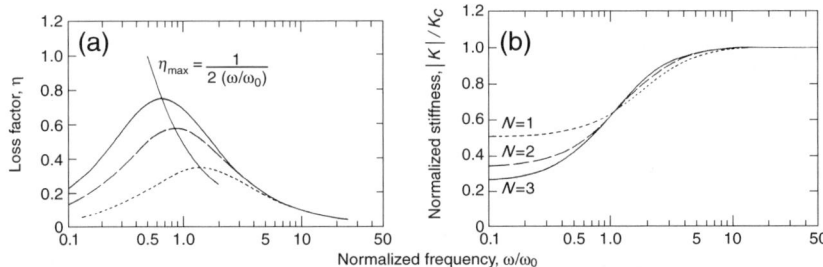

Fig. 3.5.3. (a) Loss Factor; and (b) Normalized Stiffness vs. Normalized Frequency for Damped Pneumatic Flexible Element with $N = 1, 2, 3$.

Eq. (3.5.27) and Fig. 3.5.3(b) show that the maximum stiffness occurs when $\omega/\omega_o = \infty$, with the corresponding ratio being

$$\left(\frac{|k|}{k_o}\right)_{\frac{\omega}{\omega_o}=\infty} = 1.0 \qquad \text{Eq. (3.5.29)}$$

and that the minimum stiffness occurs when $\omega/\omega_o = 0$, with the ratio being

$$\left(\frac{|k|}{k_o}\right)_{\frac{\omega}{\omega_o}=0} = \frac{1}{N+1}. \qquad \text{Eq. (3.5.30)}$$

Eqs. (3.5.22) and (3.5.23) show that any amount of damping can be achieved, but the frequency at which the maximum damping occurs decreases with the increasing damping.

To derive the transmissibility expression between absolute displacement y of the supporting structure (base) and absolute displacement u of the mass m, the damping ratio c/c_{cr} can be defined as

$$\omega_n N k_o = 2\frac{c}{c_{cr}}, \quad \text{where } \omega_n = \sqrt{\frac{k_n}{m}} = \sqrt{\frac{k_o}{Nm}} \qquad \text{Eq. (3.5.31)}$$

With these notations, the transmissibility is

$$\frac{y_o}{u_o} = \mu = \sqrt{\frac{1 + \left(2\frac{c}{c_{cr}}\frac{\omega}{\omega_n}\right)^2}{\left[1 - (N+1)\frac{\omega^2}{\omega_n^2}\right]^2 + \left(2\frac{c}{c_{cr}}\frac{\omega}{\omega_n}\right)^2\left(1 - \frac{\omega^2}{\omega_n^2}\right)^2}}, \qquad \text{Eq. (3.5.32)}$$

The parameter ω_n does not have a physical meaning for the considered system. The system behavior is determined at low frequencies by the

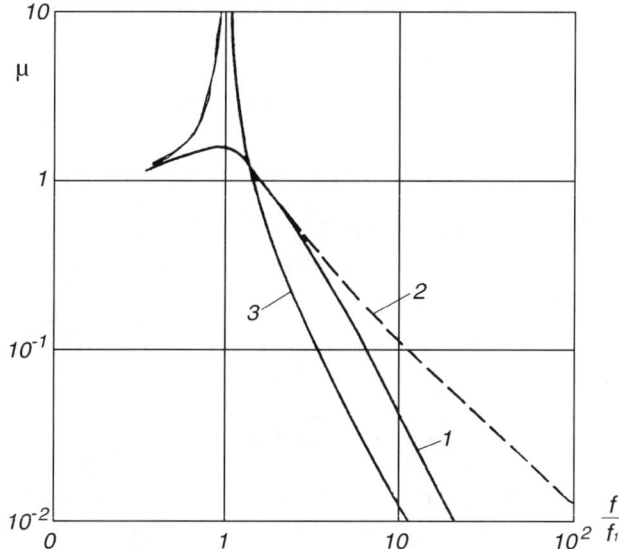

Fig. 3.5.4. Amplitude-Frequency Characteristics of SDOF Vibration Isolation Systems (1—System with an Optimally Damped Pneumatic Isolator; 2—System with Viscously Damped Spring Isolator Having the Same Resonance Amplitude; 3—System without Damping).

corresponding low frequency stiffness values, and at high frequencies, by the corresponding high-frequency stiffness values, e.g., see Fig. 3.5.3(b), and Eqs. (3.5.29) and (3.5.30). Since at the high frequencies there is practically no flow through the capillary, damping at the high frequencies approaches zero, e.g., see Fig. 3.5.3(a). Thus, damping in the pneumatic systems, as in Fig. 3.5.2, limits amplitudes at resonance but does not increase transmissibility at high frequencies, contrary to the case of vibration isolation systems with viscous damping discussed in Section 1.4.

The combined action of both stiffness and damping changing with frequency, as shown in Fig. 3.5.3, results in the following asymptotic behavior of transmissibility of the isolation system with a damped pneumatic flexible element:

$$\left|\frac{y_o}{u_o}\right|_{\omega\to\infty} = \frac{1}{\omega^2/\omega_n^2}, \qquad \text{Eq. (3.5.33)}$$

while for the system with viscous damping having the same transmissibility at resonance

$$\left|\frac{y_o}{u_o}\right|_{\omega\to\infty} = \frac{2\frac{c}{c_{cr}}}{\omega/\omega_n}.$$ Eq. (3.5.34)

Thus, effectiveness of the pneumatic isolators at high frequencies is much higher than that for isolators with viscous (or hysteretic) damping. This statement is illustrated in Fig. 3.5.4.

3.6 HYDROMOUNTS (BASICS)

It was noted in Section 2.6 that vibration isolation mounts for internal combustion engines of surface vehicles must comply with contradictory requirements—they must be soft and highly damped for large-amplitude low-frequency vibrations and be lightly damped at low-amplitude high-frequency vibrations. This contradiction is especially critical for highly vibration active four-cylinder engines. Transmissibility at high frequencies/low amplitudes can be reduced while maintaining relatively low resonance amplitudes using various techniques. One way of realizing such a characteristic is by using materials with highly pronounced amplitude dependencies of stiffness and damping (wire mesh, Section 3.3.1A; or some rubber blends, Section 3.3.2D). Another approach is to use "relaxation" isolation systems (Section 1.4.3). If in the schematic in Fig. 1.4.1(c) vibration has amplitude a and low frequency f near zero, the viscous damping force is nearly zero, the connecting ("damping") spring k_1 does not transfer any force, and the effective stiffness of the entire system is $\sim k$. At high frequencies the damping force for the same amplitude vibration is so great that the system behaves like an undamped system with two parallel springs k and k_1 (thus, stiffer than the same system at the low frequencies). Between these two extreme frequencies the system damping reaches a maximum value, as presented analytically in Section 1.4.3. If such an isolator were used as an engine mount, it could be made softer than a typical engine mount with a rubber flexible element so that the stiffening effect would not be detrimental for high-frequency isolation. It was suggested (e.g., [3.48]), essentially, to realize the relaxation isolator by using a combination rubber/fluid isolator as in

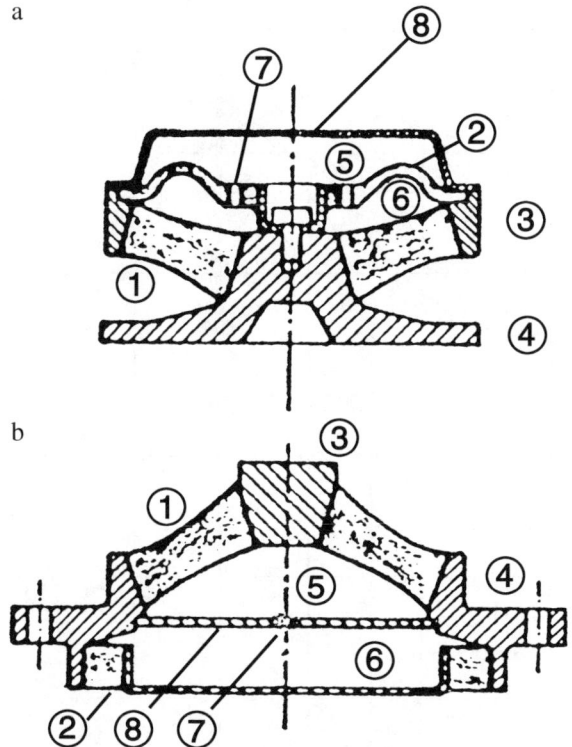

Fig. 3.6.1. Basic Designs of Hydraulic Isolators.

Fig. 3.6.1. This section gives a brief overview of the basic properties of the hydromounts, using a clear analysis from [3.48]. Recently, many modifications of hydromounts were developed and commercialized.

In Fig. 3.6.1(a), main spring 1 and damping spring 2 are separated. Main spring 1 is a solid rubber element bonded to outer ring 3 and base 4. Stamped top cover 8 attached to ring 3 encloses upper-fluid chamber 5, while lower-fluid chamber 6 is shaped by rubber elements 1 and 2. Chambers 5 and 6 are filled with antifreeze-like fluid and are connected by one or more orifices 7. The design in Fig. 3.6.1(b) combines main and damping springs in rubber ring 1 bonded to upper 3 and lower 4 attachment parts. Partition 8, having one or more fluid passages 7, separates upper chamber 5 from lower chamber 6. When static (weight) load is applied to the isolator, the fluid is pressurized and volume of fluid in lower chamber 6 increases due to compliance of rubber retaining ring 2.

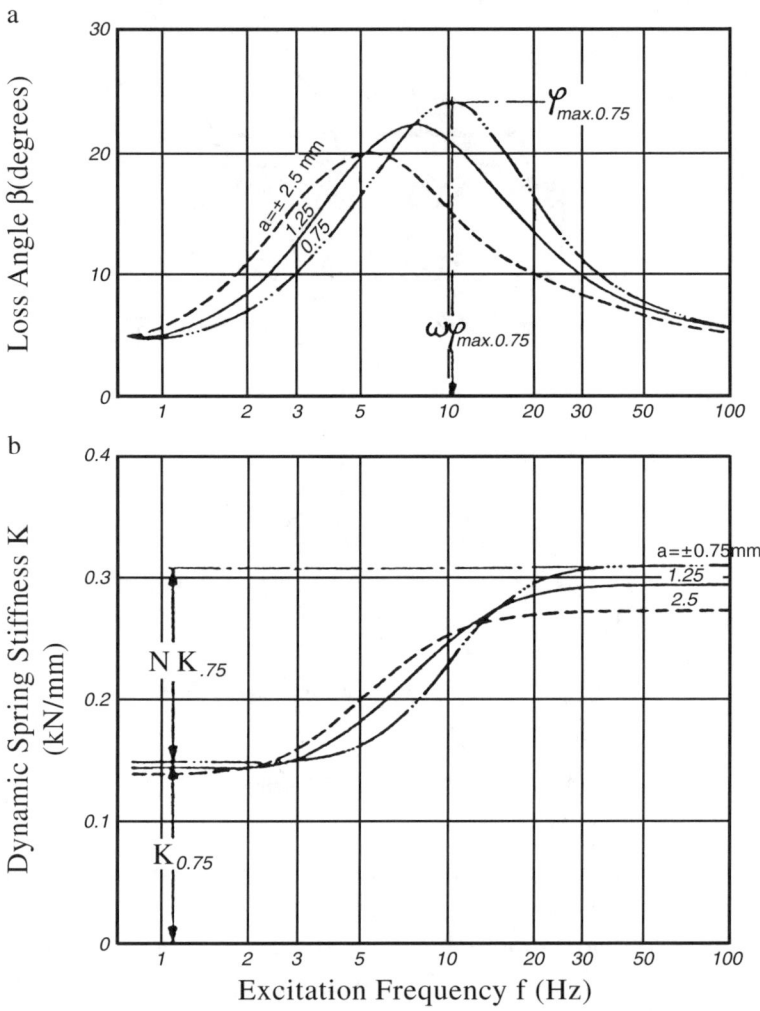

Fig. 3.6.2. (a) Loss Angle; and (b) Dynamic Stiffness of Hydraulic Isolators as Functions of Frequency.

Dynamic properties of both designs are similar. If a low frequency (e.g., $f < 3$ Hz) vibration with amplitude a is applied to top 3 of the isolator in Fig. 3.6.1(b), the fluid flows between chambers 5 and 6 with low resistance (low damping). At higher frequencies (e.g., $f > 25$ Hz), the resistance force to the fluid flow through passages 7 becomes greater than the elastic resistance of rubber spring 1. Thus the fluid pressure in chamber 5 fluctuates with excitation frequency f, but fluid pressure in

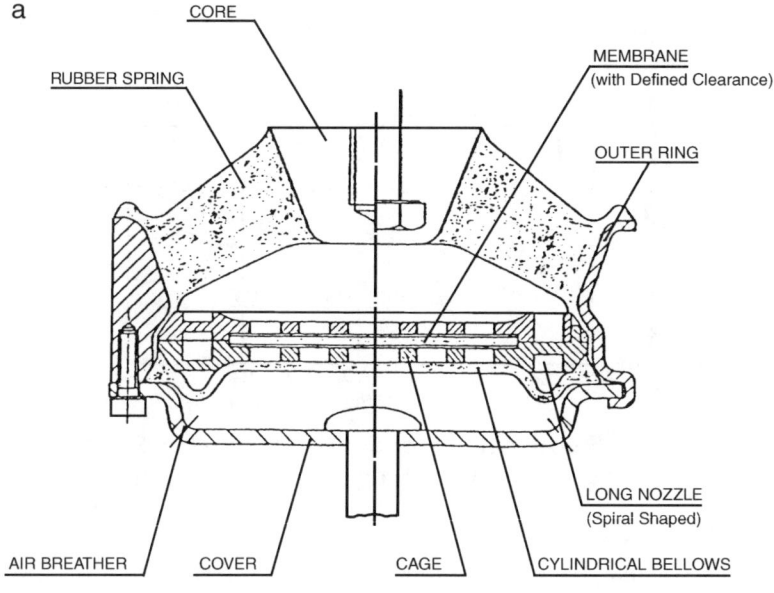

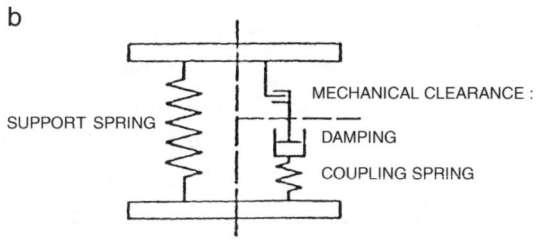

Fig. 3.6.3. Hydraulic Isolator with (a) Clearance Diaphragm and Spiral Nozzle; and (b) Its Simplistic Model.

lower chamber 6 remains almost constant. The damping maximum occurs between these two frequencies. Fig. 3.6.2(a) shows damping (loss angle) β and Fig. 3.6.2(b) shows effective stiffness K of the isolator in Fig. 3.6.1(b) as functions of the excitation frequency. Both plots are influenced by the excitation amplitude (± 0.75; 1.25; 2.5 mm amplitudes are presented). The amplitude dependency in Fig. 3.6.2(a) is explained in [3.48] as being caused by the amplitude-dependent damping in rubber spring 1; a highly elastic spring 1 would result in lower amplitude dependency.

Fig. 3.6.2(b) shows the functional difference between main spring K and damping spring NK. At $a = \pm 0.75$ mm the effective value is $N = \sim 1.2$.

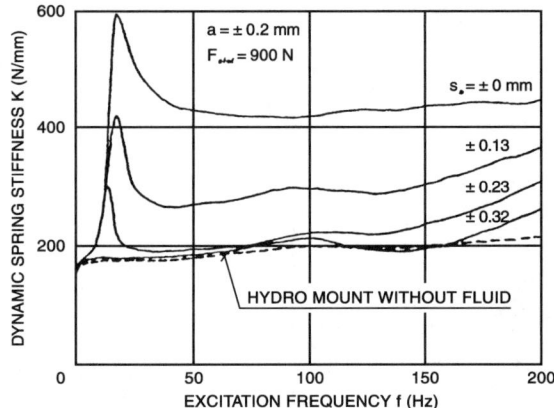

Fig. 3.6.4. Frequency Dependence of Dynamic Stiffness of Fig. 3.6.3 Isolator vs. Frequency for Various s_o.

Since both springs are combined in one elastomeric structure, it can be seen that they represent different deformation modes. Stiffness of the main spring K is represented by the vertical stiffness of spring 1. While stiffness of the damping spring NK at high frequencies also depends on the vertical stiffness of rubber spring 1, its mode of deformation at the working frequency range is not vertical but radial outward deformation.

An analysis of Figs. 1.4.5–1.4.8 leads to a conclusion that the system performance would further improve if the stiffness ratio N were frequency-dependent (large at the resonance and small thereafter). Design implementation of this requirement can be made easier if the above-noted fact of much smaller amplitudes at higher frequencies in the engine isolation systems is considered. Fig. 3.6.3(a) [3.48] shows a hydraulic isolator with amplitude and frequency-dependent parameters. This isolator is a modification of basic hydraulic isolators in Fig. 3.6.1, with added calibrated mechanical clearance realized by an encased elastomeric membrane. Fig. 3.6.3(b) shows the dynamic model of this isolator. At low frequencies/large amplitudes (resonance conditions) this small clearance is irrelevant and the fluid passes between the chambers through the long spiral nozzle that acts as an additional damped spring, which adds to the stiffness ratio N in the resonance range. At higher frequencies/low amplitudes, the membrane is moving in its space (within the clearance) and the resistance force to fluid flow through the spiral nozzle is greater than the elastic force of the rubber spring.

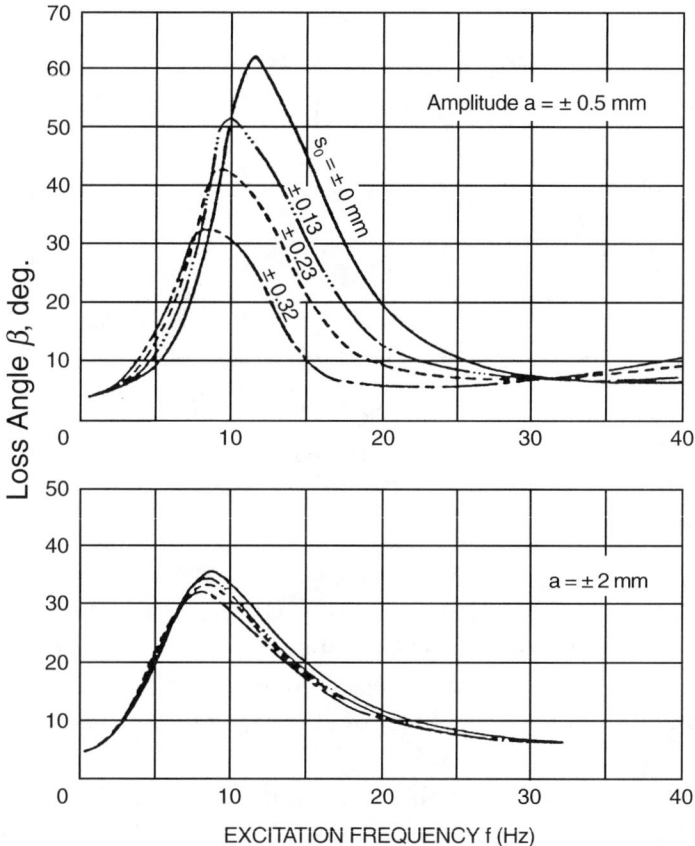

Fig. 3.6.5. Frequency Dependence of Loss Angle for Fig. 3.6.3 Isolator vs. Frequency for Various s_o and a.

Consequently, the vibration model for higher frequencies is, effectively, a system with low N.

Fig. 3.6.4 [3.48] shows dynamic stiffness vs. frequency for various clearance sizes (s_o) at amplitude $a = \pm 0.2$ mm and static (weight) force on isolator $F_{st} = 900$ N. Without clearance ($s_o = 0$), the isolator has increased dynamic stiffness at higher frequencies, somewhat similar to the behavior of isolators in Fig. 3.6.1 but showing a peak at low frequencies due to the additional damped spring effect of the spiral nozzle. As the clearance size increases above the excitation amplitude a, the plot of dynamic stiffness approaches the one for a hydraulic isolator without fluid. Fig. 3.6.5 illustrates the behavior of the Fig. 3.6.3 isolator as influenced by both a and s_o. For $a = \pm 0.5$ mm, the isolator with no clearance has the highest

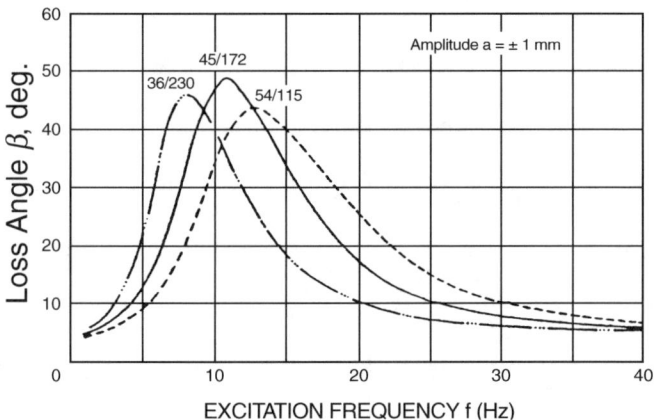

Fig. 3.6.6. Frequency Dependence of Loss Angle for Fig. 3.6.3 Isolator vs. Frequency for Various Nozzle Parameters.

damping because the isolator is damped for the entire stroke. On the other hand, for $a = \pm 2$ mm, although the changes in the clearance size are always $s_o \ll a$, they do not have a noticeable effect. Fig. 3.6.6 shows the way to shift the maximum damping frequency. The first numbers (36, 45, 54) refer to the nozzle cross-sectional area, while the second numbers refer to its length.

Realization of high damping values in hydraulic isolators (engine mounts) results in noticeable reduction (25–40%) of the low frequency "shake" of both the passenger seat and steering wheel, and engine, as well as reduction of noise transmission (up to 5 dBA) credited to the use of hydraulic engine mounts [3.48]. On the negative side, such isolators are much more expensive and heavy than rubber-metal isolators and often have maintenance problems (fluid leakage). The hydraulic mounts can be "tuned" by design to realize frequency-dependent dynamic characteristics optimized for a given vehicle. Accordingly, there is a great variety of hydromounts on the market, e.g. see [3.49]. During the tuning process, performance in a certain frequency range could be degraded to improve performance at other frequencies. Elastomeric mounts exhibit stiffness variation due to tolerances on their durometer. Hydromounts, whose performance directly depends on rubber parts and their stiffness and damping, have also other sources of parameter variation. Those sources include dimensional variations of critical hydraulic passages and temperature variations. It was also noted in [3.50] that while a hydromount exhibits a theoretically predictable performance for a sinusoidal input, its performance for complex inputs having superimposed harmonics is hardly predictable. As a result, a significant effort is extended to the development of adjustable or servo-controlled (active) hydraulic isolators, e.g. see [3.49], [3.51], [3.52], whose characteristics can be tuned after installation. Needless to say, such isolators are even more expensive, heavy, and less reliable.

REFERENCES

[3.1] Liang, C. and Roberts, C.A., 1993, "Design of Shape Memory Alloy Springs with Applications in Vibration Control," *ASME J. of Vibration and Acoustics*, Vol. 115, pp. 129–135.

[3.2] Rivin, E.I., 1994, "Damping of NiTi Shape Memory Alloys and Its Application for Cutting Tools," *Materials for Noise and Vibration Control*, ASME NCA-Vol. 18/DE, pp. 35–41.

[3.3] DeBra, D.B., 1992, "Vibration Isolation of Precision Machine Tools and Instruments," *Annals of the CIRP*, Vol. 41/2, pp. 711–718.

[3.4] Crede, Ch.E., 1951, Vibration and Shock Isolation, John Wiley & Sons, N.Y.

[3.5] Poimenov, I.A. and Legenia, B.I., 1996, "Analysis of Flexible Element of Vibration Isolator," *Vestnik Mashinostroeniya*, No. 1, pp. 14–16 [in Russian].

[3.6] Gavrilenko, I.P. and Filkin, I.M., 1987, "Design of Variable Shape Coil Springs with Constant Natural Frequency Characteristic," *Vestnik Mashinostroeniya*, No. 5, pp. 34–38 [in Russian].

[3.7] Wienand, S., 1970, "Progressive Schraubenfedern für Kraftfahrzeuge [Progressive Coil Springs for Vehicles]," *Draht-Welt*, No. 10, pp. 546–550 [in German].

[3.8] Gurti, G. and Montanini, R., 1999, "On the Influence of Friction in the Calculation of Conical Disc Springs," *ASME J. of Mechanical Design*, Vol. 121, pp. 622–630.

[3.9] Handbook for Disc Springs, 1997, *Adolf Schnorr GmbH*.

[3.10] DeBra, D.B., 1984, "Design of Laminar Flow Restrictors for Damping Pneumatic Vibration Isolators," *Annals of the CIRP*, Vol. 33/1.

[3.11] De Gaspari, J., 2002, "Hot Stuff," Mechanical Engineering, No. 12, pp. 32–35.

[3.12] Crede, Ch.E., 1961, "Application and Design of Isolators," *Shock and Vibration Handbook*, Vol. 2, McGraw-Hill, N.Y.

[3.13] Harris, B., 1980, "Little Known Facts Affecting Teflon Fabric Bearing Life," *SAE Paper* 800676.

[3.14] Rivin, E.I., 1999, Stiffness and Damping in Mechanical Design, Marcel Dekker Inc., N.Y.

[3.15] Aeroflex International, *www.vmc-kdc.com*

[3.16] Holownia, B.P., 1974, "Effect of Carbon Black on the Elastic Constants of Elastomers," *J. Instn Rubber Ind.*, Vol. 4, No. 8, pp. 157–161.

[3.17] NR/ENR High Damping Blends, 1998, *Guthrie Latex, Inc.*

[3.18] "BROMO XP-50 Rubber. Compounding and Applications," 1992, *Exxon Corp.*

[3.19] Lemieux, M.A. and Kilgoar, P.C., 1984, "Low Modulus, High Damping, High Fatigue Life Elastomer Compounds for Vibration Isolation," *Rubber Chemistry and Technology*, Vol. 57, pp. 792–803.

[3.20] Rivin, E.I., "Nonlinear Flexible Connectors with Streamlined Resilient Elements," *U.S. Patent 5,954,653*

[3.21] Byrne, C.A. and Zukas, W.X., 1989, "Shock Absorbing Polymers: Chemical Analysis and Characterization," *Polymeric Materials Science and Engineering*, Vol. 61, pp. 560–564.

[3.22] Snowdon, J.C., 1965, "Rubberlike Materials, Their Internal Damping and Role in Vibration Isolation," *J. of Sound and Vibration*, Vol. 2, pp. 175–193.

[3.23] Davey, A.B. and Payne, A.R., 1964, Rubber in Engineering Practice, McLaren & Sons, London.

[3.24] Rivin, E.I., 1965, "Horizontal Stiffness of Anti-Vibration Mountings," *Russian Engineering Journal*, No. 8, pp. 21–23.

[3.25] Herrmann, L.R., Ramaswamy, A. and Hamidi, R., 1989, "Analytical Parameters Study for Class of Elastomeric Bearings," *J. of Structural Engineering*, Vol. 115, No. 10, pp. 2415–2434.

[3.26] Rivin, E.I., 1985, "Passive Engine Mounts—Some Directions for Future Development," *SAE Transactions*, pp. 3.582–3.592.

[3.27] Rivin, E.I. and Lee, B.S, 1994, "Experimental Study of Load-Deflection and Creep Characteristics of Compressed Rubber Components for Vibration Control Devices," *ASME J. of Mechanical Design*, Vol. 116, No. 2, pp. 539–549.

[3.28] Lee, B.S. and Rivin, E.I., 1996, "Finite Element Analysis of Load-Deflection and Creep Characteristics of Compressed Rubber Components for Vibration Control Devices," *ASME J. of Mechanical Design*, Vol. 118, pp. 328–336.

[3.29] Neidhart, R., 1969, "Special Spring Unit," *Rubbers Handbook*, Morgan-Grampian, London. pp. 87–91.

[3.30] Rivin, E.I., 1999, "Shaped Elastomeric Components for Vibration Control Devices," *Sound and Vibration*, Vol. 33, No. 7, pp. 18–23.

[3.31] Dymnikov, S.I., 1972, "Stiffness Computation for Rubber Rings and Cords," *Voprosi dinamiki i prochnosti [Issues of Dynamics and Strength]*, No. 24, Zinatne Publ. House, Riga, pp. 163–173 [in Russian].

[3.32] Freakley, P.K. and Payne, A.R., 1978, Theory and Practice of Engineering with Rubber, Applied Science Publishers, London.

[3.33] Rivin, E.I., 1983, "Ultra-Thin-Layered Rubber-Metal Laminates for Limited Travel Bearings," *Tribology International*, Vol. 16, No. 1, pp. 17–25.

[3.34] Gorelik, B.M., Kolosova, V.I., Tikhonov, V.A. and Sshegolev, V.A., 1980, "Influence of Mechanical and Geometric Parameters of Thin-Layered Rubber-Metal Elements on Their Stiffness," *Kauchuk i resina*, No. 8, pp. 40–44 [in Russian].

[3.35] Lewis, T., 1991, "The Effects of Dynamic Strain Amplitude and Static Prestrain on the Properties of Viscoelastic Materials," *SAE Tech. Paper 911084*.

[3.36] Turner, P.W., 1957, *Proceedings of "Rubber in Engineering" Conference*, Natural Rubber Development Board, London, p. 117.

[3.37] Lee, S.-H. and Lim, Y.-S., 1996, "The development and Performance Simulation of Polychloroprene High Temperature Bush Type Engine Mount," *SAE Tech. Paper 940888.*

[3.38] Gruenberg, S., Blough, J., Kowalski, D. and Pistana, J., 2001, "The Effects of Natural Aging on Fleet and Durability Vehicle Engine Mounts from a Dynamic Characterization Prospective," *Proceedings of the 2001 SAE Noise and Vibration Conference (Noise 2001 CD), SAE Paper 2001-01-1449.* Traverse City, MI.

[3.39] Booth, D.A., 1969, "Butyl (IIR)," *Rubber Handbook*, Morgan-Grampian Publishers, London. pp. 135–144.

[3.40] "Affordable High Performance Reinforced Polyurethane Shock and Vibration Mounts," 1999, Globe Rubber Works Inc., Report on U.S. Navy Contract N00167-98-C-0034.

[3.41] "Materials in Application," 1992, *EAR Specialty Composites/Cabot Safety Corporation Bulletin.*

[3.42] Fey, V.R. and Rivin, E.I., 1997, The Science of Innovation, *The TRIZ Group*, West Bloomfield, MI.

[3.43] Schettky, L.M. and Perkins, J., 1978, "The Quiet Alloys," *Machine Design*, April 6, pp. 202–206.

[3.44] Beranek, L.L. and Ver, I.L. (editors), 1992, "Noise and Vibration Control Engineering," Viley-Interscience, N.Y.

[3.45] Inaudi, J.A., Kelly, J.M., Taniwanzza, W. and Krumme, R., "Analytical and Experimental Study of a Mass Damper Using Shape Memory Alloys," *Proceedings of "Damping 93" Conference*, pp. HAA 1–20.

[3.46] Bachrach, B.I. and Rivin, E., 1983, "Analysis of Damped Pneumatic Spring," *J. of Sound and Vibration*, Vol. 86, No. 2, pp. 191–197.

[3.47] Cavanaugh, R.D. 1961, "Air Suspension and Servocontrolled Isolation Systems," *Shock and Vibration Handbook*, Vol. 2, McGraw-Hill, N.Y.

[3.48] Corcoran, P.E. and Ticks, G.-H., 1984, "Hydraulic Engine Mount Characteristics," *SAE Tech. Paper 840407.*

[3.49] Yu, Y., Naganathan, N.G., Dukkipati, R.V., 2001, "A Literature Review of Automotive Vehicle Engine Mounting Systems," *Mechanism and Machine Theory*, Vol. 36, pp. 123–142.

[3.50] Ushijima, T., Takano, K., Kojima, H., 1998, "High Performance Hydraulic Mount for Improving Vehicle Noise and Vibration," *SAE Tech. Paper 880073*.

[3.51] Gennesseaux, A., 1993, "Research for New Vibration Isolation Techniques: From Hydro-Mounts to Active Mounts," *SAE Tech. Paper 931324*.

[3.52] Muller, M., Weltin, U., Law, D., Roberts, M.M., and Siebler, T.W., 1996, "The Effect of Engine Mounts on the Noise and Vibration Behavior of Vehicles," *SAE Tech. Paper 940607*.

CHAPTER 4

Passive Vibration Isolation Means

The Chapter describes the basic design concepts of regular non-reinforced rubber and plastic mats and pads and regular machinery mounts, as well as design concepts for advanced and special purpose isolators. Significant application and production/economical advantages of the Constant Natural Frequency vibration isolators are emphasized.

Special vibration isolator designs include low vertical stiffness isolators, both isolators with metal spring flexible elements and with quasi-unstable systems. Another important group includes isolators with high vertical but low horizontal stiffness values.

An important group of self-leveling isolation mounts and installation systems is represented by a servo-pneumatic and a servo-hydraulic systems. These servo systems are maintaining the static level of the installed object without affecting the vibration isolation process.

External connections of the vibration isolated objects can detrimentally affect the vibration isolation efficiency. Some basic principles and design concepts for such connections (both piping and power transmission connections) are briefly described.

4.1 GENERAL COMMENTS

The textbook treatment of vibration isolation problems results in recommendation of very low frequencies of the vibration isolation system so that its highest natural frequency is below (preferably, at least one half of) the lowest excitation frequency to be isolated or attenuated. Usually it is impractical, and also impossible and useless, to follow this recommendation since vibration isolators would be extremely soft, causing intense rocking motions, increasing deformations of the frame parts, etc. thus violating performance specifications for the object to be isolated.

Analysis performed in Chapters 1 and 2 provides directions for attaining an adequate degree of isolation while complying with the performance requirements. However, the problem of realization of the recommendations in Chapters 1 and 2 is often not a simple one. This problem can be solved only by using properly designed and selected vibration isolation means. These include both vibration isolators for direct connection of the isolated object with the supporting structure, and vibration isolating foundations. Vibration isolators have to realize required, often relatively low, natural frequencies in the principal directions. This can be facilitated by having low dynamic stiffness in the amplitude and frequency ranges of interest as well as specified stiffness ratios of the isolators in the principal directions (optimized for typical applications). In many cases the dynamic-to-static stiffness ratio K_{dyn} must be minimized, at least in the direction of the gravity vector. In some cases vibration isolation systems with motion transformers (Section 1.4.4) may be useful. It was shown in Chapter 2 that in many applications for precision objects, the specified stiffness values can be increased by enhancing damping in the system.

While K_{dyn} and damping are mostly determined by the material selected for the flexible element (unless a special damper is employed), the stiffness ratios are mostly determined by the isolator design. As was shown in Chapter 1, an important characteristic of the isolator is its load-deflection or load-stiffness plot (i.e., character and degree of nonlinearity). Proportionality between the weight load on the isolator and its stiffness constants in the principal directions [constant natural frequency (CNF) characteristic] can often improve performance of the isolation system, even with stiffer isolators. The nonlinearity (CNF characteristic, snubbers for preventing excessive deformations of the flexible element, etc.) can be achieved by design. In cases where there are special requirements for vibration attenuation in the high-frequency range, other design directions can be used. One design direction is reduction of mass of the flexible element, in accordance with Eqs. (1.10.1)–(1.10.3); another design direction is incorporation of intermediate masses into the flexible element (Section 1.5.2).

A vibration-isolated foundation can be designed as an isolated supporting structure (inertia block), which allows increase of the stiffness and robustness of isolators for the given natural frequencies, and also allows reduction or elimination of inter-coordinate coupling by a judicial placement of isolators (e.g., in the horizontal plane containing C.G.).

This chapter is intended to provide, first of all, an analytical survey of commercially available or extensively tested and promising vibration isolation means. Many commercially available vibration isolators and other vibration isolation means have been produced without significant changes for the last 50–60 years and are ripe for upgrading or replacement by much better performing designs, some of which were extensively (and very successfully) used outside of North America. Instead of providing a detailed survey of the huge variety of commercially available isolators, the intention of this chapter is to describe the most important design concepts.

4.2 MATERIAL SELECTION FOR VIBRATION ISOLATORS

Selection of material for flexible elements of vibration isolators is a very important task since it determines, to a large extent, the performance characteristics of isolators. The basic characteristics of materials used for vibration isolators are addressed in Chapter 3. The following characteristics can be considered in the selection process:

- *Dynamic-to-static stiffness ratio* K_{dyn}. Lower values of K_{dyn} result in the possibility of using statically more rigid isolators to realize the required natural frequencies of the vibration isolation system. This, in turn, leads to more sturdy installations, enhances effective rigidity of beds and frames of the objects that require reinforcement from the supporting structure, and, in some cases, allows reduction in the size (weight) of the inertia block or even total elimination of the need for the inertia blocks.
- *Damping*. As was shown in Chapter 2, in many applications higher damping in the isolators can allow a beneficial change in the required natural frequencies of the vibration isolation system (increase in some cases, reduction in other cases) without deterioration in performance. While damping of elastomeric (rubber) elements can be made very high by proper selection of the base material and the blending additives, the high-damping blends may have some undesirable "side effects," such as increased K_{dyn} and/or increased creep rate. A 15–20% damping increase can be achieved by a proper selection of the flexible element geometry (use of streamlined elements such as radially loaded cylinders, spheres, toruses, etc. [4.1]). While damping increase by blending is often accompanied by increasing creep rate, use of streamlined shapes results in 20–40% reduction in creep rates.
- *Frequency and amplitude dependencies* of stiffness and damping.
- *Creep resistance* (especially for machinery mounts).
- *Temperature stability* and *temperature dependencies* of stiffness, damping, and creep rate.
- *Design/production flexibility*, especially in order to realize desirable stiffness ratios in different directions; desirable nonlinearity of stiffness; and also to accommodate various attachment requirements, various natural frequencies, various design concepts, such as isolators with intermediate masses, relaxation isolators, etc.

Table 4.1. Ratio K_{dyn}/δ for Various Elasto-Damping Materials for Different Vibration Amplitude Ranges

Elastromeric Materials		**Double Amplitude, mm**			
Rubber Type	**Hardness**	**0.01–0.015**	**0.05**	**0.1**	**0.2**
NR	41	4.6	4.3	3.3	2.6
	46	5.4	3.7	3.1	–
	56	5.4	4.2	3.6	–
	60	4.85	3.5	3.0	3.3
	61	5.9	4.4	4.4	3.0
	75	5.0	3.8	2.4	–
CR	42	4.6	3.85	3.75	3.0
	58	3.75	3.3	2.8	2.5
	74	6.2	3.1	**2.0**	**1.6**
	77	5.5	2.85	2.75	2.1
	78	5.8	**2.65**	2.2	1.85
NBR-26	42	4.0	3.6	3.5	3.3
	56	3.1	2.9	2.7	2.5
	69	**2.9**	**2.2**	**2.1**	1.95
NBR-40	45	3.45	3.7	2.8	2.05
	58	3.1	3.0	2.3	2.05
	62	3.5	2.8	2.6	2.25
	80	**2.6**	**2.3**	**1.9**	–
BR	50	**2.8**	**2.65**	2.2	–
Vibrachok wire-mesh isolating mounts					
Model	Load, N				
V439-0	400	30	9.5	3.4	**1.5**
	1,150	45	4.8	2.4	–
W246-0	870	13	4.5	3.4	1.8
	1,150	23	11.2	6.4	2.9
W246-5	2,300	27	4.1	2.9	**1.55**
Felt Unisorb		15	7	5.35	–

Since both dynamic stiffness and damping are amplitude- and frequency-dependent, selection of an optimum material should be influenced by the amplitude and frequency ranges in the vibration isolation system. In cases when the system performance is determined by the criterial stiffness/damping relationships derived in Chapter 2, these relationships can be used for optimization of material selection for isolator designs if the amplitude and/or frequency dependencies for the candidate materials are known.

Some of the most important applications of vibration isolation are associated with isolation of vibration-sensitive objects. Usually the frequency range of interest in these cases is rather limited (~0–100 Hz or less), and frequency dependency of stiffness and damping for typical elasto-damping materials in this range is not strong. Neglecting the frequency dependency, Table 4.1 gives ratios K_{dyn}/δ for various materials (elastomers, wire-mesh Vibrachok elements, and felt) for several typical amplitude ranges, from small to large amplitudes. The optimum material should have the lowest value of this ratio, in accordance with Eq. (2.2.17). For each subrange of vibration amplitudes, three smallest K_{dyn}/δ ratios are marked by bold-face numerals. It can be seen that values of K_{dyn}/δ vary widely for different materials and amplitude ranges, and that different materials should be selected for isolators that are designed to be used in different vibration amplitude ranges, with very significant differences in their performance. Similar comparisons can be made for selecting materials for vibration isolators intended for other applications rather than for isolation of vibration-sensitive objects, using other criteria derived in Chapter 2. It should be understood that the values of K_{dyn}/δ in Table 4.1 are given for specific blends/makes of the materials and could be very different for other, maybe similar, materials (e.g., rubber blends made with the same base rubber and having the same hardness but different chemical composition). However, all the materials in Table 4.1 are widely used for vibration isolation means.

Creep resistance of some materials is addressed in Chapter 3. While creep resistance of some rubber blends, even those with relatively high damping, is good (creep rate for standard test specimens around 1% per decade), it can be further reduced 20–40% by proper selection of geometry of the element [4.2], [4.3]. Since there is still no complete understanding of creep mechanism, especially in filled rubbers, and of correlation of the creep rate with K_{dyn} and damping, testing of each candidate compound is necessary if the creep resistance is of critical importance. Creep rates of wire-mesh and fibrous (e.g., felt) materials are usually low at the rated specific loads, but they should be tested in critical cases. The lowest creep have metal flexible elements. However, although creep of metal elements proper is low, the creep-like phenomena can occur in frictional connections in vibration isolators employing coil and, especially, Belleville spring units, due to micro motions in frictional joints under vibrations (see Section 2.1.2). Accordingly, if extremely low creep rates are required, special designs of metal isolators are used wherein no frictional joints (a monolithic design) or a minimal number of frictional joints are used, e.g., see [4.4].

If the vibration isolator has to consistently perform in a broad *temperature range*, the best material selection for its flexible element would be, obviously, a metal. However, while stiffness characteristics for many metals are reasonably stable in broad temperature ranges, damping of the special damping devices that are often necessary for isolators with metal springs, or even internal damping of high-damping materials, can be temperature-dependent. There are some elastomers, such as silicone rubbers, which have little-varying performance characteristics in rather broad temperature ranges. However, these elastomers may be unacceptable because of inadequate stiffness and/or damping and/or creep and/or bonding-to-metals characteristics.

One reason why elastomeric (rubberlike) materials are very popular for vibration isolators is their exceptional *design and production flexibility*. Rubber flexible elements can be designed with or without rigid (metal or hard plastic) inserts for modification of their stiffness values in three translational directions or, if required, also in three translational and three angular directions. Even without rigid inserts, a very broad range of isotropic as well as anisotropic flexible elements can be realized, e.g. as shown in Section 4.3. Designs of elastomeric flexible elements can be simplified since the required stiffness and damping characteristics can be achieved without special damping units, just by a judicial selection of the rubber blend. Also, it is not difficult to design elements with the required nonlinear load-deflection or load-stiffness characteristics, including CNF characteristics and "quasi-zero stiffness" characteristics, and to incorporate intermediate masses in the isolator designs. Low specific density of elastomers results in very small overall mass of elastomeric elements, especially when streamlined elements are used, thus reducing the danger of high-frequency wave resonances (see Section 1.10). However, all these features can be, at least to some degree, also designed into flexible elements made from other materials.

4.3 ISOLATING MATS AND PADS

Mats and pads, especially those made from rubber or plastic, are the simplest and the least expensive vibration isolation means. They are usually manufactured as relatively large flat blocks. These blocks can be used "as produced" for isolation of heavy objects, such as under foundations carrying machines or instruments. The concrete or polymer-concrete foundation block can be cast in place directly on the mat/pad coated with a waterproof paper. Often the mats and pads are cut into relatively small pieces rated for the required weight loads and placed directly under the isolated objects. A relatively small variety of the off-the-shelf mats/pads can be used to obtain a rather broad range of natural frequencies because stiffness of the mats/pads can be easily modified by cutting (changing the loaded surface area) or by stacking. Sometimes, the segments of mats/pads are attached as flexible elements to level-adjustment devices (jacks) and used as vibration isolation mounts.

4.3.1 Rubber Mats

Rubber mats are a popular means for providing elastic connection in vibration isolation systems. The simplest mats are solid rubber sheets. However, since rubber is a volumetric incompressible material (Section 3.3.2), vertical stiffness of solid rubber mats is very high, and they are used mostly in cases when vibration protection is required only in the horizontal plane (see Section 4.7).

Pads cut from solid sheets are often made from highly damped polyvinyl chloride (PVC) and polyurethane-based elastomeric materials Sorbothane™ [4.5] and EAR™ [4.6] of various thickness, usually, 2.5–40 mm (0.1–1.5 in.). These pads are available with a pressure-sensitive adhesive and are used in cases where damping enhancement is critical. Application of these and similar materials is limited by strong dependencies of their stiffness and damping on ambient temperature, by high values of $K_{dyn} \geq 3$, by relatively high creep rates, and, in the case of Sorbothane™, by low rated loads (hardness of all Sorbothane™ blends is below durometer $H = 30$ on the A-scale used for rubber materials).

Sometimes, high-damping elastomeric foam mats are used for vibration protection of light instruments, medical devices, etc.

A serious disadvantage of solid rubber mats is their large ratios $\eta_{x,y} = k_z/k_{x,y}$ between vertical stiffness coefficient k_z (in the direction normal to the weight-loaded surface) and horizontal stiffness coefficients k_x, k_y, respectively. Since the vibration isolators for majority of machinery are selected to achieve, first of all, the desirable vertical natural frequency f_z, low horizontal stiffness values result in low rocking natural frequencies $f_{x\beta}, f_{y\alpha}$. The low rocking natural frequencies lead to large amplitudes of

306 • PASSIVE VIBRATION ISOLATION

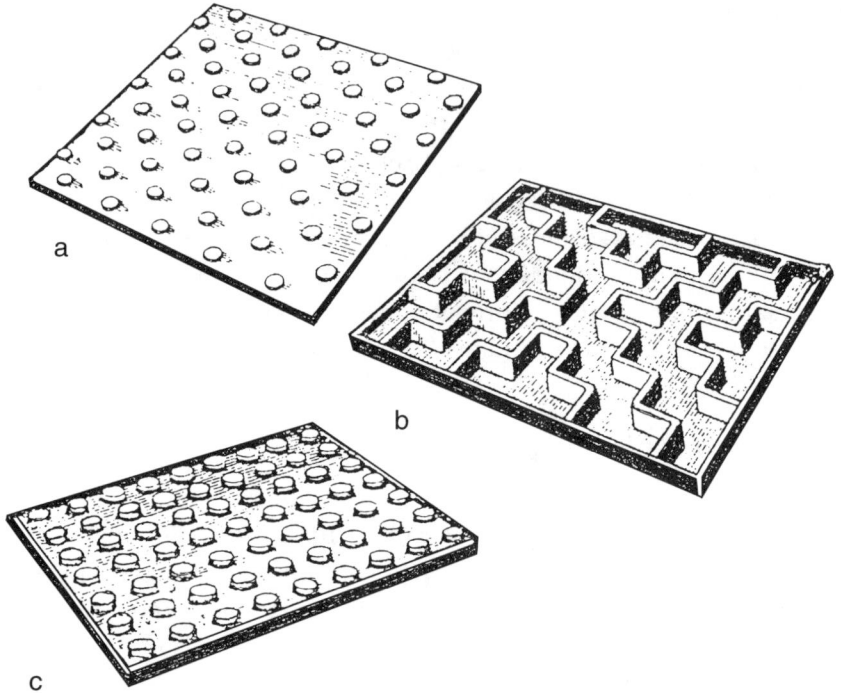

Fig. 4.3.1. (a) Light Duty; (b) Medium Duty; and (c) Heavy Duty Rubber Mats from Metalastik Ltd.

rocking oscillations of the isolated objects. This is especially important for rubber mats made from low or medium damping rubber blends and it is less important for highly damped Sorbothane™ or EAR™ mats, since the transient rocking vibrations quickly settle when these high damping mats are used. The overwhelming majority of vibration isolating rubber mats are designed to reduce the stiffness ratio $\eta_{x,y}$. As can be seen from Section 2.2.4, in many cases the typical desirable stiffness ratios should be $\eta_{x,y} = 0.5$–2.0, and also the required natural frequencies of vibration isolation systems for precision machine tools in the x- and y-directions are often quite different, $f_x/f_y > \sim 2{:}1$. Accordingly, in many cases it is desirable to have anisotropic mats in the horizontal plane with $\eta_y \ll \eta_x$.

Reduction of the stiffness ratios in rubber mats can be achieved by reducing their shape factor (see Chapter 3) in order to increase vertical deformations. A good example is presented by the rubber mats from Metalastic Ltd. (U.K.) shown in Figs. 4.3.1(a), (b), (c) [4.7]. Three types of mats are each made of eight blends of rubber with different levels of hardness (durometer), from soft to hard. Thickness of mat a is 27.6 mm, mat b is 25.4 mm, and mat c is 19 mm. The rated specific compression (weight) load range is 0.008–0.043 MPa (1.2–6.5 psi) for mat a, 0.054–0.29 MPa (8–43 psi) for mat b, and 0.2–0.72 MPa (30–110 psi) for mat c, while static stiffness constants k_z for all modifications are such that the natural frequency calculated using k_z ("static natural frequency") is $f_z = \sim 9$ Hz at the rated specific load for all variants of the mats. Thus, 24 modifications of these designs supposedly span the rated specific load range 90:1, providing the same natural frequencies with very fine differences between the rated loads for the adjacent modifications. While such coverage should not result in much higher prices since only three molds are required (one for each type in Fig. 4.3.1), its usefulness is questionable. Values of K_{dyn} between the softest and the hardest rubber blends used for these mats vary at least as 2:1 (and, thus, values of the actual "dynamic" f_z at the rated load at least as 1.4:1), and damping variations are certainly even greater. These differences are not mentioned in the catalogs for these mats thus, in critical cases, the user must perform the dynamic testing. In addition, the designs in Figs. 4.3.1(a–c) are conceptually very different. The light duty mat in Fig. 4.3.1(a) has widely spaced cylindrical studs on the upper side, each positioned in the center of a square formed by four identical studs on the lower side. As a result, the vertical load is accommodated mostly by shear and bending of the rubber base of the mat between the studs and the contribution of the stud compression is relatively insignificant. For this design, the measured $\eta_{x,y} = 3$–4 [4.7]. The design in Fig. 4.3.1(b) has relatively thin ribs only on one side, which are stabilized against buckling by their ladder-like shape. Low shape factors for the ribs result in their enhanced compression compliance and the measured $\eta_{x,y} = 4$–5 [4.7]. The design in Fig. 4.3.1(c) has relatively "stubby" studs only on one side, characterized by a higher shape factor and the measured $\eta_{x,y} = \sim 6$ [4.7]. Besides the differences in the stiffness ratios, very different breakdown between shear and compression deformation under vertical and horizontal forces may result in significant differences in damping and K_{dyn} values for mats of different designs, even those made from the same rubber blends.

Thus, the first impression is that the family of rubber mats presented in Fig. 4.3.1 provides an opportunity to fine-tune vibration isolation systems

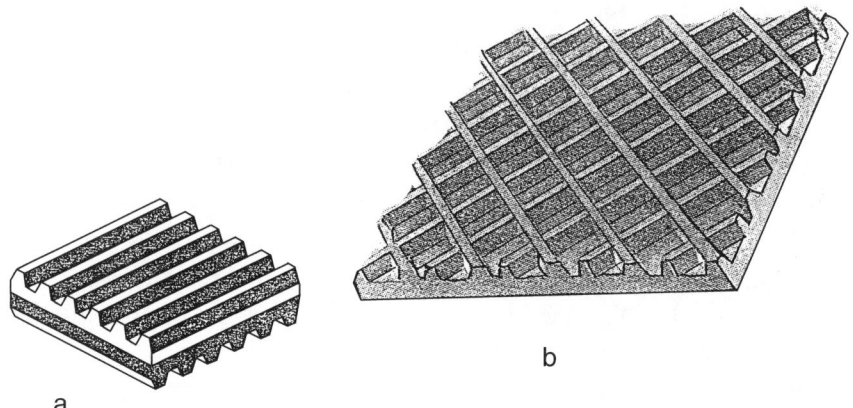

Fig. 4.3.2. (a) Double Sided; and (b) Single-Sided "Waffle" Pattern Rubber Mat.

by selecting from a large variety of similar designs with similar stiffness ratios and similar dynamic characteristics. However, this first impression is deceptive since these mats have very different values of some critical parameters and these differences are not communicated to the users. The same statement can be made for many lines of commercially available vibration isolating mats and mounts. Thus, the user often does not have much choice, but to try various products for his application.

Recently, the variety of vibration isolating rubber mats from various manufacturers was significantly reduced, possibly due to a better understanding of these factors. Also, very simple designs are produced by many current suppliers. One popular type comprises a thin rubber base with rectangular or slightly tapered cross-section ribs molded on both sides, with the ribs on one side running at a right angle to the ribs on the other side [Fig. 4.3.2(a)]. Typical dimensions: The ribs are ~5 mm wide, 3 mm high, with 7.8 mm pitch, and thickness of the base is ~2 mm, or proportionally larger (total thickness 8–13 mm). The bulk of the compressive load is accommodated by compression deformation of effective (virtual) rubber blocks generated by the intersecting ribs in the plane view. These virtual rubber blocks are $3 + 3 + 2 = 8$ mm high and are 5×5 mm^2 in the cross-section. However, bulging of these blocks under compression is limited by the rubber base at their midsections. Accordingly, compression stiffness is relatively high and the stiffness ratio of these mats is usually $\eta_{x,y} \geq 6$.

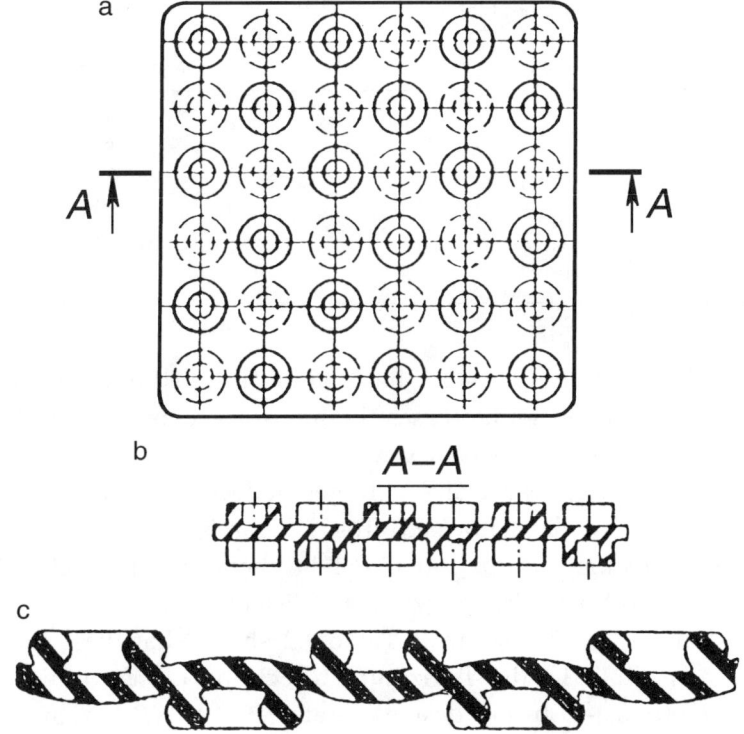

Fig. 4.3.3. Quasi-isotropic Rubber Mat KB-1: (a), (b) Design; (c)—Deformation Pattern.

Another popular design has so-called waffle pattern on one side (total thickness ~6.35 mm) [Fig. 4.3.2(b)] or on both sides (total thickness ~12.7 mm). Stiffness ratio of these mats is about $\eta_{x,y} = 5-6$.

Large stiffness ratios for the above-surveyed and other commercially available rubber mats make them suitable for vibration isolation of objects sensitive to vibration in horizontal directions. However, in order to reduce rocking vibrations of installations in which low vertical frequencies are required and/or there are internal dynamic forces acting in horizontal directions, much lower values of $\eta_{x,y}$ are required. In cases when the direction of maximum sensitivity (x) and the direction of action of the internal dynamic forces (y) are orthogonal (e.g., as in surface grinders or coordinate measuring machines), low and preferably anisotropic values of η_x and η_y are desirable. Such designs (KB-1 and KB-2) are described in [4.8], [4.9].

Table 4.2. Stiffness for Rubber Mats KB-1 and KB-2

Mat	$\eta_{x,y}@\varepsilon_z = 7\%$		$\eta_{x,y}@\varepsilon_z = 15\%$	
KB-1-1	1.5		1.4	
KB-1-2	1.1		0.6	
KB-1-3	1.3		1.0	
	η_x	η_y	η_x	η_y
KB-2-1	1.8	1.1	3.4	2.3
KB-2-2	0.6	0.7	2.3	2.5
KB-2-3	1.3	1.2	1.7	1.5

Type KB-1 mat in Fig. 4.3.3 has hollow round bosses on both its sides. This type of design increases the free surface area and results in a reduced compression stiffness while stiffness in the horizontal directions increases, especially at larger deflections in the z-direction, as it is clear from the deformed cross-section shown in Fig. 4.3.3(c). These mats are made from soft (KB-1-1), medium hard (KB-1-2), and hard (KB-1-3) rubber with values of $\eta_{x,y}$ listed in Table 4.2 for relative compressions $\varepsilon_z = 7\%$ and 15%. Table 4.2 shows that these mats have stiffness constants in the horizontal directions close to the vertical stiffness k_z (quasi-isotropic design). Mats KB-1-1, KB-1-2, and KB-1-3 have the rated specific loads $p_{max} = 0.08$; 0.19; 0.24 MPa (12, 21, 36 psi), respectively, with $f_z = 9$ Hz at p_{max}. All values of f_z at the rated specific loads p_{max} are measured, not calculated, and thus reflect the values of K_{dyn}. The latest modification of KB-1 mats (KB-1A) has through holes rather than recesses, in the bosses (Fig. 4.3.4). It results in ~20% reduction of stiffness in the z-direction and in ~70 % increase in horizontal stiffness, thus its $\eta_{x,y} \leq 0.5$.

Rubber mats KB-2 (Fig. 4.3.5) have higher rated specific loads than KB-1 made from the same rubber blends. For KB-2-1, KB-2-2 measured natural frequency $f_z = 10$ Hz at $p_{max} = 0.12$, 0.4 MPa (18, 60 psi), respectively, and for KB-2-3 $f_z = 12$ Hz at $p_{max} = 0.6$ MPa (90 psi); The KB-2 line is also characterized by $\eta_x \neq \eta_y$, which is rather remarkable since there are no rigid stiffness-modifying inserts in the design (Table 4.2).

4.3.2 Plastic and Fibrous Pads

Plastic or plastic-impregnated particle and fibrous pads are relatively widely applicable for mounting industrial machinery and equipment.

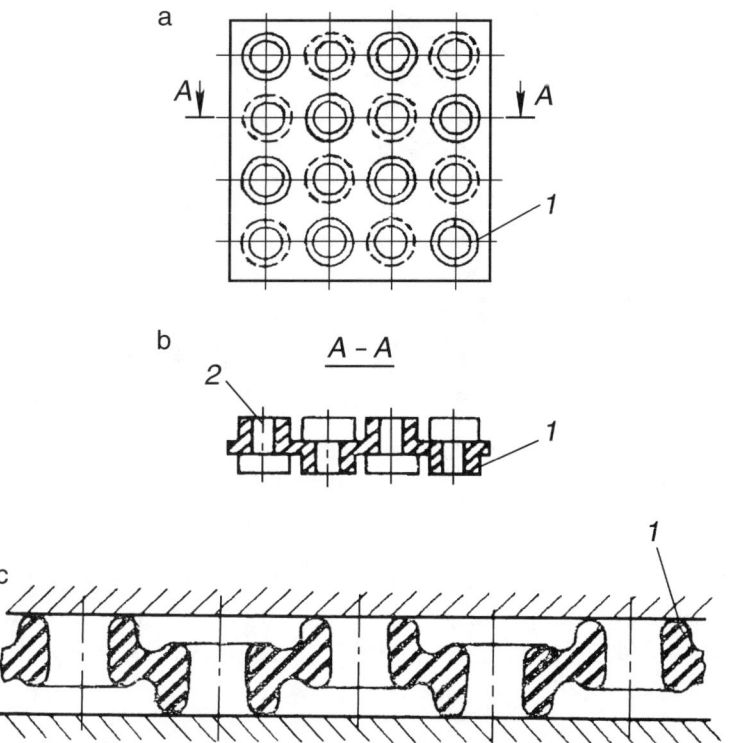

Fig. 4.3.4. Rubber Mat KB-1A with Enhanced Horizontal Stiffness ($\eta_{x,y} \approx 0.5$): (a), (b) Design; (c)—Deformation Pattern.

Non-impregnated fibrous pads (felt) are not as popular as in the past since they absorb oil and other contaminants that are always present on the shop floor. Pads made from cork slabs or from cork particles (chunks), which are pressed together while steamed are also capable of absorbing moisture, oil, etc. The felt and cork pads have to be protected by tar paper, synthetic rubber protectors, etc., unless they are installed under foundation blocks. The lowest natural frequency of cork pad-supported installations is ~10 Hz. Often, cork blocks are used for horizontal restraint and sealing of inertia blocks of vibration isolating foundations. In this application, creep is not an important parameter but resilience and impermeability (especially after coating) of cork are very beneficial characteristics.

Pure plastic pads are usually made from PVC plastics (e.g., EAR materials) and from polyurethane-based plastics (e.g., Sorbothane™ and

312 • PASSIVE VIBRATION ISOLATION

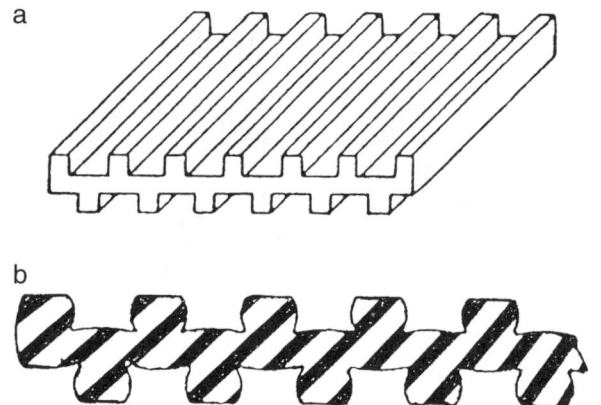

Fig. 4.3.5. Heavy-Duty Rubber Mat KB-2 Anisotropic in Horizontal Plane.

some EAR materials). Both these plastic families have elastomeric-type deformation characteristics, similar to rubber blends, and are briefly discussed in Section 3.3.3. They are highly damped but have relatively high creep rates and high values of K_{dyn}.

The cores of plastic-impregnated pads are made from particles (usually cork or wood) or natural (e.g., sisal) or synthetic fibers. The upper and lower load-carrying surfaces of plastic-impregnated pads have patterns made from the impregnating plastic (usually PVC-based) and they provide high friction for a better grip on the floor and machinery foot surfaces, as well as an additional deformation ability.

The psychological advantage of the plastic-impregnated pads is their visible sturdiness—they are usually considered desirable for mounting heavy and bulky industrial machinery. Other advantages are their proven high degree of oil resistance and high horizontal stiffness limiting rocking amplitudes of the installed objects under internal excitations. Such pads are now largely displaced by elastomeric pads and mounts, which can be readily designed to satisfy an extremely broad range of specifications such as loads, stiffness/natural frequencies, stiffness ratios, damping, resistance to environmental factors, etc. It can be seen from Section 3.3 that non-rubber natural and man-made materials often have somewhat mediocre dynamic and creep characteristics.

Synthetic rubber-impregnated cork particle pads have relatively high damping ($\delta \approx 0.55$) and $K_{dyn} \approx 3$. They cannot be used in critical

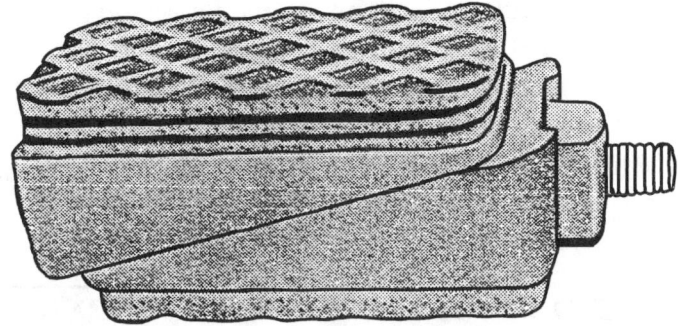

Fig. 4.3.6. Wedge-Based Leveling Mount Lined with Air-Loc Isolation Pad on Top Surface and Shim Pad on Bottom Surface.

applications since their creep rate is very significant, up to 20% per decade [4.7]. Similar properties are attributable to plastic-impregnated wood particle pads. They are available in blocks 1,200 × 1,200 mm, with a wide range of thickness—from 5.3 to 76 mm for isolation pads and from 0.5 to 3.2 mm for shims. The rated compression load is $p_{\max} = 0.35$ MPa (~50 psi); $K_{dyn} = 2.5$–4.5, $\delta = 0.5$–0.7, depending on vibration amplitude; $\eta_{x,y} = 1$–1.4; creep rate about 15% per decade [4.7].

Air-Loc® pads are marketed for installing machine tools and other machinery [4.7]. Two types of Air-Loc® pads are available: isolation pad, 9/19 in. thick, and shim pad, 3/16 in. thick. The former comprises a lamination of woven sisal fiber mats and layers of cork. The sisal fiber mats define the shape of the pad and provide its strength while cork layers impart vibration absorption properties. The laminate is impregnated with PVC, thus integrating it into a solid block (sheet) and providing resistance to water, oil, fungi, etc. The pad contains 70% PVC (by weight) and 30% of sisal and cork. In the process of impregnation, ribbed patterns are generated on both the top and bottom surfaces (Fig. 4.3.6). These patterns strengthen the frictional grip between the installed object and the floor. The shim pad consists of a single sisal mat impregnated with PVC and having the same surface patterns.

The rated pressure on Air-Loc pads is $p_{\max} = 0.8$–1.0 MPa (120–150 psi). At this pressure, pad stiffness is $k_z \approx 2$ MPa/mm; $\eta_{x,y} \approx 10$; $K_{dyn} = 3$–5; $\delta = 0.6$–0.8, depending on vibration amplitude [4.7]. Since vertical natural frequency of an object installed on such pads is quite high, $f_z \approx 40$ Hz, isolation criterion $\Phi_p = f_z/\sqrt{\delta} \approx 46$, these pads are used mostly for objects requiring relatively stiff connection with the floor structure while being sensitive to horizontal vibrations (e.g., lathes, cylindrical grinders, etc.). For grinders, two layers of Air-Loc pads are sometimes used ($f_z = 28$ Hz, $\Phi_p \approx 32$).

Air-Loc pads are usually attached to leveling devices (screw jacks, wedge mounts as shown in Fig. 4.3.6).

4.4 VIBRATION ISOLATORS WITH RUBBER FLEXIBLE ELEMENTS

While isolating mats and pads are very versatile means for isolating various units of equipment, their versatility prevents them from possessing the optimal performance characteristics necessary for diverse specific applications. There is a better chance to realize the optimal characteristics for various applications by selecting from a huge inventory of discrete vibration isolators available on the market. By far the largest group of such vibration isolators has elastomeric (rubber) flexible elements, since such designs can be tailored for a very broad range of performance specifications. The basic groups of vibration isolators with rubber flexible elements are:

- *Machinery mounts* for shop floors characterized by high rated loads, by a need for leveling devices, by exposure to environmental factors such as mineral oil, UV light, ozone, which damage rubber.
- *Engine and machinery mounts* for road, off-road, marine, and flying vehicles. This group is characterized by medium-to-high rated loads and by exposure to relatively high temperatures and to high overloads, as well as to oil and fuel spills. The mounts for mass-produced vehicles (cars and trucks) usually have finely tuned (for each individual application) stiffness constants, stiffness ratios, and damping. These mounts usually have metal (steel) armature constructed/assembled in such a way that they provide "captivity," that is, they prevent separation of the engine or some other installed object from the supporting structure in an event of catastrophic failure of the rubber element.
- *Instrument mounts*, used mostly for shock isolation of various instruments on vehicles (e.g., avionics on aircraft) and for portable vibration-sensitive devices, such as computer hard drives, computer

boards, etc. Often, computer subsystems are designed with vibration protection means integrated with them; this technique is outside the scope of this book.
- *Special vibration isolators*, such as constant natural frequency (CNF) isolators, highly anisotropic isolators, isolators having a combination of rubber flexible elements with other design components, very low stiffness isolators, etc.

This section addresses the basic design concepts for groups 1–2, with group 4 being addressed in other sections of this chapter.

4.4.1 Machinery and Engine Mounts

Machinery mounts are widely used for installing production, processing, and auxiliary machinery, such as machine tools, stamping presses, meat grinders, injection molding machines, fans, air-conditioners, etc. While in many cases these vibration isolating mounts are used for protecting precision machinery and equipment from floor vibrations or for protecting the environment from dynamic loads generated inside the machine, often such installation is used for objects that are not excessively vibration-sensitive and do not generate intensive dynamic exertions. In such cases the main purpose of the installation (often not clearly understood) is elimination of the need to attach machine to the floor by anchoring, grouting, etc. This effect results in substantial savings, especially when machinery is frequently relocated. It is caused by reduction of dynamic forces, however visibly insignificant, between the machine and the floor, which effectively reduces friction between the two and may cause "walking" of the machine. Another factor is enhancement of the gripping action between the machine and the floor due to high friction at the rubber/floor interface. It is not always appreciated that installation on vibration isolators protects the environment from accidental dynamic effects even in "dynamically quiet" machines (start/stop/reversal events, machining of odd-shaped parts causing unbalanced forces and highly variable intensive cutting forces, "crashes" in computer-controlled machines, etc.), and protects even not highly precision objects from accidental excessive vibratory and shock motions of the floor (the accidental dynamic effects in adjacent machines, accidental drops of heavy objects on the floor, rough operation of material-handling equipment, etc.). It was also observed that installation of large groups of machinery on vibration isolators in the shop results in a noticeable reduction of wear in joints, especially of bearings, and of noise (see Chapter 2).

Obviously, stiffness reduction of vibration isolators results in better isolation (see Chapter 1). However, the use of low stiffness isolators causes larger vibratory, usually rocking, motions of the isolated objects. Such motions develop due to internal excitations (e.g., stop-start and reversing events for moving heavy structural parts in the machine) and due to external excitations, especially for on-board objects on all kinds of vehicles (surface vehicles and construction/agricultural machinery, surface ships and submarines, aircraft, etc.). The special case is isolation of power transmission units, both on-board and stationary ones, wherein the isolators are subjected to torque-reaction forces.

For the stationary installed machinery, the "rocking" movements of the machine are unpleasant and sometimes disturbing for the operators, and in vibration-sensitive machines they may negate the intended purpose of the vibration isolating installation if the transient motions in the work zone do not decay before the production operation (cutting, measurement, etc.) starts (Section 2.2.6). These motions can be controlled by increasing damping in the isolators; by designing optimized stiffness constants into the isolators and/or isolation system [high stiffness or large distance between the isolators in the linear and/or angular direction(s) in the plane of the objectionable dynamic exertions while there is low effective stiffness in the direction(s) of high vibration sensitivity; see also Section 4.7]; by reducing severity of the dynamic exertions inside the object (e.g., by reducing transient accelerations); and, if the goal of reducing the rocking action is still not achieved, an overall increase of stiffness constants/natural frequencies or introduction of an inertia mass.

For objects installed on-board in vehicles, the minimum acceptable stiffness of the isolation system is determined by allowable displacements of the isolated object. Maximum displacements of the on-board objects are caused by shock loads, rocking of ships or surface vehicles due to road unevenness; for engines, also by changes in the output torque. Destruction of the vibration isolators as well as structural connecting lines to the object is usually prevented by installing snubbers (see Section 1.9) or by using nonlinear isolators (Section 4.5). Usually, the stiffness values and the packaging architecture are selected in such a way that the object displacements caused by the acceptable rocking and torque changes do not result in contact with the snubbers. The frequencies of load changes on the machinery isolators during rocking on surface ships are on the order of 0.1 Hz, with the amplitude of change about 0.5 of the weight-induced loads. Dynamic loads on rough roads (or in off-road vehicles) may cause maximum dynamic load amplitudes on the isolators up to two times the weight-induced loads.

The allowable displacements of the isolated object are often determined by its connections with other equipment units. Especially critical are kinematically-connected mechanisms. For turbine-generator installations, the allowable displacements relative to the foundation are determined by stiffness of the plumbing (piping) connecting the installation with the outside units. The allowable displacements depend not only on the type of the connection, but on the direction of displacement. For example, for coaxial mechanisms, some angular displacement that do not cause radial misalignments, can be tolerated to a larger degree than a radial misalignment. It is desirable that the stiffness center (Section 1.6.3) of the vibration isolation system in such cases be located on the axis of rotation.

If there is a choice, it is better that individual units kinematically connected between themselves not be installed directly on vibration isolators. These units should be assembled into blocks (by rigid frames or by concrete bases), and the blocks, in their turn, be installed on isolators. The blocks should be connected with outside equipment units by flexible power transmission couplings ([4.10], Section 4.12) and/or by conduit (pipes, etc.) couplings (Section 4.11). If by some reason the individual units must be directly installed on vibration isolators, without heavy bases, they have to be connected with other units by even more compliant flexible couplings. Without heavy bases, the stiffness of vibration isolators for specified natural frequencies is lower, thus stiffness of the connections should also be reduced.

Both analytical study (see Chapter 1) and practical experience (e.g., see Fig. 2.2.20) affirm significant advantages of CNF machinery mounts (described in Section 4.5) vs. constant stiffness mounts. However, the latter still dominate the market, at least in the Western countries, and many of their design features described below have universal applications. Typical machinery mounts for industrial applications on the market are quasi-constant stiffness vibration isolators with rated loads in the range of

1,000–25,000 N (220–5,500 lbs), although some isolator models are available for higher rated loads. By loading these mounts within their rated load range, vertical natural frequencies $f_z = 10\text{--}30$ Hz are usually realized. In cases where lower natural frequencies are required, pneumatic (Section 4.9) or metal spring (Section 4.6) isolators are often used. However, vibration isolators with rubber flexible elements for very heavy loads (when rocking of the machine is not a critical issue) and for mounting machinery on naval surface ships and submarines are used for f_z as low as 5 Hz [4.11] (Figs. 4.4.11 and 4.4.12). Stiffness ratios for the majority of industrial machinery mount models (with a few exceptios) are identical in both horizontal directions (isotropic mounts), with $\eta_{x,y} = 4\text{--}6$; damping is usually in the range of $\delta = 0.3\text{--}0.6$.

A very important feature of machinery mounts is a leveling device that: compensates for level variations (unevenness) of the floor within the machine footprint, which can be as large as 10–15 mm (~1/2 in.), and establishes and maintains the fine leveling of the machine, often horizontal within angular deviations in the range of $1\text{--}2 \times 10^{-5}$ rad or less. While the majority of machinery mounts have only height adjustment and provide angular compensation only due to angular compliance of the rubber element, some mount designs allow for pivot-leveling in angular directions to compensate for non-flatness of the floor (see below).

Some machinery mount designs allow measuring and monitoring of the weight load acting on the mount. This feature may be very useful for the fine leveling of the machine. The leveling procedure performed in a conventional way (using spirit or electronic levels) can be a very labor-intensive and time-consuming operation, often requiring a partial disassembly of the machine (e.g., lifting a massive table). A high degree of leveling can be quickly attained if the weight distribution between the mounting points were recorded by the machine manufacturer at the production assembly stand of the machine, while the latter is assembled and set up in a specified configuration (the specified optional units attached, moving parts positioned in the specified locations, etc.). The reconstruction of this weight distribution at the user facility using the weight-sensing mounts would take much less time and would guarantee a more accurate leveling than the conventional technique would. Although monitoring of the weight loads on each isolator is very beneficial, it is still rarely used on the commercially available vibration isolators for industrial machinery.

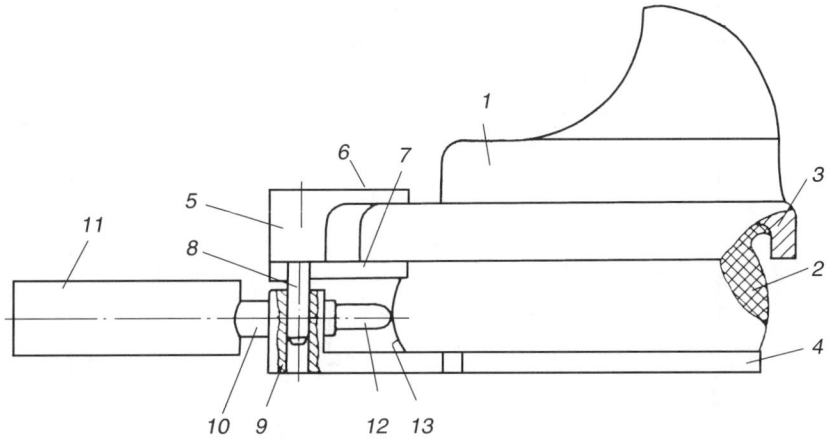

Fig. 4.4.1. Gage for Measuring Weight Load on Rubber-in-Compression Isolating Mounts.

The first information about the successful use of load monitoring is very old [4.12]. A large (weight 400 t) high-precision gear hobbing machine for 5,000 mm (~16 ft 8 in) diameter gears for submarines (max. pitch error 0.004 mm) was installed on 70 wedge mounts equipped with strain gages. By monitoring load signals from each mount, the mount levels could be adjusted continuously to assure the minimum allowable distortion of the machine bed.

Due to the incompressibility of rubber, compression of the rubber flexible elements of machinery mounts described below is always accompanied by "bulging" on the free surface(s) of the rubber flexible element. Since the free surface(s) of the rubber flexible element are usually easily accessible from the outside, measuring of the "bulge" allows evaluation of the weight load on the mount. This concept was suggested in [4.13]. The "mount load gage" in Fig. 4.4.1 allows one to measure the weight load transmitted from machine foot 1 through top cover 3 of the mount to rubber flexible element 2. The gage comprises top guide 5 with upper 6 and lower 7 guiding protrusions. Pins 8 are fastened to top guide 5 and slide into guiding holes made in bottom guide 9. Sleeve 10 of dial gage indicator 11 is fastened to bottom guide 9, and dial gage stylus 12 touches side (free) surface 13 of rubber flexible element 2. Thus, the gage is aligned with hardware components (top cover 3 and bottom cover 4) of the isolator and measures bulging of free surface 13. Obviously, electronic sensors can be used instead of dial gage 11. While this measuring concept

320 ● PASSIVE VIBRATION ISOLATION

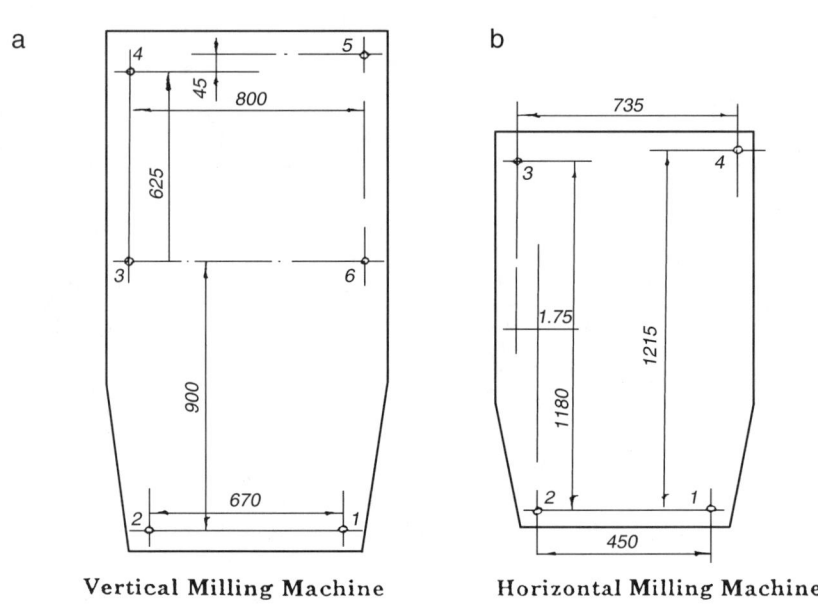

Fig. 4.4.2. Influence of Leveling on Weight Distribution Between Mounting Points.

Machine	Condition	1	2	3	4	5	6
a	Not levelled	4.0	20.9	5.0	3.7	17.7	5.9
a	Levelled	9.25	9.1	11.5	10.1	8.9	9.9
b	Not levelled	9.1	12.2	9.8	6.6		
b	Levelled	11.3	11.5	7.3	7.5		

is applicable to any isolator with its rubber flexible element loaded in compression, it is the most effective for vibration isolators that have a broad load range, such as the CNF isolators described in Section 4.5.2. On the latter, the repeatability of the weight measurement was adequate, within the 3–5% range. The gage can be used for maintenance purposes (intermittent readings of weight loads using one gage for several isolators), or for continuous monitoring (a permanent gage for each mount). Fig. 4.4.2 [4.13] shows weight load distribution between the mounting points of a vertical, Fig. 4.4.2(a), and horizontal, Fig. 4.4.2(b), knee milling machines obtained by using the gage in Fig. 4.4.1. It can be seen that the weight load distribution before leveling can generate large twisting and bending moments that distort the machine frame.

The Lod/Sen vibration isolating machinery mounts from the Vibro/Dynamics Corp., have their leveling bolts equipped with strain gages. This arrangement allows continuous monitoring of both weight load information (which can be used for deciding on the need to re-level the machine for maintenance purposes as well as for monitoring the leveling process) and for dynamic load information (which can be used for monitoring the production process [4.14]).

Rubber flexible elements of vibration isolating mounts for industrial machinery are usually shaped as cylindrical discs or rectangular blocks. Often, the central portion of the flexible element is hollow (e.g., a ring-shaped cylindrical element). The hollow shape allows increase in the overall dimensions of the mount without increasing k_z, thus enhancing its resistance to tilting. The larger size is also important in order to allay psychological concerns of the machine operators who feel better if a large machine is installed on reasonably large mounts. Another purpose of the hollow design of the rubber flexible element is to make possible a modification (increase) of stiffness of the mount in the horizontal direction. Often, the cylindrical or rectangular load-carrying rubber element has some auxiliary parts for attaching itself to the metal parts of the mount, thus integrating it into a monolithic structure.

The most popular in the West is the vibration isolating machinery mount LM, which was initially introduced by Barry Controls Co. in the 1950s. After expiration of the patent, this mount was copied by several other manufacturing companies. The flexible element of the LM mount (Fig. 4.4.3) is made from neoprene-based rubber blends. The load-carrying part of the flexible element is quasi-cylindrical; the whole rubber element is shaped in such a way that it integrates the load-carrying part with the leveling device of the vibration isolator, as shown in Fig. 4.4.3. The central opening in the load-carrying part accommodates the round outside projection of the load-distributing flange in order to constrain deformability of the mount in the horizontal direction, thus resulting in reasonably low values of $\eta_{x,y} = 2-5$ (measured in [4.7]). Originally, LM isolators were produced in seven sizes ($D_1 = 40-315$ mm, $h = 15-100$ mm), each size available in two-to-five rubber blends having hardness

322 • PASSIVE VIBRATION ISOLATION

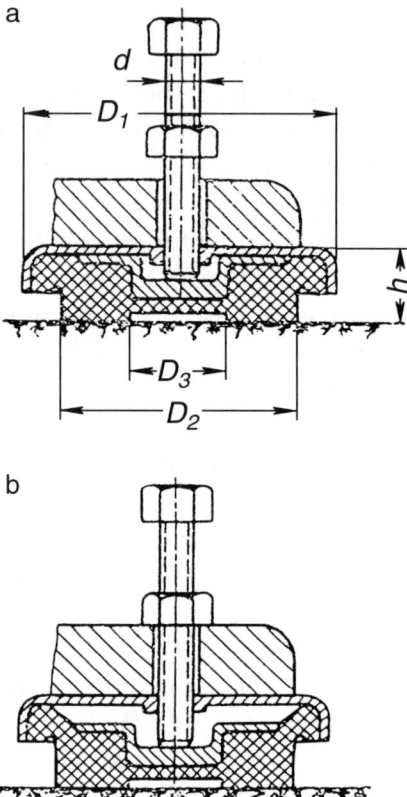

Fig. 4.4.3. Machinery Mount LM: (a) Before; and (b) After Leveling (*Barry Controls Corp.*).

(durometer) $H = 40–82$; the creep rate for these mounts is 3–4% per decade, depending on the blend of rubber [4.7], [4.15]. The range of rated weight load per mount is 1,200–120,000 N (~250–25,000 lbs) and the nominal minimal vertical natural frequencies (calculated from the static stiffness data) of the object installed on the identical mounts when the mounts are loaded with the rated load, $f_{z_{min}} = 6–10$ Hz ($f_{z_{min}} = 6–6.5$ Hz only for the largest mounts with height $h = 60–100$ mm; for the mounts with the rated load below 40,000 N or 9,000 lbs, $f_{z_{min}} = \sim10$ Hz). It was found by testing that while the natural frequencies for the mounts made from the softest rubber blends ($H = 40–50$) are within ~20% of the catalog data ($K_{dyn} = \sim1.5$), higher hardness blends have K_{dyn} up to 4.0, thus dynamic stiffness of the mounts with high durometer rubber is much different from static stiffness plots given in the catalogs, and actual natural frequencies are significantly higher than those advertised. Since many applications of vibration isolators for industrial machinery are regulated by criteria correlating f_n and δ (derived in Chapter 2) and damping (δ) of rubber usually increases with increasing H, the influence of increasing actual natural frequencies on the isolator performance is somewhat compensated by the increasing damping. However, since the isolators for each mounting point are selected on the basis of their "catalog stiffness" and the actual stiffness can be up to 2.5 times different, this factor greatly contributes to development of highly undesirable intra-coordinate coupling in the "natural coordinate frame" of the vibration isolation system (Section 1.3).

The manufacturer's recommendations on the selection of mounts for installation of various machines are based on loading of each mount and on "load-natural frequency" plots provided in the catalog. This leads to a wide scatter of the actually recommended natural frequencies. These recommendations (corrected by actual testing of LM mounts in [4.7]) are as follows: for grinders recommended $f_z = 12$–22 Hz and $\Phi_{pz} = 17$–22 Hz for the load per mount $> 3,000$ N (700 lbs), and $\Phi_{pz} = 20$–28 Hz for the load per mount $< 3,000$ N. For lathes and milling machines $f_z = 15$–27 Hz, $\Phi_{pz} = 20$–26.5 Hz for the load per mount $> 3,000$ N, and $\Phi_{pz} = 25$–34 Hz for the load per mount $< 3,000$ N. Thus, while the scatter of the actual natural frequencies is $\sim\pm 35\%$, the scatter of the actual isolation criterion Φ_{pz} is only ± 15–20%. Since the above recommendations have proven to be successful, they provide some practical validation for the concept of "isolation criterion".

As the variety of rubber blends for the rubber isolating mats was significantly reduced for recent offerings (Section 4.3.1), both Barry Controls Co. and its imitators do not currently use such a large variety of rubber blends of different durometers, and thus of different K_{dyn} and δ. Recent offerings of LM mounts have usually only one or two durometers of rubber per each size.

Numerous other designs of vibration isolating mounts for industrial machinery only insignificantly differ from the above-described LM mounts, usually by the shape of the flexible rubber element and by design of the level-adjustment mechanism, e.g., as shown in Fig. 4.4.4(a) (Nivofix mount from Paulstra Hutchinson Co. [4.16]) and Fig. 4.4.4(b) (Micro/Level mount from Vibro/Dynamics Corp. [4.17]). The latter design is characterized by a rectangular rubber flexible (resilient) element.

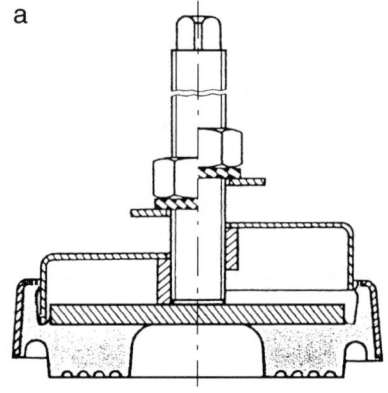

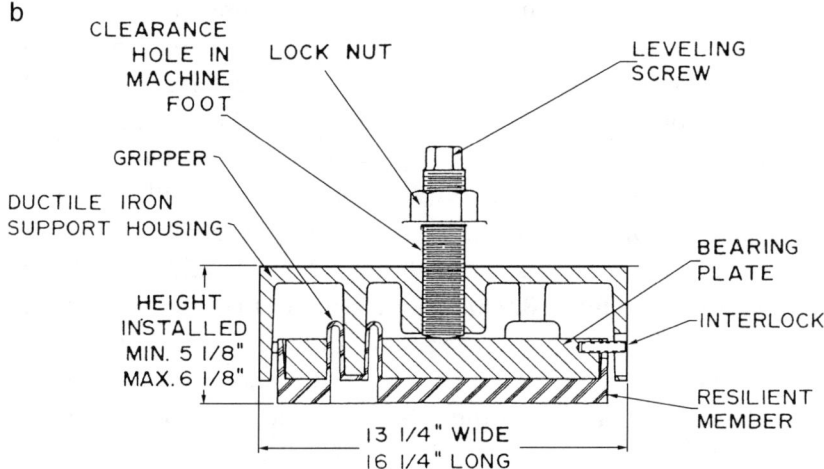

Fig. 4.4.4. (a) Nivofix Mount; and Micro/Level Mount (*Paulstra Hutchinson Co.* and *Vibro/Dynamics Corp.*).

Practically all machinery mounts with rubber flexible elements produced in industrialized countries are characterized by constant or quasi-constant stiffness within the load range of each mount.

The design of the leveling system may become important for performance characteristics of the vibration isolating mounts, since the horizontal stiffness of the leveling device may be comparable with the horizontal stiffness constants of the isolator itself and thus contribute to intensity of the rocking motions. It was found that use of "stay bolts" for leveling, e.g., as in Fig. 4.4.5(a), reduces the horizontal stiffness of the

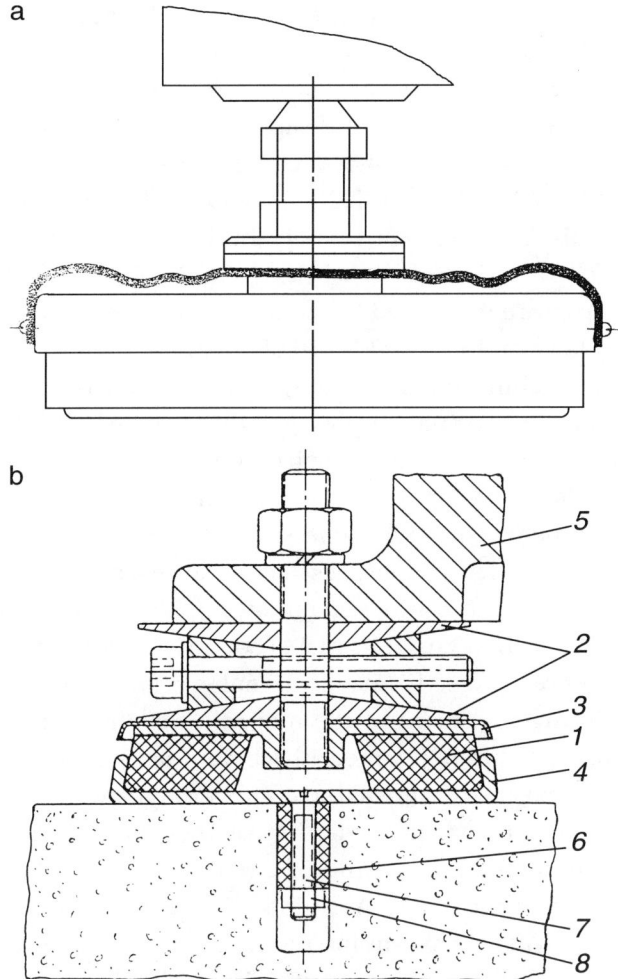

Fig. 4.4.5. (a) "Stay" Bolt Leveling Device; (b) Wedge-Based Leveling Device (1—Rubber Flexible Element; 2—Wedges; 3—Oil Spill Cover; 4—Bottom Cover; 5—Machine Foot; 6—Rubber Anchor) (SEDU Mount, *SAGA Co.*).

mount significantly, approximately in half. A better connection between the isolated object and the flexible element can be achieved by using wedge-based leveling devices, e.g., as in Fig. 4.3.6 and in Fig. 4.4.5(b). Such devices, however, often increase the effective height of the mount and thus reduce the effective rocking stiffness of the isolation system. They are also more expensive.

The majority of vibration isolating mounts for industrial machinery have only height-leveling devices. However, in cases of installation in tight corners, Fig. 4.4.6(a) [4.18], or on very uneven floors, sometimes pivoting leveling devices are used. Typical designs of such devices are shown in Fig. 4.4.6. The design in Fig. 4.4.6(a) can accommodate ±10–15° tilt and max. load 3,500–10,000 N (800–2,200 lbs); Fig. 4.4.6(b) [4.19] is designed for ±8° tilt (stainless-steel ball in polyamide socket), max. load up to 7,000 N (1,500 lbs); and a heavy duty design in Fig. 4.4.6(c) [4.19], that has a solid steel structure weakened by slots to provide for flexing, allows tilt ±2° with max. load up to 45,000 N (10,000 lbs).

Leveling of machine tools, presses, turbines, and other machines is usually performed by manual torqueing of the leveling bolts at each mount and by simultaneous monitoring of spirit or electronic levels placed on the machine bed. With the weight loads per mount exceeding ~1,000 lbs, this is a hard and time-consuming process. If the heavy machine has many mounting points, the leveling process can take several hours or, for very heavy and/or precision machines, even days. A Hydro/Level modification of heavy duty Micro/Level mounts [4.14] comprises an auxiliary hydraulic jack that can be pressurized from an external hydraulic pump and perform the required lifting/alignment of the mount. After the hydraulic-assisted lifting and alignment is performed, the bolts that are relieved from the weight loads can be easily adjusted to contact the respective machine feet and tightened. The hydraulic actuation of the mounts was claimed to reduce the alignment/leveling time of a 1.15×10^6 kg (2.5×10^6 lbs), gigantic press by three days. A solid-state electrically-operated assist device, not requiring a hydraulic pump, is now under development (prototype-tested) at Wayne State University. It should be noted that such "power assist" leveling systems not only save time but also reduce residual strains in the wedge mechanism after leveling thus alleviating the danger of vibration-induced changes mentioned in Section 2.2.2.

In cases of very intensive dynamic forces in horizontal direction(s) or the possibility of a collision between material handling vehicles and stationary machines, anchoring of machines to the floor might be desirable. Conventional anchor bolts require cementing into holes made in the concrete floor. This protocol necessitates cutting off the protruding anchor bolts if the machine is relocated. Utilization of incompressibility of rubber allows greater simplification of the anchoring/deanchoring procedures. A rubber anchor is shown in Fig. 4.4.5(b) as a part of a

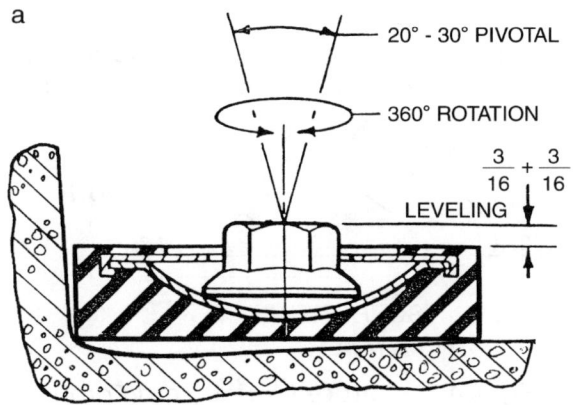

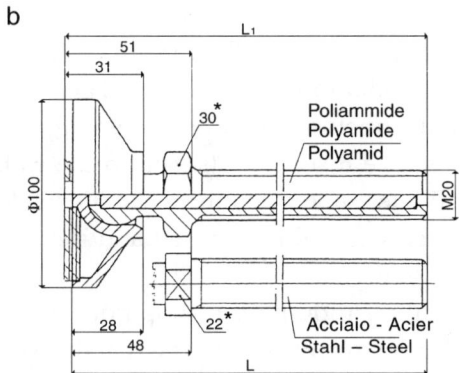

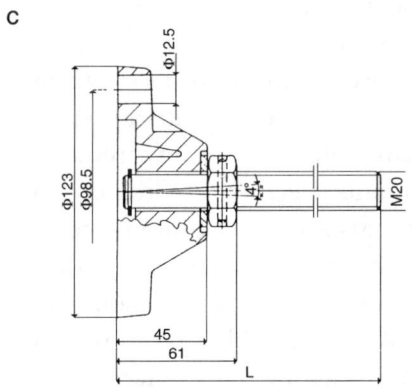

Fig. 4.4.6. Pivoting Angular Leveling Devices: (a) Medium Duty, Large Tilt Angle (*TechProducts Corp.*); (b) Medium Duty, Medium Tilt Angle; (c) Heavy Duty, Small Tilt Angle (*MarBett Plastic Engng.*).

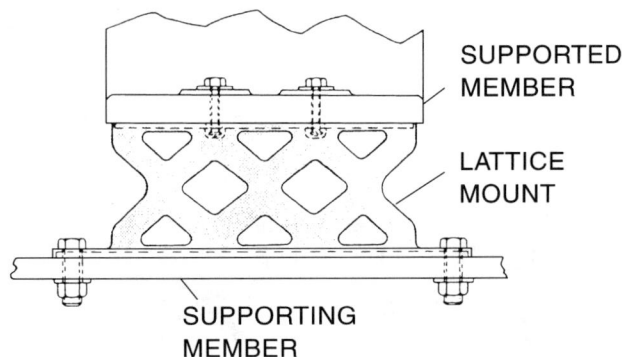

Fig. 4.4.7. Lattice Mount (*Lord Corp.*).

vibration isolating mount. Rubber sleeve 6 fits in a hole drilled into the concrete floor. After installation, bolt 7 is turned and causes nut/flange 8 to move up in the axial direction, thus compressing rubber sleeve 6. The sidewise bulging of sleeve 6 secures its grip to the walls of the hole and provides a very reliable anchor connection. For relocation, the axial compression of rubber sleeve 6 is released and the anchor can be easily withdrawn from the hole.

The vibration isolating mounts for shop-floor production/processing machinery provide for relatively high stiffness in the weight application (z) direction. However, there are many production/processing machines (sifters, screens, foundry jolt tables, rubber mills, looms, etc.) and auxiliary equipment (fans, air conditioners, chillers, vibratory conveyors, etc.) that generate intense dynamic loads at relatively low frequencies. In many cases, such units can be mounted either directly or after attachment to a cast iron, steel, or concrete plate, on low-frequency vibration isolating mounts. Since metal mounts (Section 4.6), require special oil or Coulomb friction dampers, they also need maintenance (oil refills, cleaning, sealing). On the other hand, mounts with rubber flexible elements usually do not require maintenance if the rubber blend used is resistive to the expected environment. Also, the rubber mounts provide for much better isolation of higher harmonics of the excitation forces, and their wave resonances are much less pronounced (Section 1.10).

In order to design a high-deflection (low natural frequency) rubber flexible element, a combination of several modes of deformation is often employed. For example, a lattice mount in Fig. 4.4.7 [4.20] combines

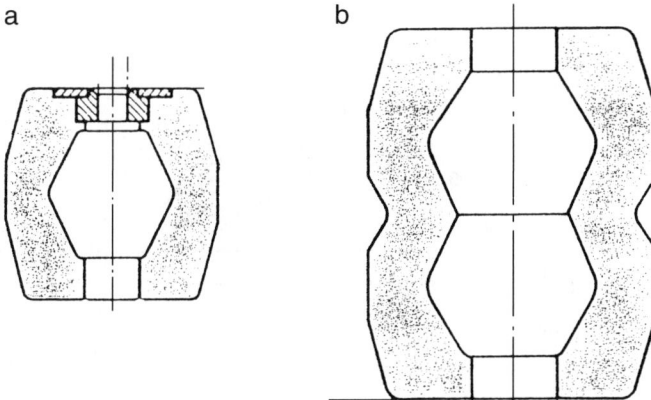

Fig. 4.4.8. Rubber Flexible Element of Barrel-Shaped Low-Frequency Vibration Isolators Evidgom (*Paulstra Hutchinson Co.*): (a) Single-Barrel Mount; (b) Double-Barrel Mount.

compression and bending. These mounts are listed in the catalog for $P_{max} = 300-4{,}000$ lbs (1,350–18,000 N). Deflection at P_{max} is 0.62 in. (15.7 mm) for medium-duty mounts ($P_{max} = 300-850$ lbs) and 1.0 in. (25.4 mm) for heavy-duty mounts ($P_{max} = 1{,}300-4{,}000$ lbs), about 15% of the mount height. Thus, vertical natural frequencies calculated from the static stiffness data are $f_z \approx 4$ Hz for the former and $f_z \approx 3.2$ Hz for the latter.

Similar deformation patterns have Evidgom mounts from Paulstra Hutchinson Co. [4.16], Fig. 4.4.8. The single-barrel mounts, Fig. 4.4.8(a), are produced for $P_{max} = 600-20{,}000$ N (130–4,400 lbs) with relative deflections at P_{max} being ~25% of the mount height or 10–35 mm ($f_z \approx 5-2.7$ Hz). The double-barrel mounts in Fig. 4.4.8(b) have $P_{max} = 50{,}000-140{,}000$ N (11,000–31,000 lbs) and relative deflection at P_{max} is ~22% of the mount height or 50–60 mm ($f_z \approx 2.2-2$ Hz). The barrel-like shape of Evidgom mounts provides for a symmetric buckling process, thus ensuring the stability necessary for such large deflections.

These vibration isolators are used for low-speed fans mounted on low-profile above-ground inertia blocks (plates), for low-speed reciprocating compressors, for driver cabin suspensions of heavy construction and agricultural machinery, etc.

Low natural frequencies together with the advantageous CNF characteristics, at relatively small outlines of the isolators, can be economically achieved using streamlined rubber elements (see Section 4.5).

330 ● **PASSIVE VIBRATION ISOLATION**

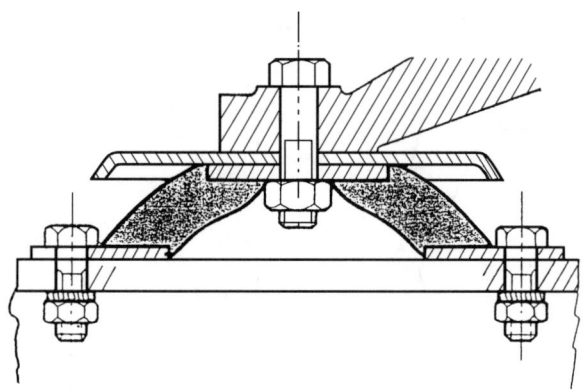

Fig. 4.4.9. Isotropic Beca Vibration Isolator
(*Paulstra Hutchinson Co.*).

Compression/bending deformation patterns realized in vibration isolators in Figs. 4.4.7 and 4.4.8 result in increasing vertical deformations as well as increasing stiffness in both horizontal directions in the isolators shown in Figs. 4.4.8(a), (b), and in one (lateral) horizontal direction in the isolator shown in Fig. 4.4.7. Further elaboration of such design concepts allows the design of an isotropic mount that has about equal stiffness constants in the x-, y-, z-directions, e.g., like the Beca isolator in Fig. 4.4.9 [4.16]. Such isolators are made for $P_{max} = 40$–12,500 N with relative vertical deflections at P_{max} being ~15% (vertical deformations 2–6 mm, natural frequency from the static stiffness data $f_z = 11$–6.5 Hz).

Sometimes, there are cases wherein relatively low natural frequency isolation of unidirectional (usually, vertical) dynamic excitations is desirable while, at the same time, the object should maintain alignment with the supporting structure in the perpendicular direction(s). Such cases are typical for installation of diesel engines with vertically oriented cylinders, reciprocating compressors, etc., on ships, trucks, railway cars, as well as for suspension elements for railway cars and other vehicles. To comply with such requirements, the isolators should have low stiffness in one (vertical) direction and much higher stiffness in at least one perpendicular (horizontal) direction. Such characteristics are easily achieved in isolators with rubber flexible elements loaded in a combination of shear and compression in three principal directions, e.g. like in Fig. 4.4.9. Variation of shear and compression stiffness constants of the

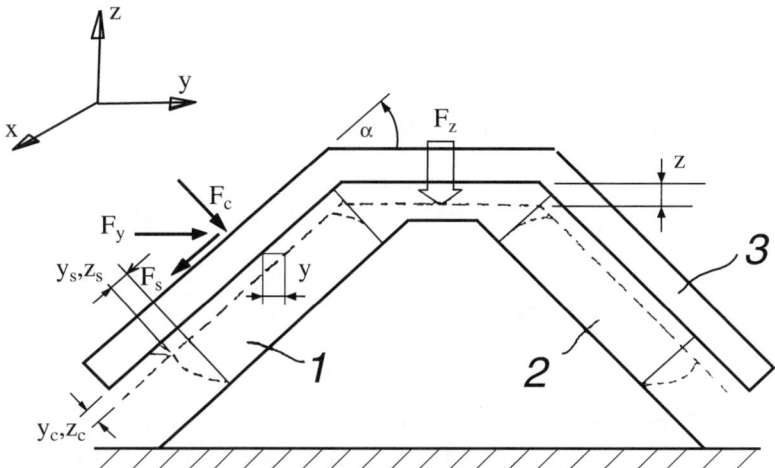

Fig. 4.4.10. Wedge-Shaped Vibration Isolator with Rubber Elements Loaded in Compression and Shear.

flexible elements by changing their shapes and their inclinations would result in variations of stiffness ratios of the isolator in these directions.

In cases where anisotropy of the horizontal stiffness constants in the x- and y-directions is required and is achieved by a combination of shear and compression, wedge-shaped isolators are used as in Fig. 4.4.10. The stiffness constants in all three directions are calculated as follows. The isolator in Fig. 4.4.10 is symmetrical relative to the x-z plane. Accordingly, magnitudes of the forces acting on the right flexible element (2) are the same as the forces shown acting on the left flexible element (1). The shown forces on element 1 are induced by one-half of the vertical force F_z, the other half being accommodated by element 2. The vertical force F_z causes vertical displacement z of cover 3 of the isolator. This displacement is due to both compression (z_c) and shear (z_s) deformations of element 1,

$$z_c = z \cos \alpha; \quad z_s = z \sin \alpha. \qquad \text{Eq. (4.4.1)}$$

These deformations are caused by the respective forces F_{z_c}, F_{z_s},

$$F_{z_c} = k_c z_c; \quad F_{z_s} = k_s z_s, \qquad \text{Eq. (4.4.2)}$$

where k_c, k_s are compression, shear stiffness constants of element 1, respectively. Obviously,

$$(1/2)F_z = F_{z_c} \cos \alpha + F_{z_s} \sin \alpha, \qquad \text{Eq. (4.4.3)}$$

or

$$F_z = 2(k_c z \cos^2 \alpha + k_s z \sin^2 \alpha) \qquad \text{Eq. (4.4.4)}$$

and the vertical stiffness of the isolator is

$$k_z = F_z/z = 2(k_c \cos^2 \alpha + k_s \sin^2 \alpha) = 2k_c[\cos^2 \alpha + (k_s/k_c)\sin^2 \alpha]. \qquad \text{Eq. (4.4.5)}$$

Stiffness in the x-direction (longitudinal direction) is due to pure shear deformation of both elements 1, 2 in Fig. 4.4.10, or

$$k_x = 2k_s. \qquad \text{Eq. (4.4.6)}$$

In the y-direction (lateral direction) force F_y is resisted by shear and compression of both elements 1 and 2. Application of force F_y may also cause rotation about the x-axis. For typical isolators with inclined flexible elements the rotational stiffness is very high and the rotational deformations can be neglected. Then, the reactions of elements 1 and 2 are identical, each reacting to $(1/2)F_y$, and

$$y_c = y \sin \alpha; \quad y_s = y \cos \alpha; \qquad \text{Eq. (4.4.7)}$$

$$F_{y_c} = k_c y_c; \quad F_{y_s} = k_s y_s; \qquad \text{Eq. (4.4.8)}$$

$$(1/2)F_y = F_{y_c} \sin \alpha + F_{y_s} \cos \alpha; \qquad \text{Eq. (4.4.9)}$$

$$F_y = 2(k_c y \sin^2 \alpha + k_s y \cos^2 \alpha); \qquad \text{Eq. (4.4.10)}$$

and the lateral stiffness is

$$k_y = F_y/y = 2(k_c \sin^2 \alpha + k_s \cos^2 \alpha) = 2k_c[\sin^2 \alpha + (k_s/k_c)\cos^2 \alpha]. \qquad \text{Eq. (4.4.11)}$$

For typical vibration isolators that have inclined or tapered rubber flexible elements $k_s/k_c \leq 0.2$, thus approximately

$$k_z \approx 2k_c \cos^2 \alpha; \quad k_y \approx 2k_c \sin^2 \alpha; \quad \text{and} \quad k_z/k_y \approx \cot^2 \alpha. \qquad \text{Eq. (4.4.12)}$$

In many cases there is no need for anisotropy in the horizontal plane (different stiffness constants k_x and k_y). In such cases isolators with tapered (conical) compression/shear elements are used. In the case of a conical isolator, $k_x = k_y$.

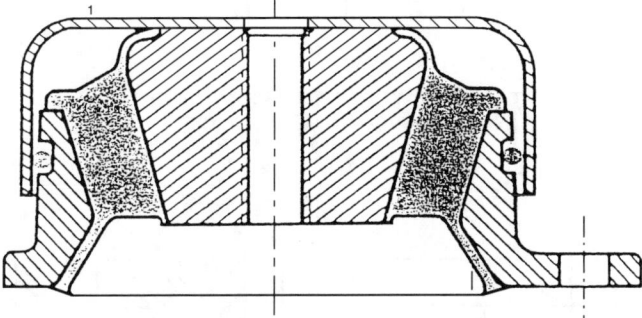

Fig. 4.4.11. Stabiflex Vibration Isolator (*Paulstra Hutchinson Co.*).

A typical isolator of such type (Stabiflex from Paulstra Hutchinson) is shown in Fig. 4.4.11 [4.16]. Since the taper angle of the annular taper rubber flexible element in Fig. 4.4.11 is small (i.e., angle α is close to 90°), the axial (vertical) deformation of the isolator is mostly shear, while the radial (lateral) deformation is mostly compression. In accordance with Eq. (4.4.12), the axial stiffness of the isolator in Fig. 4.4.11 is two to three times lower than its radial stiffness ($\eta_{x,y} = 0.3$–0.5). These isolators are made for axial loads $P_{\max} = 420$–$10{,}000$ N (93–$2{,}200$ lbs), with axial deformation at $P_{\max}$ being 3.5–8 mm (larger deformations for greater $P_{\max}$ values), or "static" natural frequencies (natural frequencies calculated from static load-deflection plots) $f_z = 8.5$–5.6 Hz. Wedge-shaped isolators with inclined compression/shear rubber flexible elements, like one shown in Fig. 4.4.10, have very different k_x and k_y; such isolators are used for suspensions of rail and, sometimes, surface vehicles as well as their engines.

Many designs of machinery mounts and other types of vibration isolators that are used on surface vehicles as well as on surface ships and submarines and on aircraft use rubber flexible elements, although frequently wire-mesh elements are used due to their large vibration amplitudes, especially on the flying vehicles. These isolators have several special features: First, the overwhelming majority of such isolators have the so-called captive design, preventing a physical separation between the isolated object and the supporting structure in the case of destruction of the flexible element, such as in the design in Fig. 4.4.12. This type of destruction may occur in rough seas or during a military attack on the seagoing vessels; during a "roll-over," during extreme accelerations during a collision or other catastrophic event on surface vehicles; during extreme maneuvers as

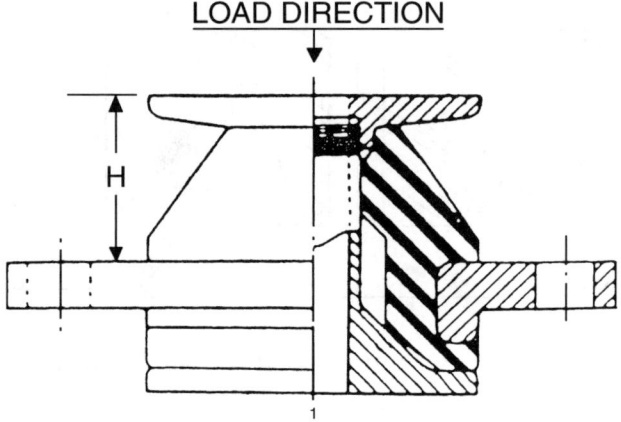

Fig. 4.4.12. Onboard Vibration Isolator 7E450 (EES line).

well as accidents on the aircraft; etc. The captive design prevents heavy objects from causing additional destructive effects. Although usually only one or two isolators must have the captive design, safety considerations lead to use all captive isolators on board vehicles and vessels.

Another typical feature of such isolators is the presence of snubbers (see Section 1.9) to prevent co-impacting of metal structural parts of the isolators in cases of excessive (shock) accelerations. Specially designed isolators with hardening nonlinear characteristics (such as the CNF isolators in Section 4.5, the isolator shown in Fig. 4.4.12, or wire-mesh isolators) are "self-snubbing" due to the progressive increase of stiffness at large deflections.

Generally, natural frequencies of vibration isolation installations on moving vehicles and vessels are noticeably lower than for the typical stationary installations, with $f_z \approx 5-15$ Hz. Typical vertical natural frequencies are $f_z \approx 3-5$ Hz for driver cabins on construction and agricultural vehicles; $f_z \approx 8-12$ Hz for engine mounting on on-road and off-road vehicles; $f_z \approx 5-15$ Hz for machinery mounting on submarines, etc. These low values can be explained by the high intensity of excitations, especially for surface vehicles. These excitations are generated both by interactions with road irregularities and by the powerful power-train system, and may get amplified due to relatively low stiffness, damping, and structural natural frequencies of the supporting structures. The latter factor is of high importance for machinery isolation on ships, especially on quiet submarines. A significant improvement of isolation in a broad

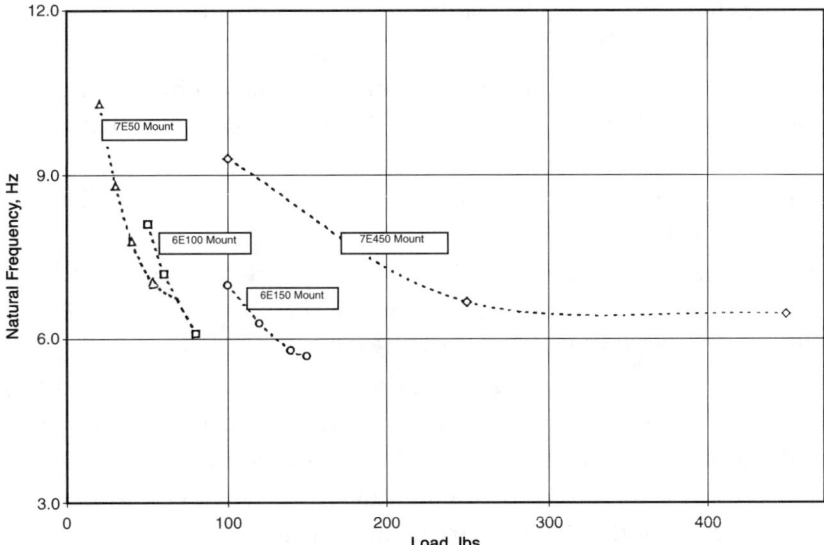

Fig. 4.4.13. Load-Natural Frequencies Characteristics of EES Line of Low-Frequency Onboard Isolators.

frequency range can be achieved by enhancing damping in the isolators; however, high-damping isolators are still a rarity.

A typical isolator (mod. 7E450) that is a part of the standard isolator lines for naval ships (including submarines) is shown in Fig. 4.4.12 [4.11]. This mount belongs to the EES line of five low natural frequency mounts (vertical load ranges from 50–100 lbs for 6E100 to 700–2,000 lbs for 6E2000). These loads can be applied only in the vertical (axial) direction. These mounts have the quasi-conical shape of the flexible element (somewhat rounded for heavy-duty mount models). The dynamic stiffness of the flexible element in both the vertical (z) and horizontal (x, y) directions increase with increasing axial loading, to a larger degree for the z-direction stiffness. As a result, two heavy-duty mounts have essentially CNF characteristics, albeit in narrow load ranges ($f_z = 5.3$–5.7 Hz in the load range $P_z = 700$–$2,000$ lbs for 6E2000 isolator; $f_z = 6.6$–6.3 Hz for a smaller 6E900 isolator in the load range $P_z = 400$–900 lbs). The smaller mounts in the EES line have f_z decreasing with increasing P_z, although less steeply than for the linear mounts. Load-natural frequency f_z of some EES isolators in the rated load ranges are shown in Fig. 4.4.13. The stiffness ratio for the EES mounts is $\eta_{x,y} = k_z/k_{x,y} = 1$–$2.5$. The smaller, doughnut-shaped segment of the flexible element functions as a snubber for high-impact shock loads.

The "captive" design of the EES mounts is achieved by making flanges of the inner metal sleeve with larger diameters than that of the hole in the main metal flange of the isolator. Thus, if the rubber element is destroyed, the metal parts remain entangled and the installed object cannot "fly away."

The mount design shown in Fig. 4.4.12 has the load range $P_z = 100-450$ lbs and weighs 4.5 lbs (2.04 kg), while the largest mount 6E2000 of this line weighs as much as 28.6 lbs (13 kg).

While machinery in stationary locations (e.g., in industrial buildings) as well as onboard a ship or of a flying apparatus is usually, with very few exceptions, installed on vibration isolators selected from a standardized (by a government agency or by the manufacturer) line, vibration isolating mounts for automotive engines are usually designed individually for each vehicle model/family. Their desired stiffness constants in three principal directions are initially obtained analytically, usually by using special software packages, e.g. [4.21], [4.22]. Unfortunately, the assumptions on which the analytical studies are based are not always clearly formulated, see Section 2.6. This, together with variations of the stiffness constants with rubber durometer scatter, deviations of the design characteristics from ones predicted by the calculations, as well as specifics of dynamic characteristics of the supporting structures not properly considered, etc., lead to the need for a lengthy "tuning" process during which numerous modifications of the isolators are made and tested until the satisfactory NVH (noise, vibration, harshness) characteristics of the vehicle are obtained. This tuning process may be shortened by using CNF isolators and/or high-damping isolators.

Onboard as well as other base- and frame-mounted equipment, such as engines, gearboxes, radiators, accessories, control boxes, instruments, etc., need to be securely attached so that vibration isolation, misalignment compensation, and shock protection are achieved. Typical mounts for such applications are shown in Fig. 4.4.14. These isolators have relatively low transverse (shear) stiffness that provides for misalignment compensation in the case of non-rigid frames. These designs also provide captivity and shock protection. The center-bonded isolator in Fig. 4.4.14(a) (Lord Corp.) has a preloaded rubber element that is used in compression with the preload amount determined by the internal sleeve of a calibrated

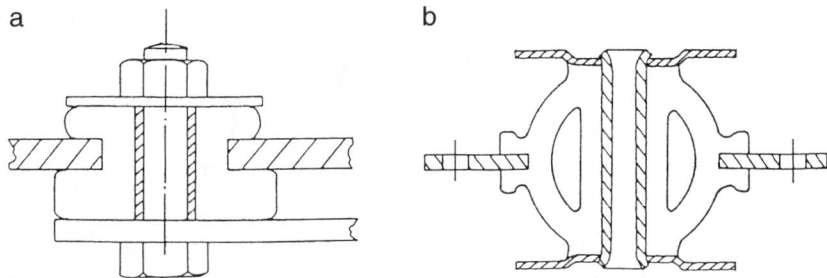

Fig. 4.4.14. Captive Vibration Isolators: (a) Center-Bonded Isolator (*Lord Corp.*); (b) Captive Isolator from *Cementation (Muffelite) Co.*

length. The stiffness ratios are $\eta_x = \eta_y = \sim 5-10$. A similar isolator in Fig. 4.4.14(b) (Cementation Muffelite Co., U.K.) has lower axial stiffness (combination of compression, shear, and bending) and higher transverse stiffness (combination of shear and compression). Judicious placement of a cavity inside the rubber element results in anisotropic characteristics in two horizontal directions, $\eta_{x,y} = \sim 0.5-2.0$.

4.5 CONSTANT NATURAL FREQUENCY (CNF) VIBRATION ISOLATORS

4.5.1 General Comments; Variable Stiffness Isolators

Understanding of dynamic processes in mechanical systems with several degrees of freedom is very important for the proper design and optimization of vibration control devices, of which vibration isolation systems are very important representatives. Analysis of such systems often is a difficult task, even with linear flexible elements. Any nonlinearity immensely increases the complexity of the analysis. Consequently, in most cases the flexible elements are assumed to be linear for analysis purposes and, partly due to the psychological inertia of the designers, these elements are intentionally designed with linear (*static*) load-deflection characteristics. Due to many design constraints, the resulting isolators often have some static nonlinearities, which may vary significantly (usually unintentionally) even within one family of elastic elements (isolators), e.g., see Fig. 4.4.13. Flexible elements made from elastomers and other polymeric materials, as well as from wire-mesh and other fibrous materials, may also have strong *dynamic* nonlinearity (dependence of effective stiffness and damping on amplitudes and frequencies of the vibratory process, Chapter 3). The dynamic nonlinearity is very difficult to consider analytically. The author suspects (and hopes he is wrong) that the wide use of low durometer rubber blends that have relatively low degrees of dynamic nonlinearity (but also low damping) is partly due to a desire to keep the system as close to linearity as possible.

While the detailed dynamic analysis of a nonlinear vibration control (isolation) system might be very useful, it is usually possible only for specific cases, and even then it is not very reliable due to the complexity of and incomplete information on the vibratory environments. However, it does not preclude the possibilities of generalized approaches to the optimization of vibration isolation systems considering both static and dynamic nonlinearities. Consideration of dynamic nonlinearities is made possible by formulating requirements for vibration isolation in various cases as criterial correlations between stiffness (natural frequency) and damping. This type of approach is explored in many situations described in Chapter 2. One example of the benefit of this approach is illustrated by Table 4.1, which provides guidance for selecting optimal materials for vibration-sensitive equipment isolators, depending on the prevailing vibration amplitudes. Another example is given in Section 2.6 for the selection of optimized materials for flexible elements of engine mounts.

The generalized/global approach to selecting optimal static nonlinearity of vibration isolators is based on analysis of inter-coordinate coupling in vibration isolation systems as discussed in Section 1.3. This analysis clearly demonstrated significant advantages of the static nonlinear load-deflection characteristic, whereas the stiffness, preferably in all principal directions, is proportional to the weight load along the principal loading axis of the isolator. Since such a characteristic results in constancy of the natural frequency f_z in the direction of application of the weight load from the isolated object on the isolator within the specified load range, such vibration isolators are known as "constant natural frequency" or CNF isolators [4.23]. The CNF isolators have many positive application-important features, in addition to their de-coupling capabilities, discussed in Section 1.3. Some of these positive features are as follows [4.23]:

1. Low sensitivity to production tolerances,
2. Possibility of use as building blocks for variable stiffness vibration isolators,
3. Simplification of vibration isolation system,

4. Self-snubbing property,
5. Reduction of size and cost of vibration isolators,
6. Possibility of standardization of isolators.

The following elaboration of these points refers in some cases to realized designs of CNF isolators discussed in detail in Section 4.5.2.

1. Low sensitivity to production tolerances is a unique feature of CNF isolators regardless of the design features of their flexible elements, such as coil springs or rubber flexible elements, etc. The survey of designs of these CNF isolators provided in Section 4.5.2 shows that the rated (nominal) natural frequency of a CNF isolator is determined by its geometry. For small loads on the isolator, $P < P_{\min}$ where $P_{\min} = m_{\min}g$ is the lower limit of its weight load range and $m_{\min}$ is the lower limit of the mass of the object generating the weight load, the flexible element usually can be assumed to be linear, Section 1.9. The CNF characteristic starts at $P_{\min}$, at which point deformation of the flexible element is $\Delta_{\min}$. Since at this point the flexible element can still be considered as linear, the natural frequency in the weight load application direction is, per (1.4.33)

$$f_{z_o} = \frac{1}{2\pi}\sqrt{\frac{k_z}{m}} = \frac{1}{2\pi}\sqrt{\frac{k_z g}{P_{\min}}} = \frac{1}{2\pi}\sqrt{\frac{g}{\Delta_{\min}}}, \qquad \text{Eq. (4.5.1)}$$

where k_z is stiffness of the flexible element on the linear (constant stiffness) segment of its load-deflection characteristic. If $\Delta_{\min}$ is expressed in cm, then the convenient formula based on (1.4.33) can be used.

$$f_{z_o} \approx \frac{5}{\sqrt{\Delta_{\min}}}. \qquad \text{Eq. (4.5.1')}$$

When the weight load increases within the CNF load range $P_{\min} - P_{\max}$, the natural frequency is practically constant and is equal to

$$f_{z\,\text{nom}} \approx f_{z_o}. \qquad \text{Eq. (4.5.2)}$$

Thus, the nominal natural frequency associated with the CNF isolator is determined by deformation $\Delta_{\min}$ of the flexible element at which the load-deflection characteristic becomes nonlinear.

340 • PASSIVE VIBRATION ISOLATION

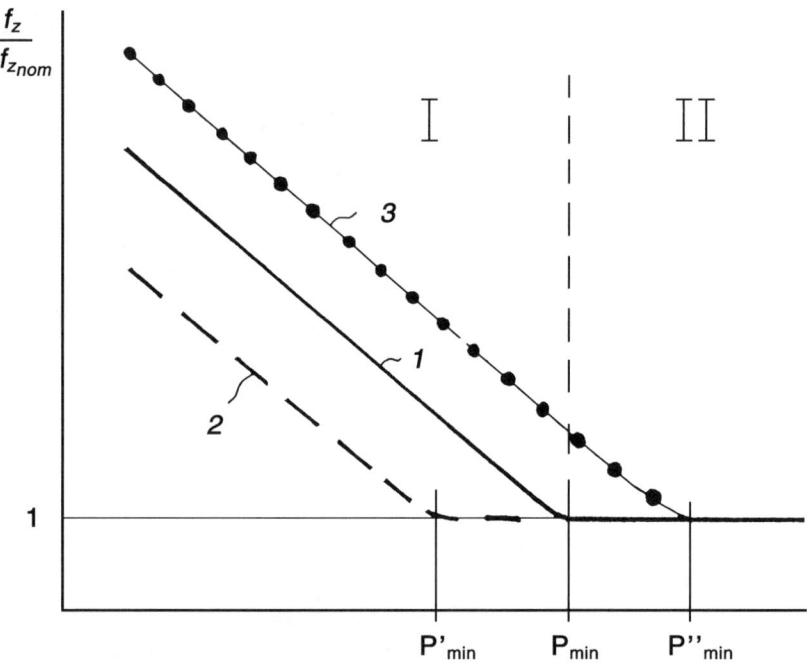

Fig. 4.5.1. Sensitivity of Constant Stiffness (I) and CNF (II) Vibration Isolators to Production Variations of Flexible Elements.

For a CNF isolator, which is manufactured without any deviations from the nominal design dimensions (e.g., coil diameters of different coils for a conical coil spring), the schematic load-natural frequency characteristic is presented by line 1 in Fig. 4.5.1. At low weight loads, the isolator has a linear load-deflection characteristic (no touching between the coils), and its natural frequency f_z decreases with the increasing load. As the deformation of the spring reaches Δ_{min} (at load P_{min}, when the large-diameter coils start touching), the CNF region begins. If the spring is wound on a slightly larger mandrel, the spring would be softer and its linear region would be represented by line 2, indicating lower values of f_z for the same load in the linear range than for the linear region of line 1. The deformation Δ_{min} occurs at a lower weight load P'_{min}; the CNF characteristic thus begins at a lower load but *at the same* Δ_{min} and $f_{z_o} = f_{z_{nom}}$ is determined by the value of Δ_{min}. This value depends only on the wire diameter defining Δ_{min}, not on the coil diameters. The deviation of the generatrix of the spring from its required shape (see Section 4.5.2) would, of course, result in deviations of the line $f_z/f_{z_{nom}}$ from line 1. Similarly, for a smaller mandrel the spring would be stiffer, its linear region represented by line 3 would indicate higher f_z values in the linear range than those of line 1, and the CNF characteristic would begin at a higher weight load P''_{min} but again *at the same* Δ_{min} *and* $f_{z_o} = f_{z_{nom}}$ determined by the value of Δ_{min}.

The dimensions of CNF isolators with rubber flexible elements are determined by the mold and usually are quite consistent. However, due to inevitable inconsistencies of the rubber blending components and their composition, as well as curing parameters (such as temperature and time), hardness (durometer) of the rubber part can vary within ±5 units of durometer scale (±17% in stiffness; Section 3.3.2). Again, line 1 represents the isolator with a "perfect" (nominal) rubber durometer; line 2, with the reduced durometer, and line 3, with the elevated durometer. For all these variations, the natural frequency remains the same while the weight load range shifts.

The shifts in the load range of CNF isolators are not significant for most applications since the load range is usually very broad ($P_{max}/P_{min} \approx 1.5:1-2:1$ for spring CNF isolators, $P_{max}/P_{min} = 5:1-20:1$ for CNF isolators with rubber flexible elements while P''_{min}/P'_{min} usually does not exceed ~1.35); here P_{max} is the maximum allowable weight load on the isolator. This conclusion about the robustness of the natural frequency values provided by CNF vibration isolators was validated by monitoring the mass-produced (~700,000 units a year) vibration isolators shown in Fig. 4.5.5 for two years of production. Variation in the natural frequency had not exceeded ±3%, while variation in the rubber durometer was observed within ±7 units of durometer (stiffness variation ±25%). The observed small variation in the natural frequency could also be explained by variations of K_{dyn} for the rubber flexible elements due to production tolerances.

2. Nonlinear load-deflection characteristic of a CNF flexible element means that its stiffness increases with increasing load along the z-axis. If two CNF isolators 1 and 2 are assembled as shown in Fig. 4.5.2(a) [4.23], so that the load along the z-axis can be adjusted by internal preload applied by preloading member

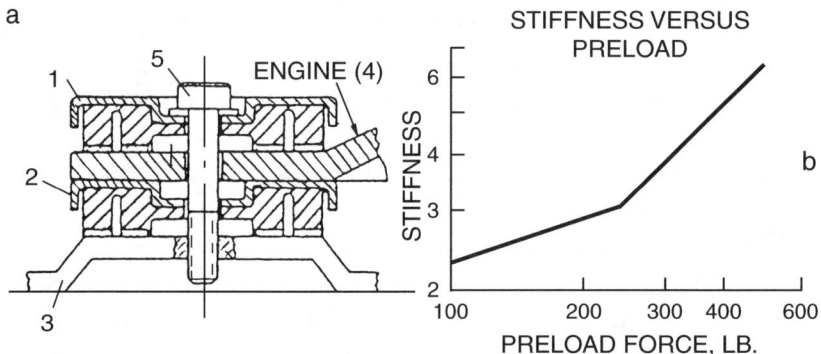

Fig. 4.5.2. (a) Variable Stiffness Vibration Isolator; and (b) Its Stiffness-Preload Characteristic.

(bolt) 5, independent of the weight load applied by the isolated object 4, then the effective stiffness of the connection between object 4 and supporting surface 3 can be varied by varying the preload force applied to both isolators 1 and 2. Change of the preload force results in changing stiffness k_1, k_2 of isolator 1, 2, respectively, with the increasing preload force corresponding to increasing stiffness, and the decreasing preload force corresponding to decreasing stiffness. Fig. 4.5.2(b) shows experimental results for the connection of two small CNF isolators of a somewhat simplified design, shown in Fig. 4.5.2(a) and similar to Fig. 4.5.4. Variation of the effective stiffness in the range of ∼3:1 have been observed. Fig. 4.5.3 shows a similar embodiment wherein small rubber spheres were used as nonlinear flexible elements. Variation of the natural frequency f_x from 20 Hz to 60 Hz was observed, corresponding to ∼9:1 stiffness ratio. An even broader stiffness variation range is feasible if the CNF vibration isolators with a broader load range are used. It is obvious that preload-adjustment means other than bolts, including some controllable electromechanical or hydromechanical actuators, can be used in this device. The embodiments in Figs. 4.5.2 and 4.5.3 are naturally of the captive kind (see Section 4.4.1).

The "building blocks" 1, 2 in Fig. 4.5.2(a) do not need to be identical. They may have different degrees of nonlinearity, different nominal natural frequencies, or one of them may even have a linear load-deflection characteristic.

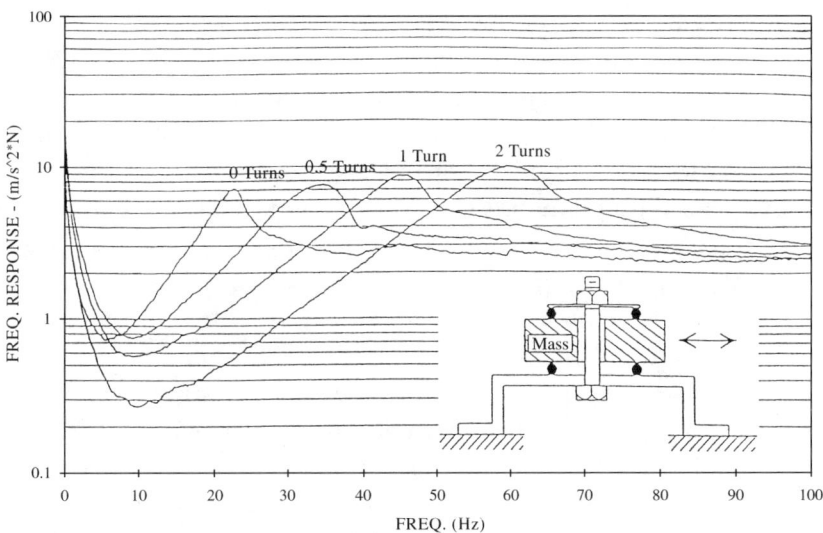

Fig. 4.5.3. Use of Preloaded Rubber Spheres for Tuning Vibration Isolators and Other Vibration Control Devices.

Use of the variable stiffness vibration isolators can be very helpful for experimental optimization of vibration isolation systems with complex excitations, with compliant supporting surfaces, etc. (e.g., see Section 2.7), as well as for adaptive and other actively controlled vibration isolation systems and dynamic vibration absorbers. Within the CNF load range $P_{min}-P_{max}$, stiffness of the variable stiffness isolator is proportional to the preload force Q. Obviously, this combination isolator is not a CNF isolator, but a locally linear isolator with an adjustable constant stiffness.

3. For some objects that have three mounting points in the horizontal plane, the weight loads acting on each point are identical and three identical vibration isolators can be used for the object. However, usually the nominal (computationally determined) weight loads are quite different. Leaving aside the issue of uncertainty of the computed weight load distribution addressed in Section 1.3, the actual loads on various mounting points can vary in a very broad range, e.g., see Fig. 4.4.2. Thus, one object may need to be installed on vibration isolators that are different for different mounting points. This fact creates logistical

difficulties in ordering, storage, and actual installation of the mounts. There is even greater danger in mixing up the isolators when the object has to be relocated/reinstalled.

Since CNF isolators usually have large load ranges, the same isolators (providing for the object's required natural frequency f_z) are used for all mounting points. In addition to their much better performance (see Section 1.3 and Fig. 2.2.20), the ordering, storage, and maintenance of such isolators are greatly simplified. The CNF isolators for different f_z can be color-coded, etc.

4. Constant stiffness isolators or weakly nonlinear vibration isolators are designed to provide necessary stiffness and natural frequency values for "normal" displacement and acceleration amplitudes of the external excitations. Momentary extreme amplitudes of external displacements and accelerations may significantly exceed the normal amplitudes, usually by a factor of two or more, thus causing quasi-proportionate large deflections of the isolators. For example, for most of the time vibratory accelerations of the structural parts and seats of trucks and construction/agricultural machinery do not exceed 0.1–0.2g, but in cases of extreme bumps and potholes on the road, accelerations can reach 2g. These deflections may result in destruction of the flexible elements and/or undesirable metal-to-metal impacts.

Snubbers are stiff auxiliary flexible elements, usually made from hard rubber, that engage in cases of excessive vibratory and shock exertions (Section 1.9). While they prevent the destructive interactions, their engagement may result in transmission of excessive and sometimes damaging vibratory inputs to the isolated objects. The effective stiffness of a snubber is the same both for rather frequently occurring slight excesses in the vibratory exertion and for extreme shocks, which may never occur.

The CNF characteristic is the result of a nonlinearity of the hardening type, both within the working load range $P_{min}-P_{max}$ and outside this range, at $P > P_{max}$. Thus, it naturally provides a *graduated* snubbing effect. When amplitudes of the external exertions only slightly exceed the normal amplitudes, the stiffness increase is also small, thus not exposing the object to excessive vibratory inputs. The more severe the perturbation, the stiffer the CNF isolator. Thus, there is no need for a special snubber—the snubbing action at the extreme exertion is as stiff as those for the special snubbers, but the most frequent and relatively small excesses in the excitation intensity are handled more gently in relation to the isolated object.

Typical CNF isolators can accommodate very high overloads. Thus, the O-ring based isolator was subjected to 100 times its rated compression axial load P_{max}, and after this overload was removed, promptly returned to its initial condition [4.11].

5. Conventional (quasi-constant stiffness) rubber flexible elements for vibration isolators are designed as "lines" or "families" comprising geometrically similar elements, which are small for low rated loads and progressively increase in size for units with higher rated loads. The dimensions can be approximately characterized by shape factors $S \leq 2-3$ (Section 3.3.2A) and specific compressive loads 0.3–0.7 MPa (45–100 psi). Rubber flexible elements with CNF characteristics usually have broad load ranges and their shape factor increase with the increasing load. The isolator in Fig. 4.5.5 in Section 4.5.2 has initial $S_{P_{min}} = \sim 0.5$ and $S_{P_{max}} = \sim 2.5$, and specific compressive loads changing accordingly, since rubber elements with higher S can accommodate greater loads. One CNF flexible element may replace the entire family of constant stiffness flexible elements. Analysis of available CNF isolators has shown that a CNF isolator is usually slightly ($\sim 20\%$) larger than the smallest constant stiffness isolator in the comparable (by the rated loads and by the lowest f_z) line/family, but is significantly smaller than the largest constant stiffness isolator in the line.

Due to the fact that the largest isolators in the line are the most expensive, replacement of 3 to 10 isolators in a line of constant stiffness isolators by one CNF isolator is, usually, very cost-effective.

Another source of financial gains is reduction of inventory of CNF isolators as compared with constant stiffness isolators. The "line of one" CNF isolator requires only one mold for the rubber flexible element and only one set of stamping dies for the metal parts. Three CNF isolators OB-33 whose load-natural frequency characteristics are shown as lines 2, 3, 4 in Fig. 4.5.9 in Section 4.5.2 (range of $f_{z_{nom}} = 11-30$ Hz) were produced in the same mold (using some simple inserts) and with the same metal parts (cover plates). Last but not the least is the scale of production, since for the same number of isolators produced, there is a much smaller number of different CNF isolator models.

6. Implementation of CNF vibration isolators opens the possibility of a significant overall reduction in the inventory of vibration isolators, since relatively small isolators have very broad ranges of rated loads and can be fabricated for required values of rated natural frequencies $f_{z_{nom}}$ with very narrow scatter. After the development of the CNF machinery mount in the former USSR in the early 1960s (shown in Fig. 4.5.5), it became so popular that in 15 years about 10 mln units were produced and put to use. Although only one isolator model OB-31 with $f_{z_{nom}} = \sim 18-20$ Hz was produced, about 90% of the industry needs have been satisfied due to the advantages of its CNF characteristics, its very broad load range, and relatively high damping. However, there was still a perceived need for lower (10 Hz, 15 Hz) and, sometimes, higher (30 Hz with high damping) $f_{z_{nom}}$ isolators, to be used in the remaining 10% of industrial machinery cases as well as for other applications. This led to development of a State Standard [4.24] for the parametric series of CNF isolators. The suggested series of the nominal natural frequencies $f_{z_{nom}}$ in this standard are: 10, 12.5, 15, 16, 20, 25, 31.5, and 40 Hz, with damping of the isolators $\delta \geq 0.5$. This parametric series was materialized by designing and testing prototype CNF isolators OB-33, whose basic characteristics are close to the Standard's requirements (see Fig. 4.5.9 in Section 4.5.2).

Use of the CNF characteristics of streamlined rubber flexible elements (see Section 3.3.2B), makes the establishment of an inventory of highly effective isolators for diverse applications even more realistic. Use of such rubber flexible elements would also allow expansion of the above parametric series toward the lower natural frequencies, starting at 3–5 Hz.

4.5.2 Designs of CNF Isolators

Various commercially available vibration isolators have CNF-like characteristics. In some cases the CNF or quasi-CNF characteristic is

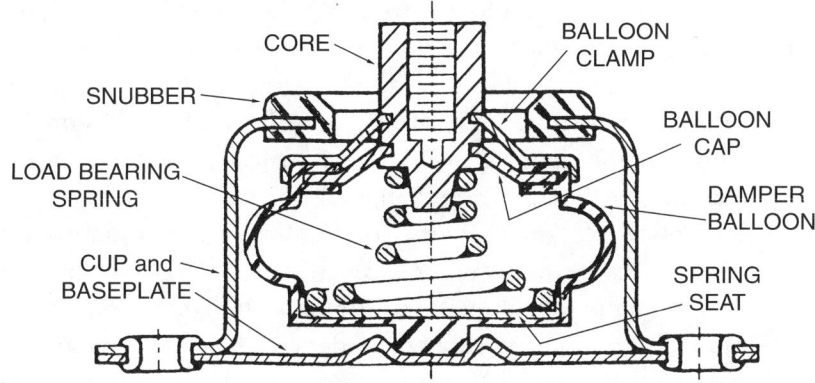

Fig. 4.5.4. Constant Natural Frequency Isolator with Shaped Coil Spring.

unintentional and occurs naturally due to intrinsic nonlinearity of the material used in the isolator design. Such cases are represented by some models of rubber isolators (e.g., Figs. 4.4.12 and 4.4.13), to some extent by wire-mesh isolators (Section 3.3.1 and Section 4.6.3), and by plastic or composite pads (Section 3.3.3 and Section 4.3.2).

The first intentionally designed CNF vibration isolators seem to be isolators that have nonlinear coil springs (see Section 3.1.1) as their flexible elements. The isolator in Fig. 4.5.4 [4.25] was designed for mounting onboard instruments in various surface and flying vehicles. The rated load ranges for these isolators are from 3–6 N to 100–150 N ($\sim$1–35 lbs) with the CNF load range $P_{max}/P_{min} = \sim 2{:}1$ (and $f_{z_{nom}} = 9{-}10$ Hz) for smaller isolators and $\sim$1.5:1 (and $f_{z_{nom}} = 7{-}8$ Hz) for larger isolators. The isolators have pneumatic damping (energy dissipation due to flow of air through a calibrated orifice in and out of the rubber shell/damper balloon enveloping the spring). The resulting damping is equivalent to $\delta = 0.2{-}0.7$, depending on the isolator size. Similar (conical) springs are sometimes used in car and truck suspensions, since the achievable load ranges are close to variations of the weight load on suspension springs (1.5–2:1) between empty and loaded vehicles in typical applications.

CNF isolators with elastomeric flexible elements have much broader CNF load ranges than those that have shaped coil springs. This feature is especially important for basic isolators that can be used for installation of diverse objects. Most of the known/available designs employ rubber flexible elements that accommodate the weight loads by compression and are intended for relatively high payloads, $P_{min} > {\sim}500$ N (110 lbs), unless streamlined rubber elements, with which both the upper and lower values of the rated loads are practically unlimited, are used. One reason for using the compression deformation mode is sensitivity of the compression stiffness to the shape factor S, which can be made load-dependent by design. Since conventional (e.g., rectangular or axially loaded cylindrical) rubber elements have quasi-linear load-deflection characteristics in compression up to large deformations (see Section 3.3.2), the nonlinearity has to be intentionally introduced into rubber CNF isolator designs, with the exception of streamlined rubber elements. The nonlinear nature of the load-deflection characteristic is usually achieved by incorporating into the design a continuous change (increase) of the shape factor S with the increasing compression load.

This type of change in S can be achieved by a judicious design of the flexible element comprising two or several components that are initially separated by clearances, so that their free (non-loaded) surfaces have unrestrained bulging at small compression loads. With the increasing load, the bulging free surfaces start contacting one another and/or rigid components built into the flexible element. With a further increase in the compression load, the contact areas enlarge, thus further limiting the bulging surface areas.

Another direction for designing CNF rubber flexible elements is by using streamlined unconstrained or constrained, if necessary, rubber bodies (Section 3.3.2B).

The first mass-produced CNF vibration isolator for machinery with a rubber flexible element was isolator OB-31 [4.26], [4.8] (Fig. 4.5.5). Its flexible element is a monolithic rubber block molded in a relatively simple mold cavity, but the core of this element is composed of two quasi-independent rubber rings, 1 and 2. These rings are bonded to top 3 and bottom 4 metal covers. Top cover 3 has lid 5 so that there is a calibrated clearance (gap) Δ_1 between the inner surface of lid 5 and the outer surface of rubber ring 1. Another calibrated gap Δ_2 is between the inner surface of ring 1 and the outer surface of ring 2. The weight load from object 6 is transmitted by leveling nut 8 to leveling bolt 7, and by washer/flange 9 to top cover 3 and to the rubber block. At low weight loads, all four side surfaces of rings 1 and 2 are free to bulge, thus resulting in the shape factor $S = {\sim}0.5$. With an increasing weight load, the bulge on the outer surface of ring 1 touches lid 5, and the bulge on the inner surface of ring 1 makes

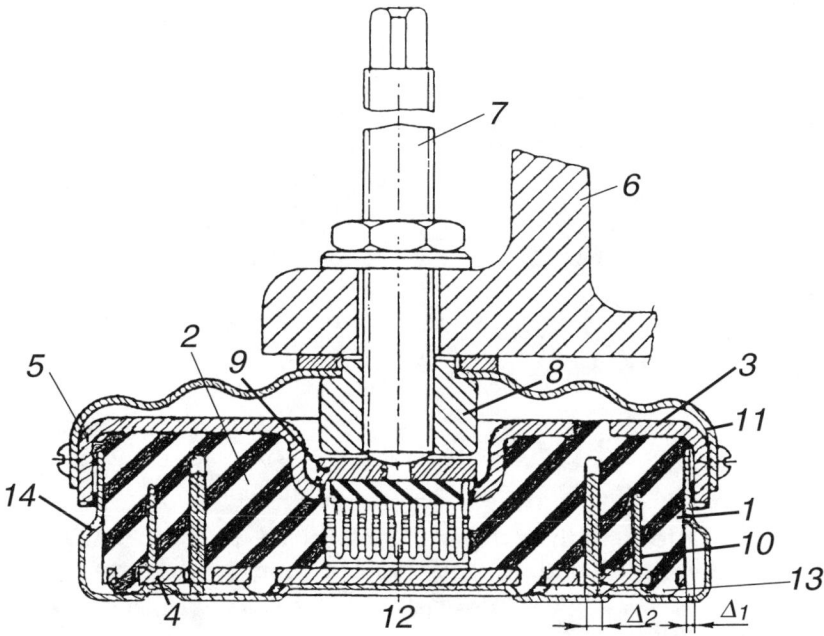

Fig. 4.5.5. CNF Rubber-Metal Vibration Isolator OB-31.

contact with the bulge on the outer surface of ring 2. With the increasing weight load, gaps Δ_1 and Δ_2 are gradually filled with the bulging rubber (Fig. 4.5.6), thus further increasing S and stiffness of the isolator. Fig. 4.5.6 depicts deformation stages of the isolator in Fig. 4.5.5 obtained by compressing a sector cut from the isolator and placed, with petrolatum lubrication, between transparent Plexiglas plates that always maintain the initial angle of the sector. Before the tests, a grid was painted on the side rubber surface of the sector.

Optimization of the design dimensions during the prototype development of the isolator resulted in the load-natural frequency f_z characteristic 1 in Fig. 4.5.7. The natural frequency $f_z = 17.3$ Hz ± 15% is maintained within the rated load range 4,200–40,000 N (930–9,000 lbs), or $P_{max}:P_{min} = \sim 10:1$. Some additional features of the CNF isolator in Fig. 4.5.5 are worth mentioning.

Flange 9 has rubber "brush" 12 molded to it. Brush 12 is immersed in viscous fluid 13 (whose reservoir it also plugs), thus generating a three-dimensional damping effect for the isolator. This increases the effective damping of the isolator by 15–20%, from $\delta = 0.4$–0.6 for the selected

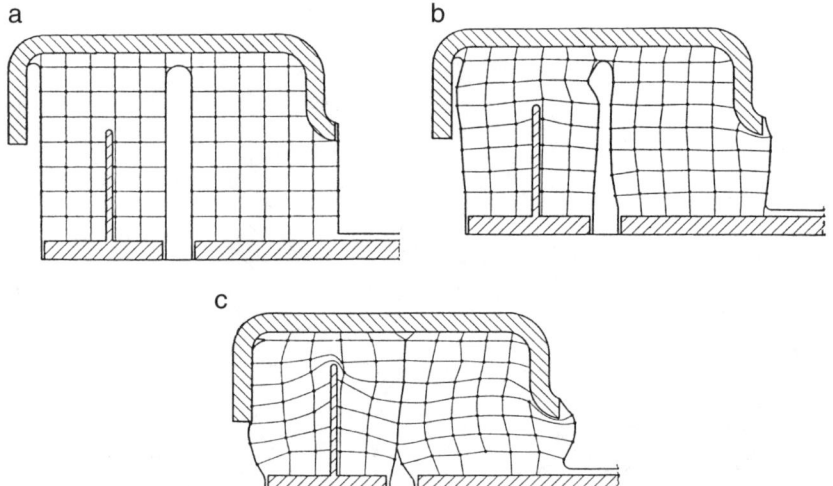

Fig. 4.5.6. Deformation Patterns of Isolator OB-31 (Without Damper): (a) No Weight Load; (b) Low Weight Load; (c) Deformation Pattern at $P_{min} < P < P_{max}$.

low-creep rubber to $\delta = 0.5$–0.7, thus allowing expansion of the isolator's application domain in accordance with the criteria derived in Chapter 2. The interaction between the damper and the rubber flexible element is quite complex and results also in some modification of the load-natural frequency characteristic, as shown by line 2 in Fig. 4.5.7. Line 2 can be described as $f_z = 20$ Hz $\pm$ 15% within the rated load range 3,200–40,000 N (700–9,000 lbs), or $P_{max}:P_{min} = {\sim}13{:}1$, and the deviations from the CNF characteristic in the important range of smaller weight loads are much less pronounced. Since smaller units of equipment usually need vibration isolation with higher natural frequencies, the isolator was successfully used in the load range 2,000–40,000 N (440–9,000 lbs), or effectively with $P_{max}:P_{min} = {\sim}20{:}1$.

Bottom cover 4 has thin metal ring/rib 10 attached (welded) to it. The presence of ring 10 provides resistance of the flexible element to lateral forces, especially due to compression of rubber between ring 10 and lid 5 (see also Fig. 4.5.6). The presence of this rib results in a relatively low $\eta_{x,y} = k_z/k_{x,y} = 2.5 \pm 20\%$ across the rated load range, as compared with values of $\eta_{x,y} = 5$–6 for the majority of the commercially available vibration isolators, Section 4.3.1, while $\eta_{x,y} = 0.5$–2.0 is the optimal range for isolation of many types of precision objects.

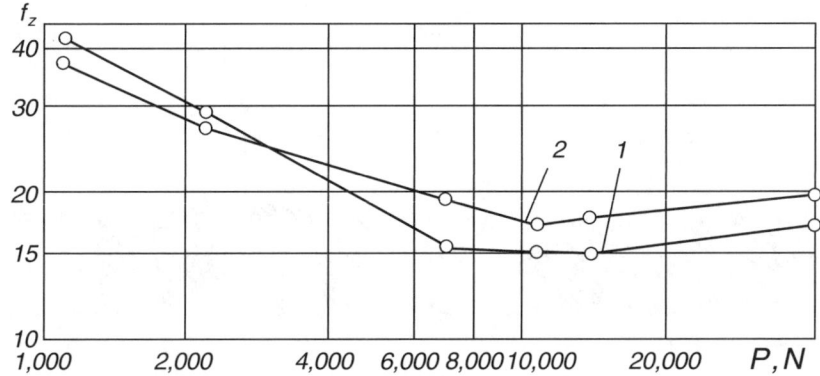

Fig. 4.5.7. Frequency Characteristic of CNF Isolator OB-31 (Line 1—No Damper; Line 2—with Damper).

Spring 11 attaches leveling nut 8 to top cover 3 while allowing its vertical displacements during the leveling process. Rubber friction rings 13 on bottom cover 4 are generated in the same mold as the whole flexible element, with the rubber flowing to form rings 13 through holes in the bottom cover 4. Metal parts 3 and 4 can be fabricated with very loose tolerances since the critical dimensions Δ_1, Δ_2 are defined by the mold.

While the above-noted uniformity of $\eta_{x,y}$ values throughout the rated load range is an important positive feature of the OB-31 design, sometimes there is a need for enhanced lateral stiffness in one (x) direction. This can be achieved by using insert 14 in gaps Δ_1, Δ_2 of the OB-31 isolator, [4.27]. The width of the insert in the direction perpendicular to the plane of the drawing is ~0.25D, where D is diameter of the isolator. The stiffness ratios with the insert are $\eta_x = 1.4$, $\eta_y = 1.9$.

The latest modification of this isolator (OB-32) is shown in Fig. 4.5.8. In this design, the connection between the leveling nut and the top cover is provided by the rubber membrane. This isolator is easier to produce (fewer parts) and has a better appearance. Due to a minor modification of the rubber flexible element and to the use of a different rubber blend (high-damping/low-creep NBR rubber), the damper is not used but the load-natural frequency characteristic is close to that of line 2 in Fig. 4.5.7.

The next generation of CNF vibration isolators based on the design concept shown in Fig. 4.5.5 is isolator OB-33. This is a line of CNF isolators providing for four $f_{z_{nom}}$ complying with the Standard [4.24] in broad load ranges and having much smaller sizes. Diameter of OB-31 and OB-32 is

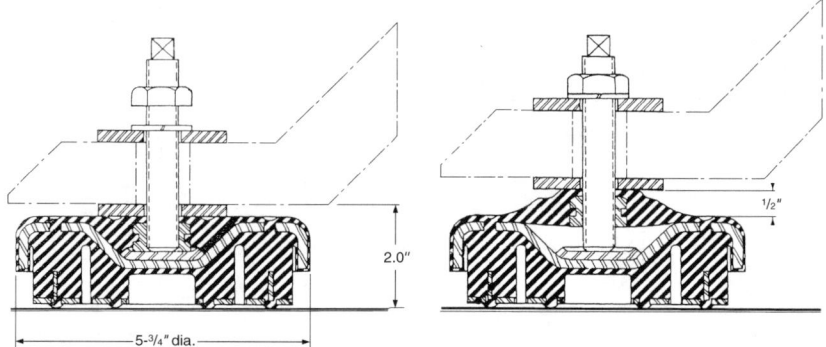

Fig. 4.5.8. Modified CNF Isolator OB-32: (a) Before Leveling; (b) After Leveling.

$D = 125$ mm (5 in.), with height (without leveling spring) $H = 35$ mm (1.4 in.). Such a large D results in a very high $P_{max} = 40,000$ N, as well as in relatively high $P_{min} = 3,200$ N. Many practical applications would benefit from shifting this load range toward lower load magnitudes. Accordingly, isolators OB-33 were designed with $D = 100$ mm (4 in.), thus reducing the loaded surface area of the rubber flexible element by ~60%. Isolators for nominal natural frequencies $f_{z_{nom}} = 10, 15, 20$, and 30 Hz were developed by using the same top and bottom cover stampings and by varying Δ_1 and Δ_2, rubber blends, and the height of the rubber flexible element. Load-natural frequency characteristics of OB-33 isolators are shown in Fig. 4.5.9. While they are shifted toward lower rated loads, their ratios $P_{max}:P_{min}$ are quite large, $P_{max}:P_{min} = 6-20:1$. For OB-33-1, $f_{z_{nom}} = 11$ Hz ± 2% at $P = 850-6,000$ N, $\eta_{x,y} = 2.25$, $\delta = 0.2$ (NR); for OB-33-2, $f_{z_{nom}} = 15$ Hz ± 2% at $P = 300-5,000$ N,

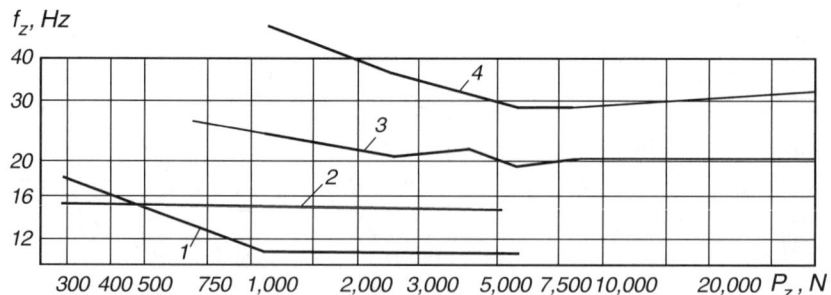

Fig. 4.5.9. Load-Natural Frequency Characteristics of OB-33 Line of CNF Isolators.

$\eta_{x,y} = 2.,25$, $\delta = 1.0$ (BR); for OB-33-3, $f_{z_{nom}} = 20$ Hz ± 5% at $P = 2{,}000-40{,}000$ N, $\eta_{x,y} = 2.8$, $\delta = 0.55$ (NBR); and for OB-33-4, $f_{z_{nom}} = 30$ Hz ± 7% at $P = 4{,}000-40{,}000$ N, $\eta_{x,y} = 3.1$, $\delta = 0.6$ (NBR).

It is interesting to note that implementation of OB-33 isolators was impaired by their smaller sizes. Operators of heavy equipment felt uneasy when a large machine was installed on several "tiny" mounts although the weight loads were safely within the rated load range. In some cases, clusters of three isolators attached to large and heavy cast-iron plates have to be used under each mounting point of the machine to be installed, for no technical reason whatsoever, except to reduce operators' anxiety.

Figs. 4.5.10(a), (b) show recently launched CNF vibration isolators from Newport Corp. [4.28]. The flexible elements of these isolators are hollow cones made from high-damping polyurethane elastomers ($\delta \approx 0.8$, $K_{dyn} \approx 4$). The hollow conical shape of the flexible element enhances its bulging on the side surfaces under compression. The bulging process is restrained by the concave internally tapered metal parts, thus resulting in a progressive increase in stiffness and in a CNF or a quasi-CNF characteristics. Fig. 4.5.10(a) shows ND01-A isolator whose load-natural frequency characteristic is represented by line 1 in Fig. 4.5.10(c). The nominal natural frequency $f_{z_{nom}} = 22$ Hz ± 5% in the rated load range $P = 315-1{,}350$ N (70-300 lbs), or $P_{max}:P_{min} = 4.3:1$; $\eta_{x,y} = 9-16$. The low-frequency ND20-A isolator [$f_{z_{nom}} = 12$ Hz ± 20% in the rated load range $P = 300-1{,}150$ N (70-250 lbs), $\eta_{x,y} \approx 1.5$] is shown in Fig. 4.5.10(b) and represented by line 2 in Fig. 4.5.10(c). These isolators are expensive since they require precision elastomeric and metal components.

The use of "ideal shape" (streamlined) rubber flexible elements is very promising for the design of CNF vibration isolators. Such elements (spheres, ellipsoids, toruses, radially-loaded cylinders) are naturally nonlinear in compression and their nonlinear load-deflection character-istics result in quasi-CNF load-natural frequency characteristics. Production of such elements requires relatively simple molds. By changing dimensions/shapes and the selection of the rubber blend, practically any natural frequency and/or rated loads can be easily realized. Preferably, such elements are not bonded to metal or other rigid components since such bonding requires cutouts and other distortions in their shapes, which may introduce stress concentrations. The latter negate or reduce the advantages of the ideal shape elements as described in Section 3.3.2.

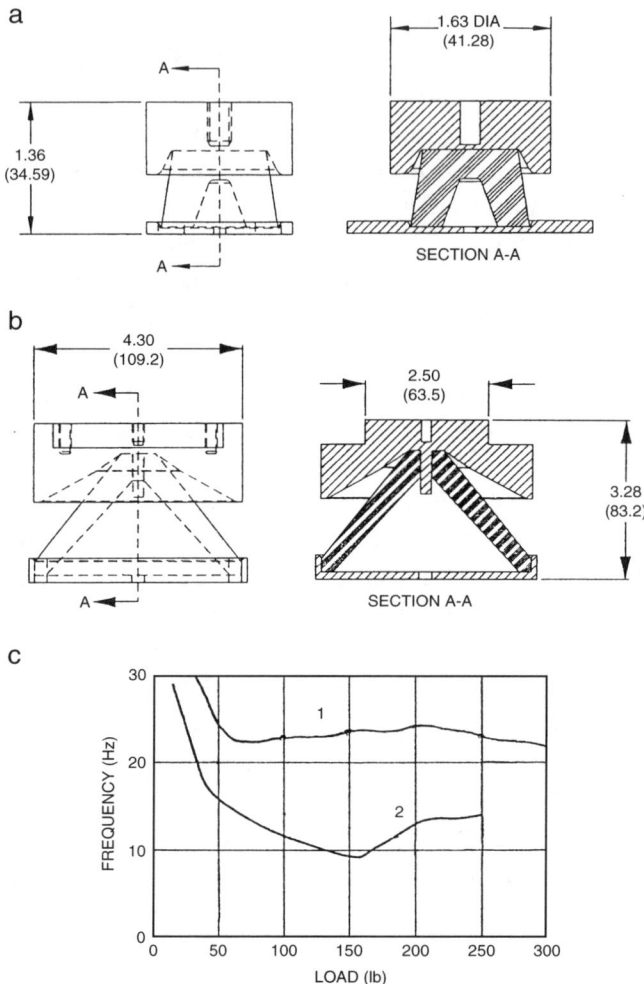

Fig. 4.5.10. Polyurethane CNF Isolators: (a) NDO1-A; (b) ND20-A; (c) Load-Natural Frequency Characteristics (1—NDO1-A; 2—ND20-A) (*Newport Corp.*).

Vibration isolators Lastosphere from Lord Corp. [4.20], (Fig. 4.5.11), comprise a rubber body of revolution having a cylindrical midsection with semi-spherical ends and loaded in compression along its major axis. The line contains four isolators with the maximum static rated load P_{max} = 2,000, 3,500, 6,000, and 10,000 lbs, and a maximum total (static + dynamic) rated load that is four times greater. The unloaded height of these isolators and nominal natural frequencies $f_{z_{nom}}$ are,

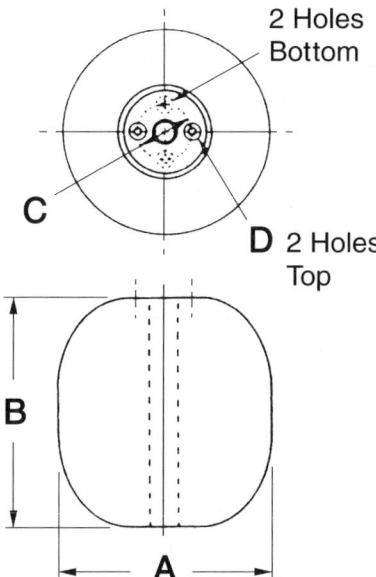

Fig. 4.5.11. Vibration Isolator Lastosphere with Streamlined Shape Rubber Flexible Element (*Lord Corp.*).

respectively: $B = 3.875$ in. (98.4 mm), $f_{z_{nom}} = 5.5$ Hz; 5.187 in. (132 mm), 5.1 Hz; 6.75 in. (171.5 mm), 4.4 Hz; 8.687 in. (221 mm), 4.0 Hz. The diameter of a non-loaded isolator $A \approx 0.97\ B$, which implies that the cylindrical section of the flexible element is rather short. The range $P_{max}:P_{min}$ varies from 2:1 for the smaller isolator to 2.5–3.0:1 for the largest ones. Deformation at the greatest rated static load P_{max} is $\Delta_{max} = (0.41–0.46)B$ and the creep/set of the isolator is (8–18)% of Δ_{max}. These isolators are used for isolation of low-speed equipment and machinery and for vehicle suspensions.

Although this isolator is close to the "ideal shape," the presence of the mounting plates at both ends may create stress concentrations, albeit not very intense. Also, the axially loaded cylindrical section of the isolator and potential discontinuities in transition from the cylindrical and semi-spherical sections may have detrimental effects on performance.

Still, the rated performance characteristics of the Lastosphere vibration isolator are exceptionally good and are proven/validated by many years of its commercial availability. No other commercially available vibration isolator loaded in compression by the weight load has a compression deformation that exceeds 15% under the rated load. But in addition to

40^+ % compression deformation under the rated static load, this isolator can be exposed to continuous dynamic/vibratory loads whose magnitudes can be four times (!) greater than already unconventionally high maximum static loads.

Even better performance can be expected from a geometrically "pure" ideal shape flexible rubber element. Fig. 3.3.8 shows the load-natural frequency characteristic of a rubber O-ring (torus) with outer diameter $D = 2$ in (51 mm) and cross-sectional diameter $d = 0.5$ in. (12.7 mm) made from elastomer with durometer $H = 30$. The nominal natural frequency $f_{z_{nom}} = 12$ Hz ± 5% is maintained in the load range 50–240 lbs, $P_{max}:P_{min} = \sim 5:1$, but this range can be extended since the O-ring was tested only up to $\sim 35\%$ relative compression. Conservatively, Δ_{max} can be at least $(0.4-0.45)d$ since the O-ring does not have stress concentrations, while the isolator in Fig. 4.5.11 has some; such an increase in the maximum static compression would result in $P_{max}:P_{min} = \sim 7-8:1$ for the rubber toruses.

Similar characteristics, with about same $P_{max}:P_{min}$, were obtained with rubber spheres used as flexible elements, e.g., in [4.29]. It was established that for very soft silicone rubber, the measured values of the vertical natural frequency are $f_z = 7.5-8$ Hz for $D = 1.5$ in; ~ 6.3 Hz for $D = 2.0$ in; ~ 5.5 Hz for $D = 2.5$ in. The last data can be compared with the smallest Lastosphere isolator also having $f_{z_{nom}} = 5.5$ Hz, but with 3.9 in. height vs. 2.5 in. for the spherical isolator. This difference might be explained also by higher K_{dyn} for rubber used in Lastosphere isolators.

It was shown in [4.2] that the load-natural frequency characteristic can be modified by varying contact conditions between the flexible element and the loading surfaces (no treatment of the loading surfaces; lubricated loading surfaces; immersion into foam; etc.). For the same contact conditions, $f_{z_{nom}}$ is inversely proportional to the root square of the diameter taking deformation under load (cross-sectional diameter d for an O-ring, sphere diameter D for a sphere, etc.). Thus, in order to reduce $f_{z_{nom}}$ in half, the diameter has to be quadrupled, e.g., $f_{z_{nom}} = 6.3$ Hz for $D = 2$ in. if $f_{z_{nom}} = 12$ Hz for $d = 0.5$ in.

O-rings or radially-loaded rubber cylinders can accommodate very large weight loads. If lower natural frequencies f_z are required, the cross-sectional diameter of the element must be increased according to the statement in the above paragraph, thus even larger forces are required for deforming the flexible element, often much greater than the weight loads of the objects to be isolated. *Although the required forces for spherical*

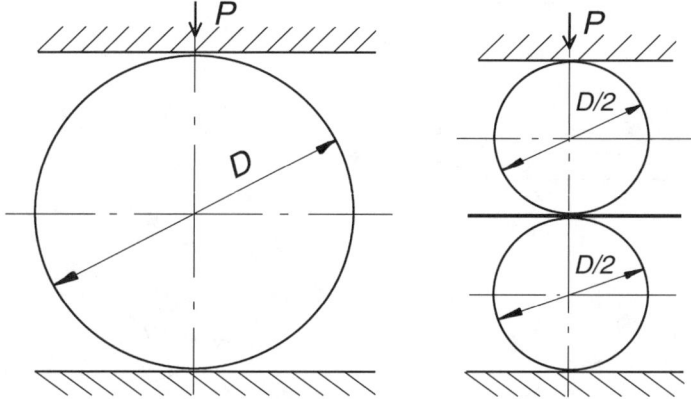

Fig. 4.5.12. (a) Straightforward; and (b) "Split Diameter" Flexible Element with Rubber Spheres.

elements are much smaller, they may become excessive if low natural frequencies, associated with the large D, are required. Also, due to such interdependence between the realized natural frequencies determining the diameter and the required forces, the volume of the flexible element can become excessive. Several approaches can be used to resolve this problem. In one approach, spheres are used instead of radially-loaded cylinders or O-rings. The force required for a given relative compression of a sphere with diameter D, is about three times smaller than the force required for the same radial deformation of a cylinder made from the same material and having diameter d and length l, so that $D = d = l$ [see Eqs. (3.3.10) and (3.3.11)].

Another approach is to use softer (lower durometer) materials. While this approach allows construction of vibration isolators for lower rated loads, it does not lead to dimensional changes since the critical dimensions (D or d) are determined by the required $f_{z_{\text{nom}}}$.

The third approach is to "split the diameter." If diameter D of the ideal shape rubber element is needed for realizing the required $f_{z_{\text{nom}}} = f_o$, then the same shape element with diameter $D/2$ would result in $f_{z_{\text{nom}}} = f_o\sqrt{2}$ and two identical elements having diameter $D/2$ each and connected in series, as in Fig. 4.5.12(b), would result in $f_{z_{\text{nom}}} = f_o$. For the same surface conditions of the loading surfaces, the critical loads $P_{\max}$ and $P_{\min}$ are proportional to the load-application area A, or to dl for cylinders and D^2 for spheres [see Eqs. (3.3.10) and (3.3.11)]. Thus, the CNF load range for the embodiment in Fig. 4.5.12(b) will be defined by loads that are four

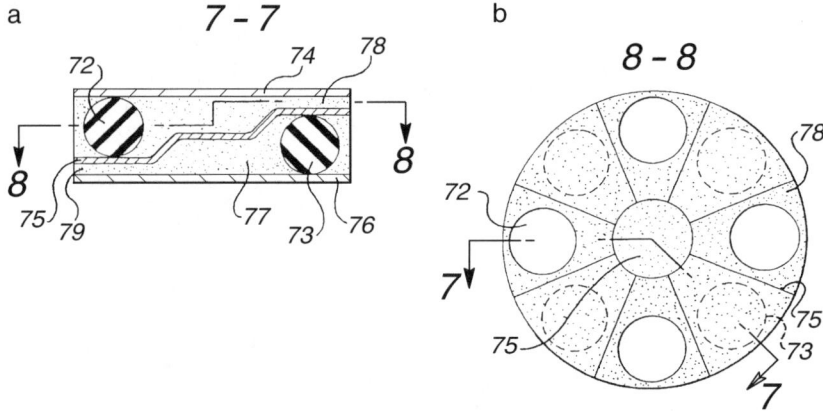

Fig. 4.5.13. Rubber Flexible Element with Staggered Ideal Shape Components.

times smaller if the elements are rubber spheres of $D/2$ diameter, or two times smaller if cylinders of $d/2$ diameter and the same length l as in Fig. 4.5.12(a) are used in Fig. 4.5.12(b). The total volume (weight) of two rubber flexible elements in Fig. 4.5.12(b) is 1/4 of the volume (weight) of the element in Fig. 4.5.12(a) if spheres are used, or 1/2 of the volume (weight) of the Fig. 4.5.12(a) element if the same length cylinders are used. This concept can be further extended by "splitting" one element into three or more smaller elements with a corresponding reduction in the isolator dimensions. In special cases, it may be desirable to use individual flexible elements in the "split" assembly, which are made from different rubber blends having different δ and/or K_{dyn}.

While the lateral dimensions of the flexible element for the required $f_{z_{nom}}$ can be significantly reduced by this approach, the dimension along the weight load application axis does not decrease; it can even increase marginally if intermediate spacer(s) between the smaller flexible elements are used. Of course, the overall height of an "ideal shape" flexible element is already much smaller than a comparable (by rated loads and achievable f_z) bonded flexible rubber element loaded in compression due to the noted differences between the allowable relative compression under the rated static load (40–45% vs. 10–15%).

An additional height reduction can be achieved by staggering the small-diameter flexible elements in the "split diameter" isolators in Fig. 4.5.12(b). In the staggered arrangement in Fig. 4.5.13 [4.30], rubber

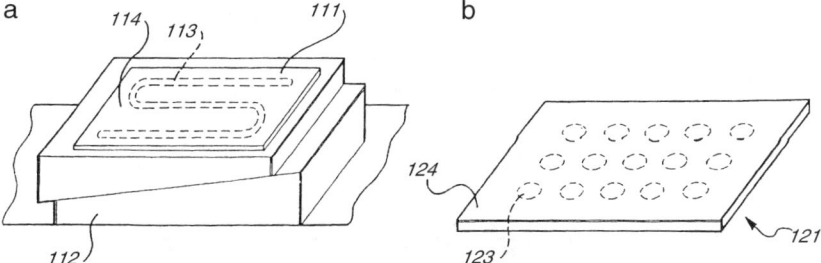

Fig. 4.5.14. Ideal Shape Rubber Mats with: (a) Cylindrical Rubber Cord; and (b) Spheres.

spheres (or other ideal shapes) 72 are placed between top cover 74 and intermediate insert 75, and rubber spheres (or other ideal shapes) 73 are placed between intermediate insert 75 and bottom cover 76. Fig. 4.5.13(b) shows the axial view of the arrangement. When an axial (weight) force is applied between covers 74 and 76, spheres 72 are compressed with the maximum compression deformation being limited by the size of clearance 78. Similarly, spheres 73 are compressed within clearance 79. Thus, the arrangement of Fig. 4.5.13 results in a series connection of spheres 72 and 73, which is the same as in Fig. 4.5.12(b) but with a reduced axial dimension. If the isolator is designed for static axial loads, without large dynamic components (the case of isolation of vibration-sensitive objects), then clearances 78 and 79 must accommodate the maximum allowable deformation of components 72, 73, e.g., (0.4–0.45)D and provide for $\sim$0.05D safety allowance. This type of arrangement results in the total height of the flexible element (assuming that thickness of intermediate insert 75 is small) to be about 1.5D rather than 2D in the case of Fig. 4.5.12(b). This difference is quite significant, especially for low $f_{z_{nom}}$ isolators with $D = 1.5$–2.5 in. (40–63 mm). More than two levels of rubber elements 72, 73 can be used [e.g., as shown in Fig. 3.3.8(a),(b), where a combination of spheres 92 and O-rings 93 is used]. A small mass of isolators with streamlined rubber elements improves their high frequency performance (per Section 1.10).

Values $\eta_{x,y}$ for ideal shape rubber elements are in the range of $\eta_{x,y} = 5$–9. These values can be reduced if the lateral deformation of the elements is restrained by rigid (metal) inserts, e.g., by inclined areas of inserts 75 for elements 72 and 73 in Fig. 4.5.13(a), provided that these inclined areas are at a properly selected distance from elements 72, 73.

360 ● PASSIVE VIBRATION ISOLATION

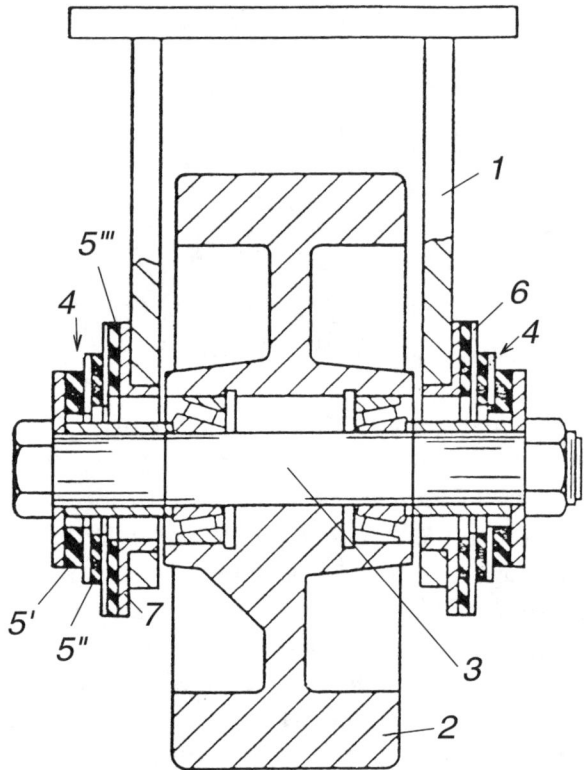

Fig. 4.5.15. Wheel Caster with Shear Disc Suspension.

Since the ideal shape rubber elements may lose some of their advantageous properties if bonded to rigid components of the isolator, the most difficult problem in designing isolators with ideal shape rubber flexible elements is maintaining a desired packaging/interrelation of rubber and rigid (metal) components of the isolator. It is proposed in [4.30] to achieve this goal by using a relatively soft foam matrix, properly positioning all rubber and metal components without a noticeable restraint in the deformation and bulging. This foam matrix is visible in Fig. 4.5.13 and can be used for any design configuration of isolators as well as CNF mats/pads employing rubber cord 113 [Fig. 4.5.14(a)] or spheres 123 [Fig. 4.5.14(b)], or other ideal shape components; foam matrix is marked 114 in Fig. 4.5.14(a) and 124 in Fig. 4.5.14(b).

A different design of rubber CNF vibration isolator using shear deformation is proposed in [4.31], [4.32] and commercialized by Lord

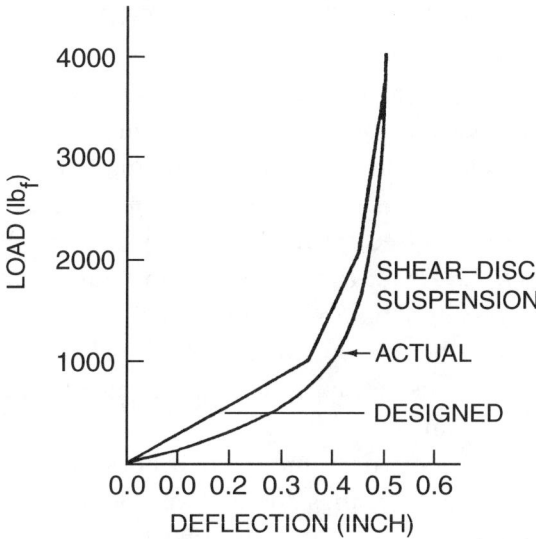

Fig. 4.5.16. Load-Deflection Characteristic of Shear Disc Suspension.

Corp. [4.20] for suspension of caster wheels for small vehicles (e.g., in-plant trailers; Fig. 4.5.15). Axle 3 of wheel 2 is connected to caster 1 via two so-called shear discs 4. Each shear disc 4 consists of two or more rubber layers 5 sandwiched between metal discs 6. The holes and outer diameters of metal discs 6 and respective rubber layers 5 are progressively smaller toward the ends of axle 3. The inner surfaces of holes in metal discs 6 will successively land on the axle surface as the shear deformation of rubber layers 5 increases with the increasing vertical (e.g., weight) load. Consequently, the shear deformation of the outer layers 5′, then 5″, etc., stops and the stiffness of the suspension increases (hardening nonlinear characteristic). The number and sizes of rubber layers 5 are determined in accordance with the desired payload range and the desired dynamic characteristics of the suspended vehicle. If the load is excessive, bushings 7 land on axle 3, thus preventing excessive shear deformation of shear discs; this feature is especially important for an in-plant trailer, which can be used in a storage mode with a supported weight that is much greater than the maximum rated payload.

Calculated and actual load-deflection characteristics of the shear disc suspension for in-plant trailers are shown in Fig. 4.5.16 [4.32]. The natural frequencies of the system are $f_z = 8.5$ Hz for load on one caster $P_1 = 250$ lbs (empty trailer); $f_z = 7.2$ Hz for $P_1 = 750$ lbs (loaded trailer); and $f_z = 9.3$ Hz for $P_1 = 1,500$ lbs (heavily loaded trailer). This CNF characteristic resulted in a more than a decimal order of magnitude reduction of dynamic loads on the wheels and on the trailer structure, and in 16–18 dBA noise reduction (noise levels without suspension can reach 110 dBA). The highest noise levels were observed from a moving empty trailer, thus the above relatively low f_z was necessary for the empty trailer conditions. If a linear suspension were used, providing $f_z = 8.5$ Hz for $P_1 = 250$ lbs, it would result in $f_z \approx 3.5$ Hz for $P_1 = 1,500$ lbs, and static deformation of the suspension would be over 0.8 in. (20 mm) at $P_1 = 1,500$ lbs. Such low f_z would not lead to further significant dynamic loads/noise reduction, but such large deformations complicate connecting (hitching) differently loaded trailers in a train.

The dynamic load reduction due to incorporation of the suspension also resulted in significant savings due to a ten-fold reduction of failures of the axle bearings and of tread wear of the wheels. Thus, the intended noise abatement has also become a money-saving project.

The shear disc CNF isolators in Fig. 4.5.15 have high anisotropy between fore-aft and lateral stiffness values, which is useful for stability of the suspended vehicle and might be desirable in some cases of vibration isolation of other objects.

4.6 VIBRATION ISOLATORS WITH METAL FLEXIBLE ELEMENTS

4.6.1 General Comments

There are two most widely used groups of vibration isolators with metal flexible elements. One group includes isolators with coil springs, usually designed to provide low natural frequencies and intended for installation either directly under the isolated object ($f_z = 3-10$ Hz) or under inertia blocks ($f_z = 1-5$ Hz). Another group employs wire-mesh/cable flexible elements ($f_z = 13-30$ Hz), sometimes combined with coil springs ($f_z = 1.7-15$ Hz). These isolators are designed for high-amplitude vibrations and for harsh environmental conditions. In addition to these large groups, metal flexible elements are used for special applications wherein high predictability and consistency of the mechanical characteristics of metals, together with the possibility of precision machining of the components, are used for special applications, e.g. [4.4].

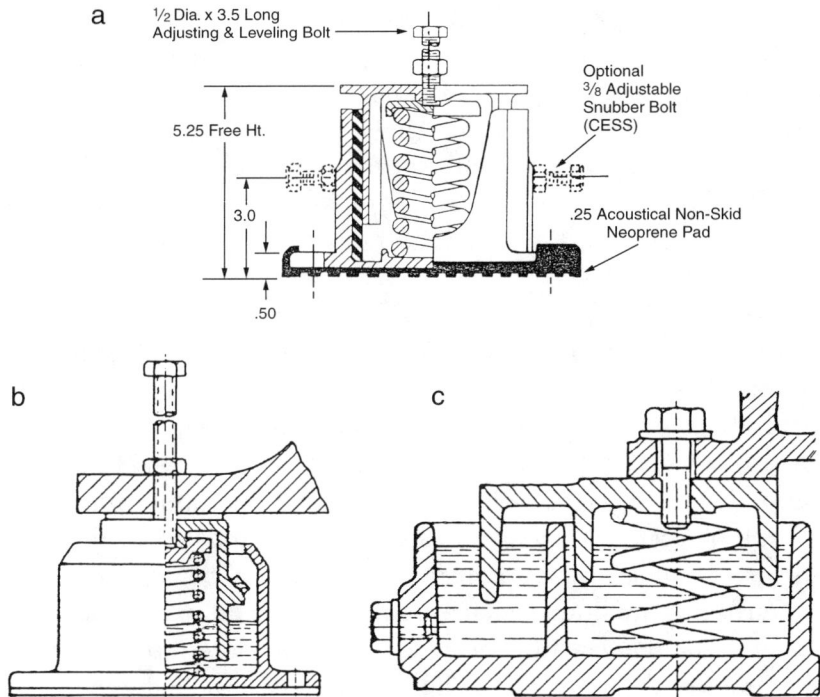

Fig. 4.6.1. Coil Spring Isolators: (a) Isolator CE-1 (*Korfund Dynamics*); (b), (c) Coil Spring Isolators with Oil Dampers.

4.6.2 Vibration Isolators With Coil Springs

With the exception of the low-load conical spring isolators shown in Fig. 4.5.4, commercially available coil spring isolators employ steel cylindrical constant pitch/constant stiffness springs and are made for large loads, up to 560,000 N (125,000 lbs) per isolator. These types of isolators are used in cases where frequent maintenance of isolators is difficult and thus, undesirable, e.g., under large concrete foundation blocks or, sometimes, under buildings. Practically, the only environmental hazard for the coil spring isolators is corrosion, which can be prevented by using a protective coating on the spring.

Coil springs have $K_{dyn} = \sim 1.0$, thus they have higher static stiffness for a specified natural frequency f_z than rubber and plastic elements. This is especially important for low f_z cases where the required stiffness is naturally low and the system may develop a rocking motion even with $K_{dyn} = \sim 1.0$. Use of materials with $K_{dyn} > 1.0$ leads to further increase of the rocking amplitudes of the isolated object. When the rocking intensity becomes unacceptable, an inertia mass must be used, which leads to an increase in the installation costs.

Another advantage of steel coil spring isolators for low f_z applications is their large allowable compression deformation as a fraction of the initial height, usually ~30–40%. Since smaller (shorter) isolators can be used, it reduces the effective C.G. height a_z and thus further adds to reduction of the rocking amplitudes.

Coil spring isolators are used for slow speed devices such as heating, ventilation and air-conditioning (HVAC) equipment, compressors, motor generators, pumps, etc. Many of these devices have unidirectional (vertical) dynamic force excitations. This fact is utilized in the design of typical isolators, like the one shown in Fig. 4.6.1(a) [4.33]. Since the excitation from which the isolator is protecting the supporting structure and/or the object has only significant vertical components, the rocking motion is restrained by an adjustable rubber snubber that comes into action at rocking amplitudes that are considered excessive by the adjuster. The lateral snubbing action normally does not interfere with vertical deformations of the spring. The isolator does not have a damper: on one hand, this would lead to deterioration of the high-frequency isolation but, on the other hand, it would also somewhat alleviate the rocking motion of the object.

Heavy steel springs exhibit intense high-frequency (wave) resonances in the spring material even at relatively low frequencies, as illustrated in Section 1.10. As a result, there may be a high transmissibility of objectionable high-frequency vibrations and noise. This effect is somewhat alleviated by elastomeric "acoustical," Fig. 4.6.1, or "sound" pads, usually attached to the bottom surface of the coil spring isolators. The effect of these pads is limited by their low thickness (usually 5–6 mm). Thicker pads cannot accommodate high specific pressures from the weight forces and would also increase the rocking amplitudes of the isolated object.

The lowest f_z (at the maximum rated load) of CE series of isolators shown in Fig. 4.6.1(a) is about 3 Hz. The rated load is adjusted by the number of springs in the isolator, from one in CE-1 isolator in Fig. 4.6.1(a) to seven in CE-7 isolators. The maximum rated load of CE-7 isolators is 10,500 lbs (47,000 N), with the footprint 6.7 × 9.75 in. (170 × 246 mm).

Passive Vibration Isolation Means • 365

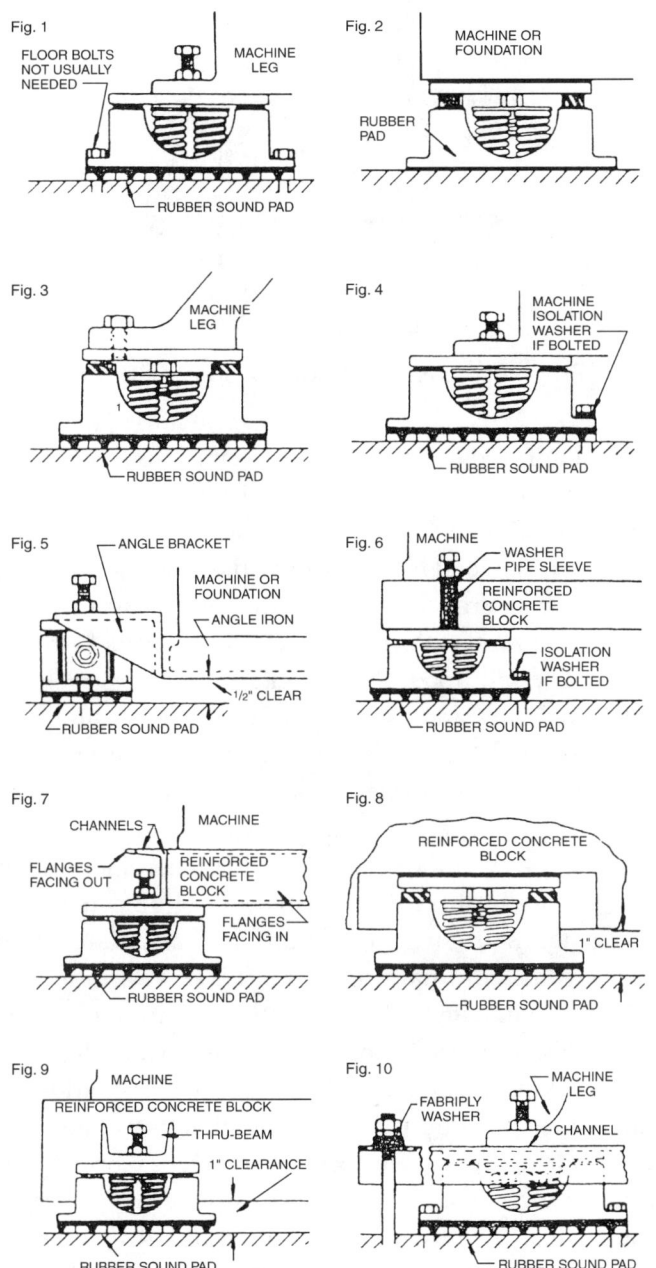

Fig. 4.6.2. Typical Installation Configurations for Coil Spring Isolators (*Korfund Dynamics*).

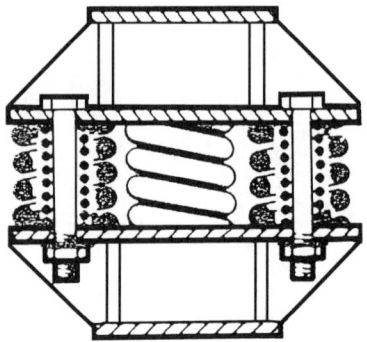

Fig. 4.6.3. Precompressed Coil Spring Isolator (*Vibrachok USA*).

Typical installation patterns of coil spring isolators are shown in Fig. 4.6.2 [4.33]. Sometimes the coil spring isolators are packaged in units like those in Fig. 4.6.3 [4.34], wherein the springs are precompressed by lug bolts to apply initial load to the springs; the initial load is slightly lower than the minimum operational load. After installing under the object/foundation, the bolts are relieved by the additional deformations of the springs under the weight load. Such packaging arrangement reduces the initial height of isolators and thus simplifies their handling.

Basic isolators with coil springs, such as the one in Fig. 4.6.1(a), have very low damping. While in some cases, such as installation of low-speed objects, it might be a beneficial feature since it improves the isolation efficiency, in many cases some damping is very desirable. High damping is important for low natural frequency isolation of objects that have high-speed rotors, since start-stop operations are associated with passing through quasi-resonance situations (Section 2.3). Vibration amplitudes in such regimes can be dangerously large unless brought down by damping in the isolation system. Damping is also important for reducing the rocking motions of the isolated object. Vibration isolation of impulse-producing machinery (Section 2.3) also benefits from higher damping in the isolators, resulting in a faster decay of transient vibrations induced by the pulses. Because of this, coil spring isolators are often equipped with viscous dampers, e.g., as shown in Fig. 4.6.1(b),(c) [4.7]. While simple and inexpensive, such dampers are prone to spills and contamination of the oil and are not maintenance-free.

When coil spring isolators are installed under inertia blocks (usually, reinforced concrete blocks), rubber blocks are often placed next to and

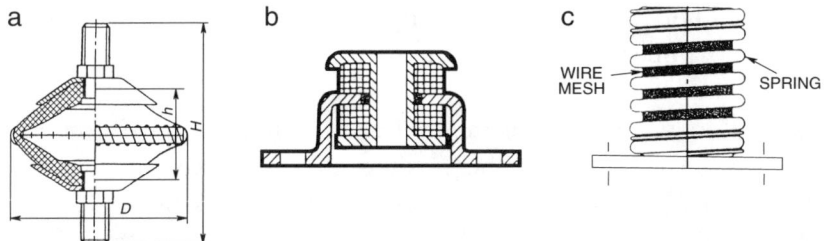

Fig. 4.6.4. Wire-Mesh Vibration Isolators: (a) "Double Bell" Isolator; (b) Typical Isolator with Wire-Mesh Pads in Compression (*Vibrachok USA*); (c) Composite Coil Spring/Wire-Mesh Isolator (*Vibrachok USA*).

parallel to the isolators and serve as dampers. In such designs there is a contradiction since the resulting damping (log decrement) is

$$\delta = \frac{k_1 \delta_1 + k_2 \delta_2}{k_1 + k_2}, \qquad \text{Eq. (4.6.1)}$$

where k_1, δ_1 are stiffness and damping of the isolators ($\delta_1 \approx 0$ for coil springs); and k_2, δ_2 are stiffness and damping of the rubber blocks. The damping effect of the rubber block is proportional to its stiffness, but since K_{dyn} for medium- and, especially, high-damping rubber is usually significantly larger than one, increasing the stiffness of rubber blocks in order to enhance their damping effect may negate one of the advantages of the coil-spring isolators—their low K_{dyn}. Thus, very high-damping (large δ_2) rubber blends have to be used, but they usually have high K_{dyn}, and bring back the same problem.

Wire-mesh dampers are sometimes placed inside the coil springs, e.g., see Fig. 4.6.4(c). While such dampers are more expensive, they reduce the size of the isolators and exhibit the highest damping at large (usually resonance or transient) amplitudes, while not significantly impairing high frequency isolation.

The main advantages of the coil spring isolators—low K_{dyn} and large allowable compression—are challenged now by elastomeric streamlined (ideal-shape) isolators, as described in Section 4.5. If low-filled rubber elements (preferably, high load capacity radially compressed cylinders) are used, their K_{dyn} can be very low, $K_{dyn} = 1.05$–1.15. Still, these rubber blends may have $\delta = 0.1$–0.2, close to the damping values for the coil spring isolators with dampers. The allowable deformation is ~40%

of the initial height, thus matching the coil spring isolators. At the same time, such rubber flexible elements can be made even smaller if configurations like those in Fig. 4.5.13 are used. The high-frequency resonances in the rubber elements occur at much higher frequencies due to the lower mass of the flexible elements. Nonlinearity of the streamlined rubber elements provides CNF characteristic as well as self-snubbing capabilities.

4.6.3 Wire-Mesh and Cable Isolators

Properties of wire-mesh and cable materials are described in Section 3.3.1A. Vibration isolators with such flexible elements, usually made from stainless steel, do not change their characteristics in a broad temperature range ($-90C°$ to $+400C°$ [4.34]) and in harsh environmental conditions. These types of vibration isolators are especially suitable in cases where there are very intense vibrations, because in such conditions these isolators exhibit very high damping and relatively small K_{dyn} (e.g., see Fig. 3.3.1 and Table 4.1). Wire-mesh and cable elements have excellent fatigue and creep durability necessary for accommodating large vibration amplitudes. On the other hand, it can be seen from Fig. 2.2.20 that installation of a precision surface grinder on wire-mesh mounts results in relative vibration amplitudes in the work zone (between the wheel and the table) far exceeding the amplitudes for the same machine installed on isolators with rubber flexible elements. Although these materials are not as versatile as elastomers, their stiffness characteristics in various directions can be, to a large extent, shaped by design. While wire-mesh isolators are available in a broad range of natural frequencies and rated loads, cable isolators are more specialized for use in cases of large required deformations wherein both vibration and shock protection are desirable. They are available in more narrow ranges of natural frequencies and rated loads.

As was stated in Section 3.3.1A, wire-mesh materials have a rather paradoxical combination of nonlinear hardening load-deflection characteristic for static loading, and amplitude dependence of dynamic stiffness typical for nonlinear systems that have a softening load-deflection characteristic. The static nonlinearity is not as steep as that required for a CNF isolator; change of the weight load in the ratio 2.5:1 usually results in the ratio of natural frequencies ~1:1.5. On the other hand, the dynamic nonlinearity is very pronounced, both for stiffness and damping. To reflect this fact, the dynamic stiffness k_{dyn} values for a "double bell" isolator mod. DK, Fig. 4.6.4(a), [4.35] are presented in the catalog table as a coefficient A (in N/m$^{0.5}$) in the expression

$$k_{dyn} = A/a^{0.5} \qquad \text{Eq. (4.6.2)}$$

This expression differs slightly from Eq. (3.3.1). For the smallest isolator, DK-I-1 (rated weight $P_r = 5$ N, $D \times h = 32 \times 18$ mm), $A = 190$; for a medium one, DK-III-1 ($P_r = 200$ N, $D \times h = 53 \times 36$ mm), $A = 1,400$; and for the largest one, DK-V-2 ($P_r = 1,500$ N, $D \times h$ 117×87 mm), $A = 12,640$. For the vibration amplitude $a = 1$ mm $= 0.001$ m, these values correspond to $f_z = 17.3, 7.4$, and 8.1 Hz respectively, while for $a = 0.5$ mm, $f_z = 20.6, 8.8$, and 9.7 Hz. These numbers show a very significant scatter of f_z depending on the size of the isolator and on the vibration amplitude. Unfortunately, other catalogs do not give such specific information on stiffness constants and their amplitude dependencies, but only on nominal values of f_z, which can be very far from the real numbers.

The steep amplitude dependence of dynamic stiffness of wire-mesh and cable isolators may significantly contribute to the undesirable intra-coordinate coupling in the natural coordinate frame (Section 1.3), since actual vibration amplitudes at different mounting points can be different, thus changing the stiffness values.

The isolator in Fig. 4.6.4(a) has top and bottom covers of a special shape enveloping wire-mesh "bells." These covers distribute compressive weight load to the bells and also restrain their lateral deformations, thus resulting in reduction of stiffness ratios $\eta_{x,y}$.

The majority of commercially available isolators with wire-mesh flexible elements employ ring-shaped wire-mesh pads loaded in compression, e.g., as shown in Fig. 4.6.4(b) [4.34]. Nominal natural frequencies of vibration isolation systems with such and similar isolators are in the range of $f_z = 13-25$ Hz.

Very high damping of wire-mesh materials, especially at large vibration amplitudes, is utilized in composite coil spring/wire-mesh isolators, like the one shown in Fig. 4.6.4(c). Presence of the coil spring allows reduction in both magnitude of K_{dyn} and its amplitude dependence, as well as realization of low natural frequencies. The presence of wire-mesh inserts (sometimes made shorter than the initial height of the coil spring and

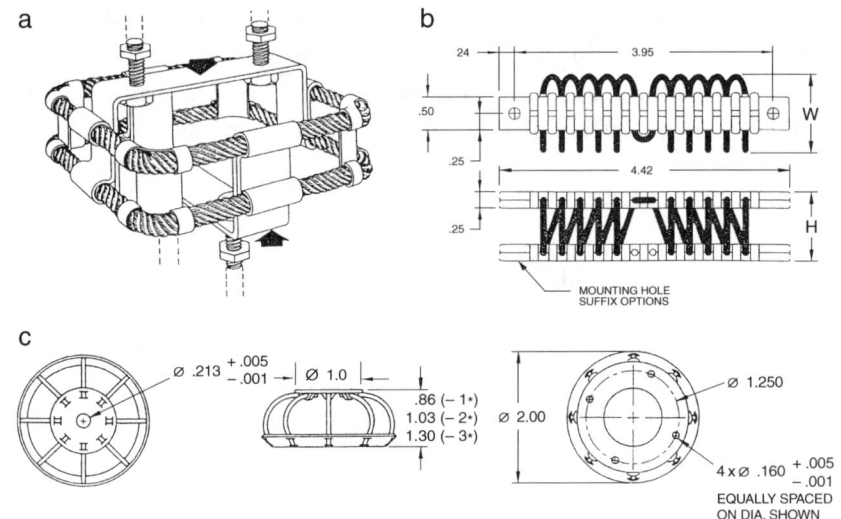

Fig. 4.6.5. Cable Isolators: (a) Basic Design; (b) Helical Coil Isolator Mod. C3 (*Aeroflex International*); (c) "Circular Arch" Isolator Mod CCA2 (*Aeroflex International*).

"coming to action" only after a certain deformation of the coil spring) results in higher damping while maintaining a compact self-contained structure. Such composite isolators are available from $f_z = 1.7$–2.1 Hz ($P_r = 600$–$100{,}000$ N) to $f_z = 8$–10 Hz ($P_r = 15$–$110{,}000$ N) [4.34].

Cable or *wire rope* isolators use stainless-steel cables that have dynamic characteristics similar to the wire mesh. However, since the cables are used as 3-D structures, Fig. 4.6.5(a) [4.36], Fig. 4.6.5(b),(c) [4.37], rather than as wire-mesh pads, static load-deflection characteristics of such isolators are very different. Instead of hardening load-deflection characteristics, cable isolators have softening load-deflection characteristics due to so-called controlled buckling of the structure (Fig. 4.6.6 [4.37]), which is especially beneficial for shock isolation. Because of this feature, cable isolators are often used in cases where both vibration and shock protection are required, e.g., for packaging fragile components into containers. The softening load-deflection characteristic results in the natural frequency decreasing with the increasing weight load even steeper than for the constant stiffness isolators, e.g. see Fig. 4.6.7 [4.38]. The weight load change from 1 to 4 lbs results in f_z change from 22 to 7Hz. This characteristic, directly opposite to the preferred CNF characteristic, may

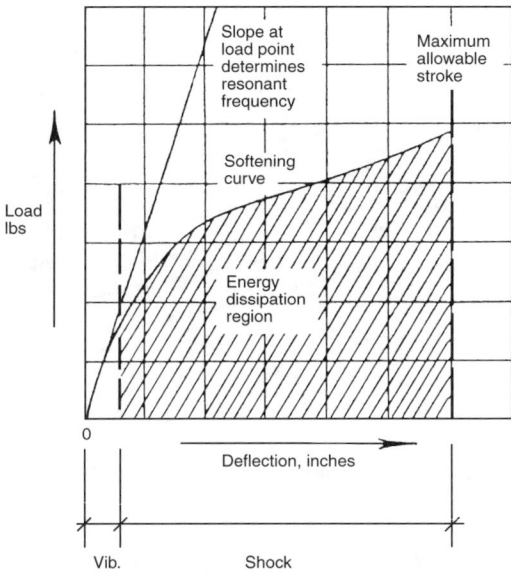

Fig. 4.6.6. Load-Deflection Characteristic of Helical Coil Isolator in Compression (*Aeroflex International*).

result in an enhanced inter-coordinate coupling in the vibration isolation system and in a need to use much softer isolators (or lower f_z) than could be used if CNF isolators were employed.

Fig. 4.6.5(a) shows one of the first designs of cable isolators for use onboard a spacecraft. The most widely used cable isolators are isolators with helically coiled cables, Fig. 4.6.5(b) [4.37]. The flexible element is a coil spring coiled from a steel cable instead of wire [4.39]. The coils are clamped by two diametrally placed pairs of steel strips with grooves accommodating the coils. Each pair is bolted together, thus ensuring maintenance of the 3-D shape. One pair of the strips is attached to the isolated object, another to the supporting structure, and the weight load is applied laterally—somewhat similar to the system shown in Fig. 3.1.3. The isolator allows compression deformation up to 50% and this can be used to realize low natural frequencies $f_z \leq 5$ Hz and log decrement $\delta = 1-2$. Off-the-shelf isolators have rated loads up to and exceeding 15,000 lbs (70,000 N). The shear stiffness is usually 0.25–0.4 of the compression stiffness ($\eta_z = 4.0-2.5$). The longitudinal stiffness is usually lower than the lateral stiffness.

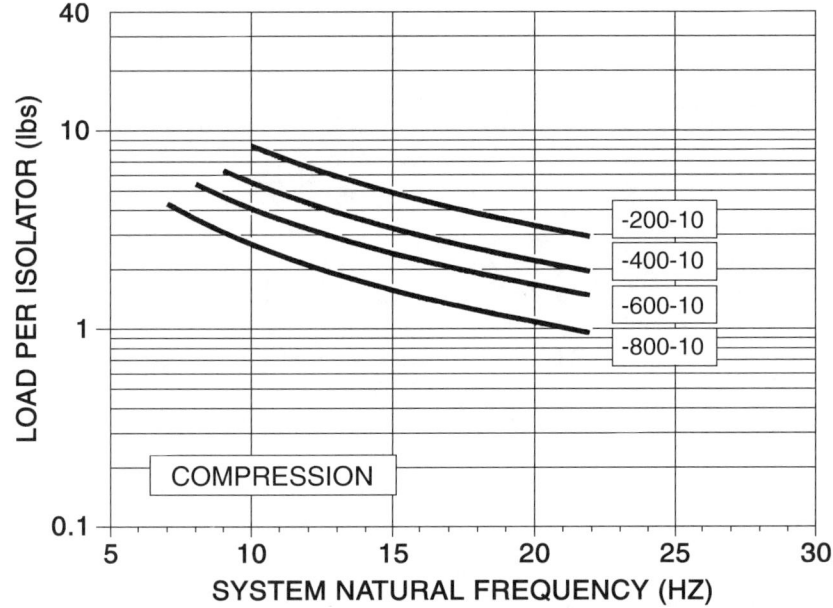

Fig. 4.6.7. Natural Frequency-Load Characteristic of Cable Isolators (*Enidine Inc.*).

"Circular arch" cable isolator, Fig. 4.6.5(c) [4.37] have higher lateral (shear) stiffness in both the x- and y-directions than other designs, about 0.4–0.6 of the compression stiffness, $\eta_{x,y} = 2.5-1.6$.

4.7 HIGH ANGULAR STIFFNESS AND ANISOTROPIC VIBRATION ISOLATION SYSTEMS

Many cases of vibration isolation installations have to deal with intense horizontal excitations generated either inside the isolated object or transmitted from the supporting structure. Examples of the former include both vibration-sensitive objects, such as coordinate-measuring machines or surface grinders isolated from the floor vibrations, or vibration producing machines, such as looms or cold headers. The latter is typical for civil engineering structures subjected to earthquake excitations whose horizontal components are usually much more intense than their vertical components. If these objects are supported by conventional vibration isolators having commensurate, albeit possibly quite different, values of stiffness constants in the vertical and horizontal directions, these horizontal excitations often result in intense rocking motions of the object—potentially associated with excessive amplitudes at the points of the object located far from the center of rocking as expressed by Eq. (1.6.5)—e.g., on higher floors of a building constructed on vibration isolators for earthquake protection or at the CMM probe. Such intense rocking motions are very unpleasant to feel and/or to observe, and they can cause significant harm. Precision coordinate-measuring machines, surface grinders, etc., have to be slowed down since the intense rocking motions do not fade enough before the next work event (probe-part contact in a CMM, new grinding pass in a surface grinder, new exposure in a photolithography stepper, etc.) commences. These motions are transmitted into the work zone and may negate the positive effects of vibration isolation. In vibration producing machines, such as looms or cold headers, intense rocking vibrations can result in loosening of structural connections, accelerated wear, etc. Large vibration amplitudes on the higher floors of earthquake-isolated buildings are associated with high dynamic forces resulting in dangerous displacements of heavy objects, or even in structural damage. The rocking motions are also undesirable for mounting sensitive gyroscopes, which may require high degree of vibration isolation in translational directions, but whose performance may degrade if the angular stiffness is inadequate.

The situation can be corrected if the intensity of the rocking motion is reduced. This type of effect can be achieved by a total decoupling in the vibration isolation system and/or by increasing angular stiffness of the isolation system in the vertical plane wherein the rocking motion is undesirable. The decoupling of the rocking modes (x-β and y-α) can be achieved by satisfying Eq. (1.3.8) (placing isolators in the plane containing C.G.) or by using the system with inclined isolators whose parameters satisfy expressions given in Section 1.6.3. The first approach (placing isolators in the C.G. plane) may require changes in the object design, often undesirable, unless an inertia block is used (see Section 4.10). The second approach (inclined isolators) is not always practical since it requires custom-made isolators with rigidly defined parameters. Even when such isolators are available, the inevitable scatter of their stiffness constants due to tolerances, as well as deviations of isolator installation angles and inaccuracies in determining the C.G. coordinates, and changing mass distribution in the object, result in some residual inter-coordinate coupling.

Increasing angular stiffness of the isolation system is, in many cases, a more straightforward, realizable, and also a more robust approach. The expression for the angular stiffness is given in Eq. (1.2.7). It can be seen that the angular stiffness in the vertical plane x-z is determined by three parameters: effective distances a_{xi} (in the x-direction) between the isolators and the projection of C.G. onto x-axis; angular stiffness $k_{\alpha i}$ of isolators; vertical stiffness k_{zi} of the isolators. If the angular stiffness of the isolation system increases, the vibratory modes x-β and y-α become less and less dynamically coupled, separating into uncoupled horizontal (x, y) and angular (β, α) modes. The latter are characterized by relatively high natural frequencies and usually are not very important for performance of the isolation system unless there are intensive angular excitations and/or high sensitivity of the isolated object to angular vibrations. Both of these situations are rather infrequent. Uncoupling of x and β (or y and α) modes enables reduction in stiffness constants k_x, k_y and natural frequencies f_x, f_y, usually critical for performance of the vibration isolation system without creating the intense rocking motions.

The first and the most versatile *method* for enhancing angular stiffness is by increasing magnitudes of $a_{x,y}$. This can be achieved without any changes in the isolators' designs, and can be performed by the users by introducing a small inertia block or beams attached to the supporting surface of the isolated object so that the distances between the isolators in the desirable direction(s) can be increased (also see Section 2.2.6). While very effective, this technique requires additional space around the isolated object and thus is not always applicable. It is usually expensive since it requires a valuable "real estate."

The second method is increasing $k_{\alpha,\beta}$ without influencing k_x, k_y, and k_z. This technique is very attractive but requires special isolator designs with an intermediate member necessary for decoupling between k_x, k_y, and k_z. A design providing for high angular stiffness of the isolator and independently adjustable k_x, k_y, and k_z was suggested in [4.40] and is shown in Fig. 4.7.1. Element 1 is attached to the object, element 2 is attached to the supporting structure. Linear stiffnesses in the y- and z-directions in this designs can be independently adjusted by assigning dimensions to rubber-metal laminates 16, 17, respectively, to obtain their desirable shear stiffness values. Angular stiffness in the α-direction can be adjusted by selection of arm lengths a, b of intermediate member 12, 13, 14, 15.

Passive Vibration Isolation Means • 375

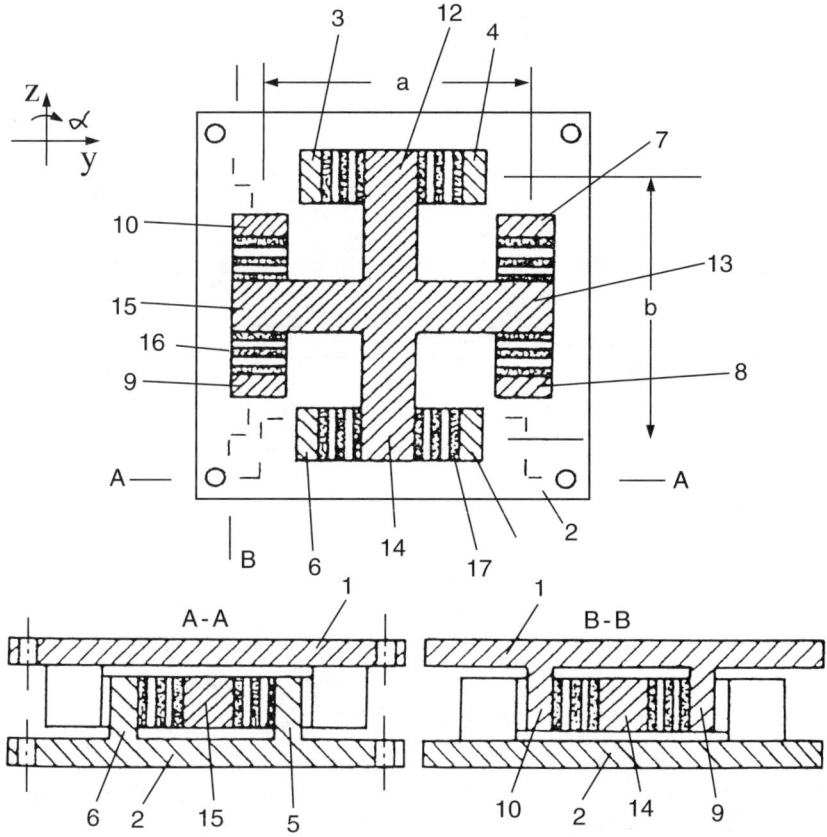

Fig. 4.7.1. Vibration Isolator with Independently Set Stiffness Values in Two Linear and One Angular Directions.

The third technique for increasing the angular stiffness of the vibration isolation system, while allowing for any desirable degree of isolation in the horizontal direction(s), is by increasing the vertical stiffness k_z relative to horizontal stiffnesses $k_{x,y}$ (anisotropic isolators).

This goal is achieved in mod. RM isolator from Barry Control Co., as shown in Fig. 4.7.2 [4.7], by connecting the top cover of the isolator carrying the object with the bottom cover attached to (or standing on) the support structure (floor) by steel rollers. The rollers provide for high k_z while allowing relative motion between the object and the supporting structure in one horizontal (x) direction. Vibration isolation in the x-direction is provided by two compression rubber blocks attached horizontally between the top and bottom covers.

376 • PASSIVE VIBRATION ISOLATION

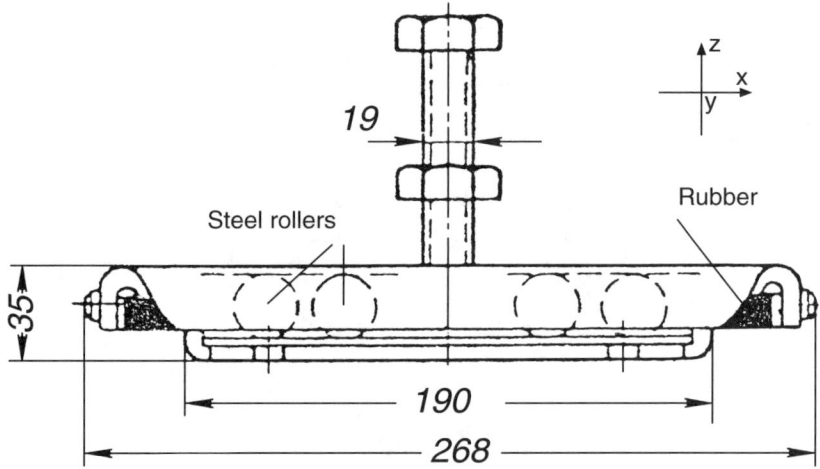

Fig. 4.7.2. Uni-Directional Isolator RM for Horizontal Direction (Dimensions in mm) (*Barry Controls Corp.*).

While achieving the goal of increasing angular stiffness of the isolation system without compromising isolation in the x-direction, this isolator has serious shortcomings. First of all, it provides isolation only in one (x) horizontal direction while maintaining a rigid connection between the object and the supporting structure in the other (y) direction due to high sliding friction between the steel rollers and the covers in the y-direction. This type of characteristic is unacceptable in many cases, e.g., for surface grinders having the highest vibration sensitivity in the y-direction while generating rocking-inducing dynamic forces in the x-direction. The RM isolator is very large and seems to be quite expensive since a good performance (low friction, uniform motion) of the rolling connection between the rollers and the covers requires high hardness of the contracting surfaces of both the rollers and the covers. This, in turn, calls for an expensive heat treatment and precision machining of the rolling surfaces. The isolator also requires protection of the rolling contacts to prevent their corrosion and contamination.

Since the thin-layered rubber-metal laminate elements have excellent characteristics as limited-travel bearings (Section 3.3.2C), it is natural to use them in designs that have high k_z and low and adjustable $k_{x,y}$ vibration isolators. The first applications of such elements were for earthquake protection and vibration isolation of buildings and other civil engineering structures, e.g., [4.41]. Such isolators can accommodate very high

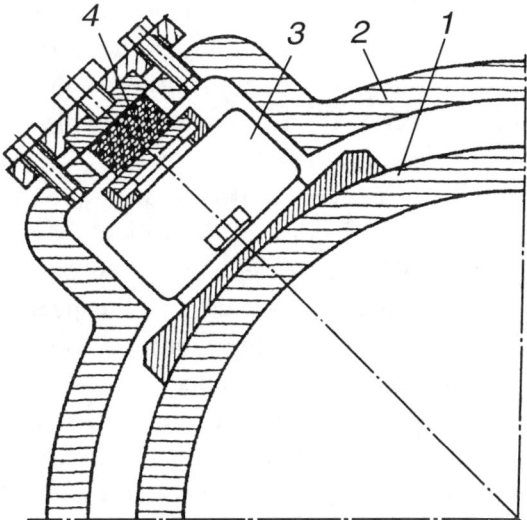

Fig. 4.7.3. Isolation of Stator of Electric Motor.

compressive forces while being rather small (see Section 3.3.2C), and provide low horizontal stiffness in all directions. Since the laminates are essentially subjected to only static weight load in the z-direction and to relatively small dynamic exertions in the x, y-directions, they can be easily fabricated by room-temperature post-vulcanization bonding rather than by the more expensive vulcanization bonding.

Another use for the anisotropic characteristics of thin-layered laminates is shown in Fig. 4.7.3 [4.42]. Dynamic excitations from electric motor 1 are isolated from outside housing 2 by conventional vibration isolator 3. Since its horizontal stiffness is inadequate to provide isolation from torsional vibrations, rubber-metal laminate 4 is installed in series with 3. While having a very low profile and not affecting the radial stiffness of isolators 3 significantly, laminate adds the desirable circumferential compliance. It results in reduction of transmission of the torsional vibrations to the housing on the frequency of rotation by 10 dB (∼3 times).

4.8 LOW STIFFNESS ISOLATORS

Reduction of isolator stiffness can be, of course, achieved by modifying the critical dimensions of the flexible element, e.g., using expressions for the stiffness constants given in Chapter 3. However, this approach results in reduction of the load-carrying capability of the flexible element and in an undesirable substantial increase in its initial height in the direction of the weight load application. If the isolators support the weight of the isolated object, a low required natural frequency f_z in vertical directions and the corresponding reduced stiffness of the isolator lead to an unacceptably large static deflection of the isolator(s) as well as to their static instability and packaging problems. Very low natural frequencies in the 1–3 Hz range, which are often desirable especially for vibration protection of humans as well as for vibration isolation of precision equipment, are associated with static deflections 3–25 cm (1.2–10.0 in.). Such deflections are hardly attainable since they require very large dimensions of the isolators, although the streamlined rubber flexible elements might be suitable for the lower end of the displacement range. Rubber isolators of conventional design loaded in compression have to be at least 6.5–10 times taller than the required static deformation, which results in highly unstable systems. The stability can be enhanced by introduction of a bulky and expensive "inertia mass" (foundation block).

Better designs of low-stiffness isolators can be realized by using non-linear effects associated with varying (usually reducing) margins of elastic stability of the appropriately designed flexible elements, and by using internal compensation devices. Although the low-frequency vibration isolation is the best candidate for using servo-controlled or "active" vibration isolators, some of the designs described below can provide for adequate low-frequency vibration isolation combined with high reliability and relatively low costs of passive vibration isolators.

4.8.1 Buckling and Stiffness

"Buckling" of an elongated structural member loaded with a compressive axial force P is loss of stability (collapse) of the structural member when the compressive force reaches a certain critical magnitude P_{cr}, which is also called the *Euler force*. Usually, the buckling process is presented as a discrete situation: *stable/unstable*. However, the process of development of instability is a gradual continuous process during which bending stiffness of the structural member is monotonously decreases with increasing axial compressive force P. The member collapses at $P = P_{cr}$, when its bending stiffness becomes zero.

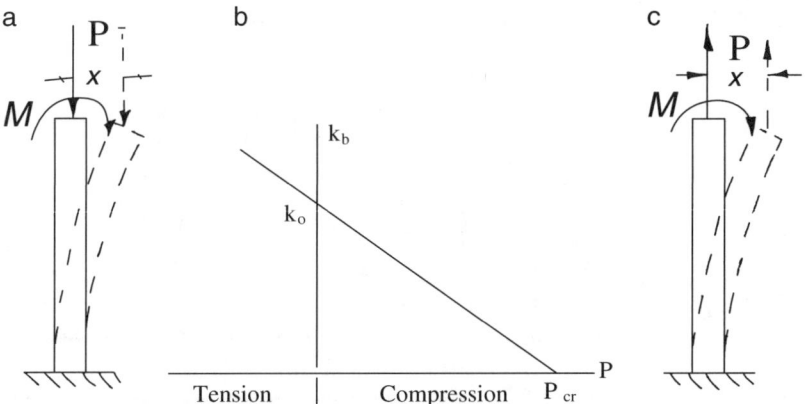

Fig. 4.8.1. (b) Stiffness Change of a Cantilever Beam Loaded by (a) Compressive; and (c) Tensile Force.

This process can be illustrated by the example of a cantilever column in Fig. 4.8.1(a). Bending moment M causes deflection of the column. If the axial compressive force $P = 0$, the bending stiffness is

$$k_o = M/x, \qquad \text{Eq. (4.8.1)}$$

where x is deflection at the end of the column due to moment M. If $P \neq 0$, it creates additional bending moment Px, which further increases the bending deformation and thus reduces the effective bending stiffness. The overall bending moment becomes [4.43]

$$M_{ef} = M/(1 - P/P_{cr}), \qquad \text{Eq. (4.8.2)}$$

and the resulting bending stiffness is approximately

$$k_b = (1 - P/P_{cr})k_o . \qquad \text{Eq. (4.8.3)}$$

Fig. 4.8.1(b) illustrates dynamics of the stiffness change for the column in Fig. 4.8.1(a) with increasing compressive force P. The stiffness-reducing effect presented by this equation is considered in many practical applications [4.44]. Buckling instability at a predetermined load can be designed by an appropriate shaping of the flexible element.

The effect of bending stiffness reduction for slender structural components loaded in compression can be very useful in cases where stiffness of a component has to be adjustable or controllable. One application of this effect for vibration isolators is proposed in [4.45].

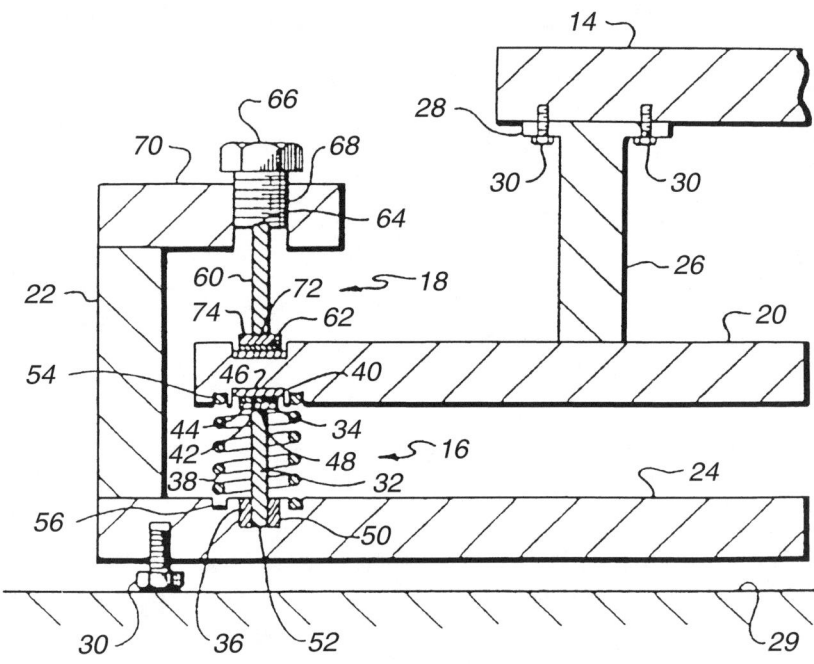

Fig. 4.8.2. Vibration Isolator for Horizontal Motion, Based on Using Buckling Effect for Stiffness Adjustment.

Fig. 4.8.2 shows a device protecting object 14 from horizontal vibrations transmitted from supporting structure 24. The isolation between 14 and 24 is provided by several isolating elements 18 and 16. Each isolator 18, 16 is a thin stiff (metal) post 60, 32, respectively. The horizontal stiffness of the isolation system is determined by bending stiffness of posts 60 and 32. This stiffness can be adjusted by changing the compression force applied to posts 60 and 32 (in series) by loading bolt 66. The horizontal stiffness can be made extremely low (even negative if the compressive force applied by bolt 66 exceeds the critical force). In the latter case, the overall stability can be maintained by springs 38. Vibration isolators built on a similar principle were also designed for six-degrees-of-freedom isolation [4.46].

In some cases, the problems associated with isolating systems characterized by very low natural frequencies can be alleviated by using so-called zero stiffness. The zero stiffness is equivalent to "Constant Force" (CF) F as a function of the load-deflection characteristic,

$$K = dF/dz = 0. \qquad \text{Eq. (4.8.4)}$$

CF systems can provide zero stiffness at one point or on a finite load interval of the load-deflection characteristic. Since the stiffness values vary with changing deflection, the CF systems are always nonlinear. Besides vibration isolation systems, CF systems are very effectively used for shock absorption devices where the CF characteristic usually is the optimal one (e.g., see [4.47]). Rubber flexible elements loaded beyond the critical compressive force P_{cr} causing buckling, often have zero-stiffness, or even negative stiffness segments in their load-deflection characteristics. Such elements can be used as vibration isolators or as correction elements for systems that have other flexible elements with positive stiffness.

Fig. 4.8.3(a) shows a radially compressed thin-walled rubber tube. When the compressive load P initially increases from $P = 0$, the load-deflection characteristic is shown as line 1 in Fig. 4.8.3(d), where $\alpha = \Delta/R$, $\beta = 12PR^2/Et^3$ (E is Young's modulus of rubber). The second stable configuration (the buckled tube) is shown in Fig. 4.8.3(c), and the corresponding load-deflection plot is shown as line 2 in Fig. 4.8.3(d). The transition between the two stable configurations in Figs. 4.8.3(b) and (c) (buckling event) starts in this case at a relative compression $\varepsilon = \Delta/R = 0.285$ and is shown by line 3 in Fig. 4.8.3(d), which is characterized by negative stiffness. The size and character (negative stiffness, low positive stiffness, or quasi-zero stiffness) of line 3 can be modified by changing R, t, and E. Many other shapes of rubber flexible elements possessing quasi-zero stiffness can be designed.

Fig. 4.8.4(a), (b) [4.47] shows design and stress-strain characteristics of two modifications of rubber mat, which is used mostly for packaging but can be used as a vibration isolation mat. One plot in Fig. 4.8.4(b) has a slightly negative stiffness in its post-buckling segment, another has a slightly positive stiffness.

Fig. 4.8.5(a) shows Dunlop buckling-type shock mount, and Fig. 4.8.5(b), (c) shows a Silentblock "Delta" shock mount and its force-deflection characteristic [4.47]. Again, these designs can be used as quasi-zero stiffness vibration isolators or as correction elements reducing the overall stiffness of the vibration isolation system.

There are numerous designs of constant force multi-component elastic systems with mechanical springs [4.48]. Their advantage relative to the rubber elements is adjustability of the load-deflection characteristics, which is more difficult for molded rubber elements. Fig. 4.8.6 [4.48]

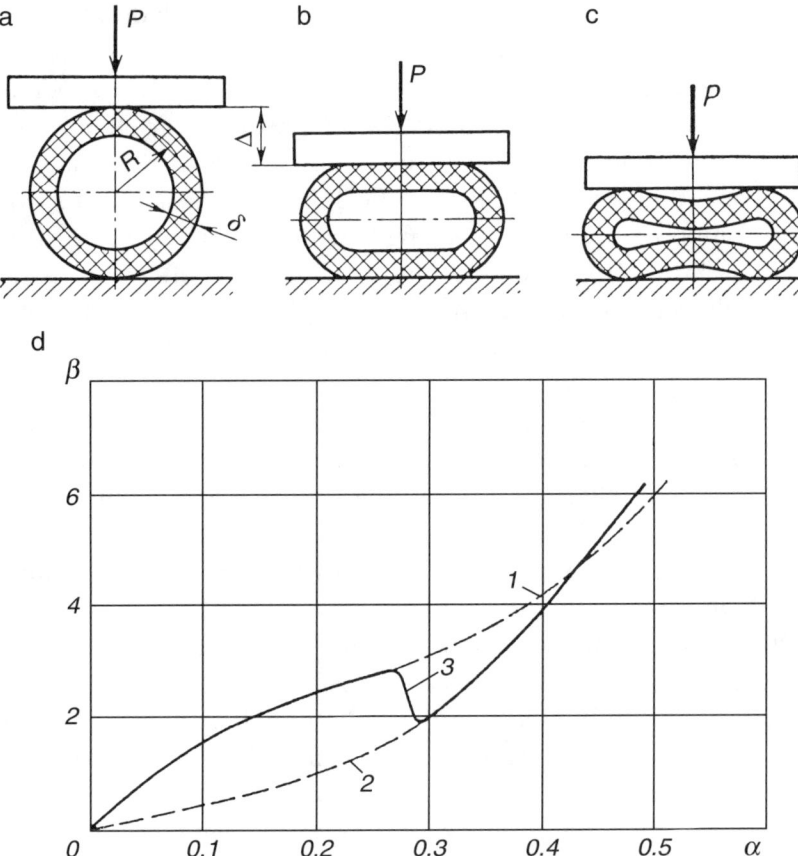

Fig. 4.8.3. Compression of Thin-Walled Rubber Tube: (a) Initial Condition; (b) First Stable Configuration; (c) Second Stable Configuration (Buckled Tube); (d) Load-Deflection Characteristics.

illustrates a CF system that has a stiffness compensator. Main spring with stiffness k_1 is connected with two correction springs each having stiffness k_2. Mass m, supported by spring k_1, vibrates in vertical direction z. The origin $z = 0$ is at the static equilibrium position, and the correction springs are horizontal at $z = 0$. If a vertical force F is applied to mass m, reactions of the springs will be F_1, F_2, respectively. The "virtual work" equation for this case is

$$(F - mg - F_1)\delta z - 2F_2 \delta L = 0. \qquad \text{Eq. (4.8.5)}$$

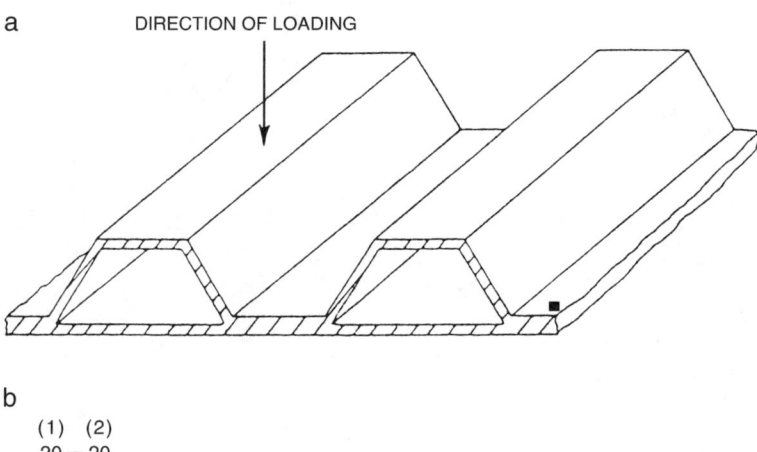

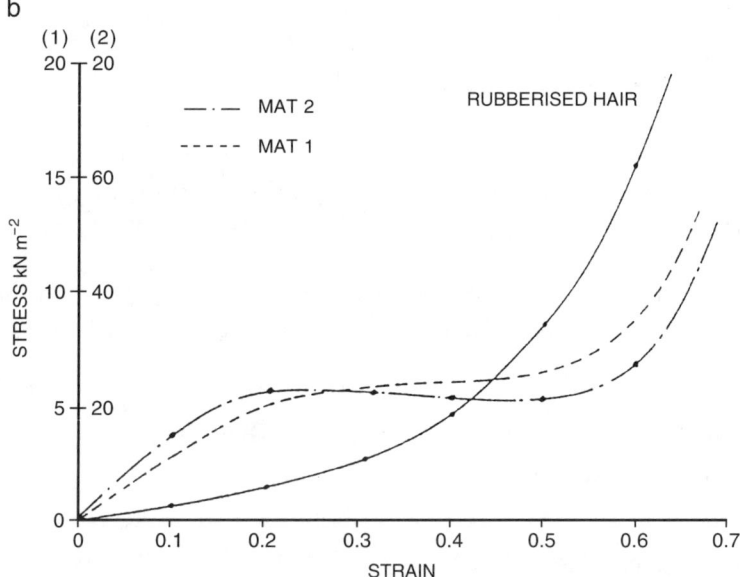

Fig. 4.8.4. Rubber Packaging Mat (a) and: Stress-Strain Characteristics of Its Two Modifications in Comparison with Rubber-Impregnated Felt Mat ("Rubberized Hair") (b).

Here F_1 and F_2 are, respectively, reaction forces of the main and correction springs,

$$F_1 = k_1(z - \Delta b), \ F_2 = k_2(L - \delta_o), \quad \text{Eq. (4.8.6)}$$

and

$$L = \sqrt{z^2 + a^2} \quad \text{Eq. (4.8.7)}$$

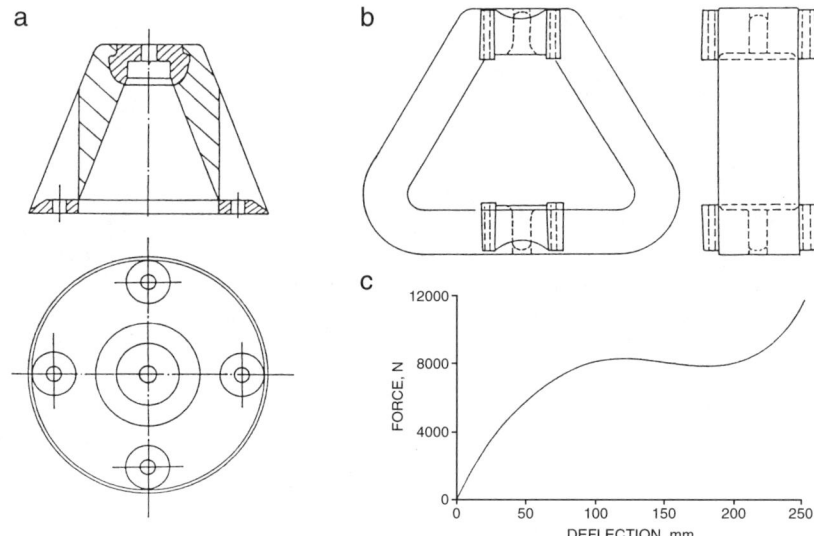

Fig. 4.8.5. Buckling-Type Shock Mounts: (a) *Dunlop*; (b) Delta (*Silentblock*); (c) Force-Deflection Plot for Delta Mount.

is length of the correction spring at deformation z of the main spring. Lengths of both springs in their undeformed conditions are, respectively,

$$L_{01} = b + \Delta b, \quad L_{02} = a + \delta_0. \quad \text{Eq. (4.8.8)}$$

Combining Eqs. (4.8.7) and (4.8.6), from Eq. (4.8.5) the system response to force F can be written as

$$F = k_1 z + 2k_2 z \left(1 - \frac{a + \delta_0}{\sqrt{z^2 + a^2}} \right), \quad \text{Eq. (4.8.9)}$$

where the condition of static equilibrium under the weight force is considered. Differentiation of Eq. (4.8.9) by z gives stiffness k_s of the isolator, which can be expressed in relative (dimensionless) terms as

$$k_c = \frac{k_s}{k_1} = 1 + 2k_{2c} \left[1 - \frac{1 + a_c}{(1 + z_c^2)^{3/2}} \right], \quad \text{Eq. (4.8.10)}$$

where $k_{2c} = k_2/k_1$; $z_c = z/a$; and $a_c = \delta_0/a$. At the static equilibrium position

$$k_c = 1 - 2k_{2c} a_c. \quad \text{Eq. (4.8.11)}$$

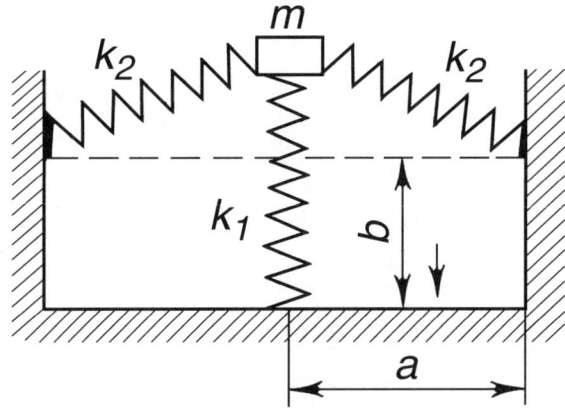

Fig. 4.8.6. Zero-Stiffness Vibration Isolator with Stiffness Compensation.

It is clear from Eq. (4.8.11) that in the static equilibrium position the value of the system stiffness depends on k_{2c} and a_c. The stiffness constant becomes zero if

$$2 a_c k_{2c} = 1. \qquad \text{Eq. (4.8.12)}$$

Plots in Fig. 4.8.7 illustrate variation of stiffness as influenced by various system parameters for a system complying with Eq. (4.8.12). The maximum displacement from the equilibrium position while the vertical stiffness does not exceed a certain magnitude k' can be derived from Eq. (4.8.10) as

$$z_1 = \sqrt{\left(\frac{1 + 2k_{2c}}{1 + 2k_{2c} - k'}\right)^{2/3} - 1}. \qquad \text{Eq. (4.8.13)}$$

It can be concluded that the system in Fig. 4.8.7 can accommodate the specified payload capacity at as low stiffness as desirable. The payload capacity is determined, in a manner similar to that of a linear system, by the static equilibrium condition, or

$$F_0 = mg = k_1 \Delta b. \qquad \text{Eq. (4.8.14)}$$

Thus, the payload capacity depends only on the stiffness constant of the main spring and its initial deformation.

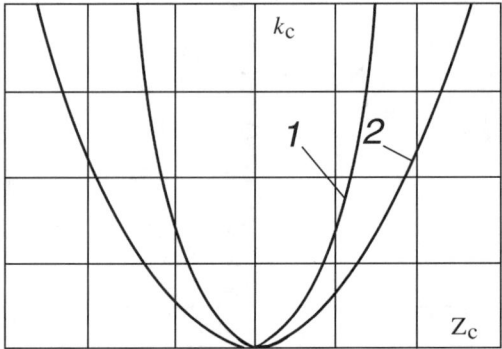

Fig. 4.8.7. Stiffness of Vibration Isolator with Quasi-Zero Stiffness (1—$k_{2c} = 4.0$, $a_c = 0.125$; 2—$k_{2c} = 2.0$, $a_c = 0.250$).

4.9 PNEUMATIC ISOLATORS

4.9.1 General Comments

Pneumatic isolators have many positive features in cases of low natural frequency isolation in comparison with coil springs also used in this frequency range. First of all, they have smaller size for a given rated load and natural frequency. For example, the static deformation of a spring under the weight load of an object supported with 1 Hz vertical natural frequency is about 10 in. (250 mm), thus requiring springs that have an initial length (height) about 30–40 in. and a loaded height of 20–30 in. A pneumatic isolator requires less than half of this height since the weight load is compensated by selection of the initial air pressure (Section 3.5).

The quasi-CNF characteristic of pneumatic isolators (Section 3.5), results in a reduction of the inter-coordinate coupling in the isolation system. This allows for use of higher natural frequencies in the system employing these isolators than in the spring system, which is especially desirable for low natural frequency systems. High-frequency resonances, addressed in Section 1.10, and strongly pronounced in steel springs, develop in pneumatic isolators at very high frequencies (in a KHz range, "organ effect" in the gas, and also at quite high frequencies in the rolling diaphragm). However, if special attention is not paid to designing the air-holding structure (e.g., roll seals) of pneumatic isolators, there is a possibility of high-frequency vibration transmission along the structural parts.

It is very important for low natural frequency/low stiffness applications that pneumatic isolators can be relatively easily equipped with self-leveling devices, which can maintain the object level within micrometers when the weight distribution between the mounting points is changing in the broad range.

While pneumatic isolators can be easily damped to a very low resonance amplitude, this high damping does not impair their high-frequency transmissibility (Section 3.5).

Pneumatic isolators are not perfect, they have their share of shortcomings. The isolators combining elastomeric and pneumatic flexible elements, as in Fig. 4.9.1, provide commensurate low stiffness values in both vertical and horizontal directions. However, they show relatively strong transmission of high frequency vibrations through their structures. Pneumatic isolators comprising a hard cylinder with the piston isolated from the cylinder by a rolling diaphragm seal show much lower high frequency transmission, but they have a relatively high stiffness and a very low damping in horizontal directions. For a typical case of the low frequency pneumatic isolation system, $f_z = 1.5$ Hz, $f_{x,y} = 2.5$ Hz, $\delta_z = \sim 3.0$ (quasi-critical damping), $\delta_{x,y} = 0.1$. Accordingly, isolation in several directions may require additional structures attached to the main pneumatic isolator, see Section 4.9.2, or introduction of coupling between the corresponding vibratory modes.

4.9.2 General Purpose Pneumatic Isolators

Fig. 4.9.1 shows pneumatic-elastomeric isolator that does not have pneumatic damping [4.49], [4.25]. The elastomeric portions of the isolator serve to contain the pressurized compressible fluid (usually air), contribute flexibility in the vertical direction, provide all the flexibility in radial (lateral) directions (resulting in $\eta_{x,y} \approx 1$), provide damping, and act as a cushioning and motion-limiting snubber in cases of severe transient/shock excitations. The elastomeric sidewall acts not only as a snubber, but as a temporary mounting pad for the payload when the isolator is uninflated or underinflated; it simplifies the mounting of heavy equipment and acts as a fail-safe in case of fluid leakage. While the isolator is designed for the natural frequency $f_z = 3$ Hz (at air pressure 20 psi) to 5 Hz (at air pressure 80 psi) when inflated, it maintains the natural frequency $f_z = \sim 10$ Hz when deflated. The three steel rings strapped around the

388 • PASSIVE VIBRATION ISOLATION

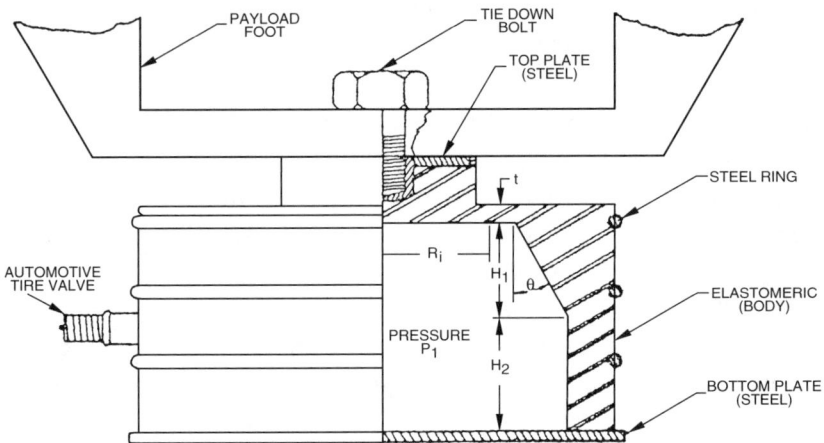

Fig. 4.9.1. Pneumo-Elastomeric Isolator.

outside of the sidewall prevent bulging, provide the desirable deflection characteristics, and maintain radial stability when the isolator is pressurized. The isolators are produced in eight sizes for the rated load (at the full pressure 80 psi or ~5.5 bar) 100–19,200 lbs (450–87,000 N) with dimensions L × W × H from 3 × 3 × 2.5 to 24 × 24 × 3.5 in. The leveling of the installed object (within ±1/4 in) is performed by varying air pressure inside the mount.

Pneumatic flexible elements can also be designed as bladders made from a fiber- (cord) reinforced rubber. The simplest design is shown in Fig. 4.9.2. The unloaded flexible element shown in Fig. 4.9.2(a) is a rubber cylinder attached to two end plates and reinforced by diagonal cords at the angle β to its axis. The bladder is pressurized by air (pressure p), and its sides bulge under the applied load as shown in Fig. 4.9.2(b). Such flexible elements work well for unidirectional loading but have very low lateral stiffness.

The Firestone sleeve-type air spring shown in Fig. 4.9.3(a) [4.47] has somewhat higher stiffness. The effective load-carrying area of such pneumatic spring is controlled by the initial circumferential dimension, the structural stiffness of the shell, and by the shape of the stationary "piston." By shaping the piston, it is possible to obtain quasi-zero stiffness over a certain limited rated range of travel (position), and progressively increasing stiffness after this range is exceeded, Fig. 4.9.3(b). Reinforcing

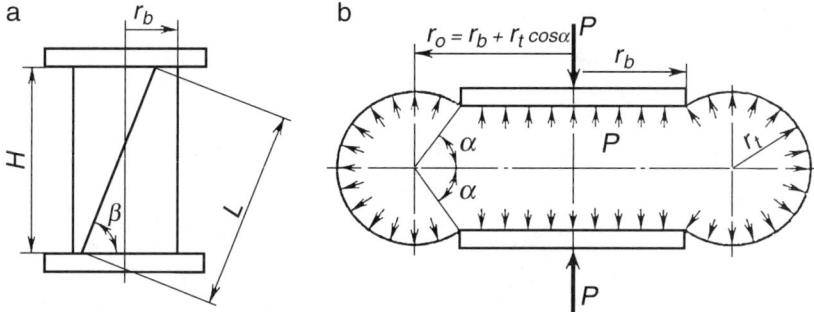

Fig. 4.9.2. (a) Unloaded; and (b) Loaded Pneumatic Flexible Element with Cord-Reinforced Rubber Shell.

the rolling diaphragm with steel rings, similar to the ones in Fig. 4.9.1, relieves some circumferential tension and allows to use higher air pressures which results in higher load ratings. Significant disadvantages of such elements for use in vibration isolation of stationary objects are their low lateral stability, which requires some lateral restraints, and their low damping.

Pneumatic vibration isolators for precision objects are usually designed as hard-wall cylinders with pneumatic damping in accordance with the embodiments shown in Figs. 3.5.2(a), (b). Typical pneumatic isolators of such configurations provide for the vertical natural frequency $f_z \geq 1$ Hz. It was shown in [4.50] that further reduction of the vertical stiffness determining f_z is difficult because of high stiffness of the diaphragm connecting the cylinder with the piston [Fig. 3.5.2(a) and Fig. 4.9.4(a)]. Changing the geometry of the rolling diaphragm to one shown in Fig. 4.9.4(b) significantly reduces its stiffness, thus f_z can be reduced to ~0.6 Hz [4.50]. Wide use of pneumatically damped pneumatic flexible elements, per sketches in Figs. 3.5.2(a), (b), is somewhat hampered by their unidirectional action. These elements usually have high stiffness and low damping in the lateral directions, as well as very low stiffness and low damping in the angular directions (tilting). This situation can be corrected by incorporating special devices into designs of damped pneumatic isolators in order to provide the required lateral mobility. One such design is shown in Fig. 4.9.5 [4.51]. The supported object (platform) in Fig. 4.9.5 is hinged to the top of the inverted pendulum whose lower end is hinged to the piston of the damped pneumatic cylinder. Stability and control of

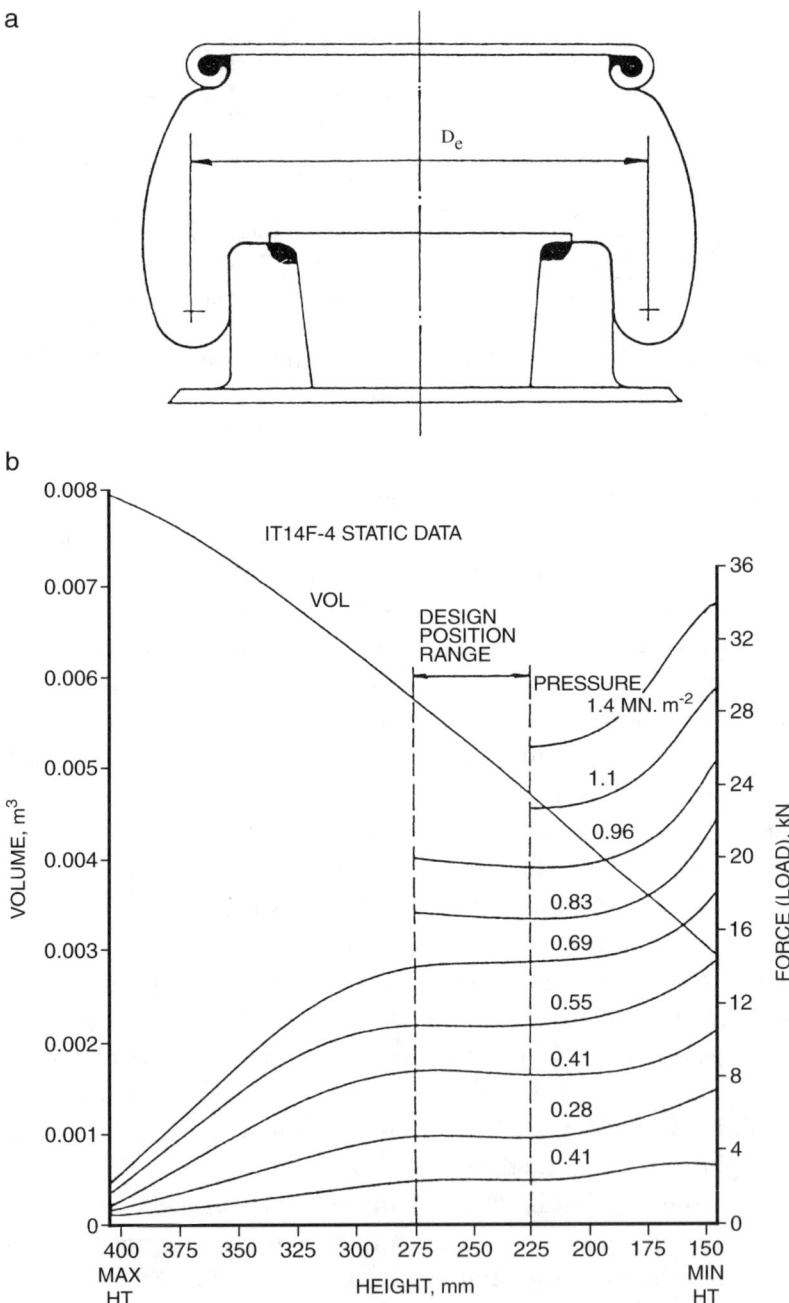

Fig. 4.9.3. (a) Sleeve-Type Air Spring; and (b) Its Performance Characteristics (*Firestone Industrial Products Co.*).

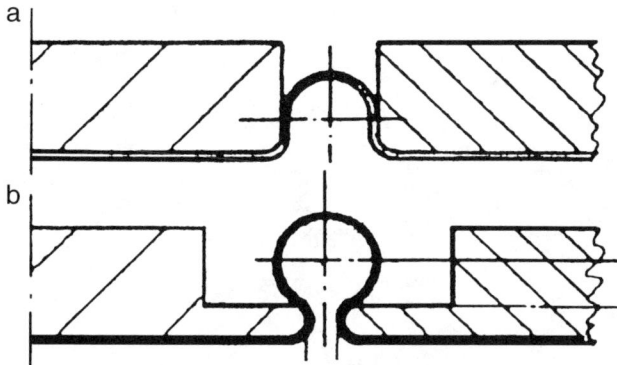

Fig. 4.9.4. Geometries of Sealing/Guiding Diaphragms for Pneumatic Vibration Isolators: (a) Conventional (More Rigid) Design; (b) Advanced (Less Rigid) Design.

lateral stiffness of the pendulum is provided by springs. Horizontal stiffness k_x of the pendulum support is

$$k_x = \frac{k_s a^2}{L^2} - \frac{mg}{L}, \qquad \text{Eq. (4.9.1)}$$

where L is length of the pendulum; m is mass of the object associated with this isolator; k_s is total stiffness of springs supporting the pendulum; a is the distance between spring attachment points and C.G. of the pendulum; and g is acceleration due to gravity. The horizontal stiffness k_x can be adjusted by varying k_s and a, provided that the right-hand side of Eq. (4.9.1) is > 0, otherwise the system becomes unstable. The horizontal natural frequency $f_x = 0.01$ Hz was reported as realized with this type of system [4.51]. Obviously, a similar set of springs can be provided for the y-direction, possibly with different tuning. The isolator in Fig. 4.9.5 also has a height control ("self-leveling") system, which is illustrated in Fig. 4.9.7.

Another design of pneumatic vibration isolator providing for lateral mobility is shown in Fig. 4.9.6 [4.52]. In this design, the object-carrying platform is attached to the piston B via three rods, which play a role of conventional pendulums providing for low resistance lateral mobility in the x- and y-directions. Other interesting features of this design are the special shape of the piston B allowing for both economic packaging of the pendulums and for significant reduction of the main chamber C, thus

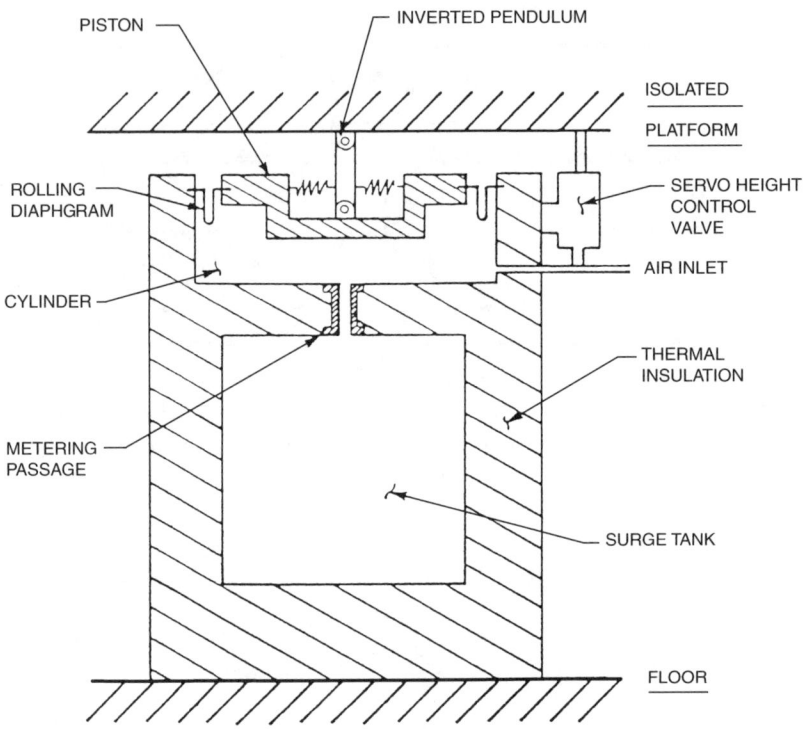

Fig. 4.9.5. Pneumatic Vibration Isolator with Inverted Pendulum for Lateral Mobility.

greatly increasing the volume ratio N (see Section 3.5), as well as use of special laminar flow air damping restrictors A instead of simple capillaries (see Section 3.5).

4.9.3 Self-Leveling Isolators

Since pneumatic isolators are used for low natural frequency isolation systems, they have low or very low stiffness and thus are sensitive to even small changes in the weight load on each isolator and/or to small deformations of the structural frames of objects, not providing significant rigidization of the structure (see Section 2.2.6). While these shortcomings can be alleviated by attaching the structural frame of the object to a rigid plate or to a rigid inertia block, such solutions result in a costly and bulky system. Another approach is to use self-leveling pneumatic isolators. This approach was developed by Barry-Wright Corp. in the 1960s and is now

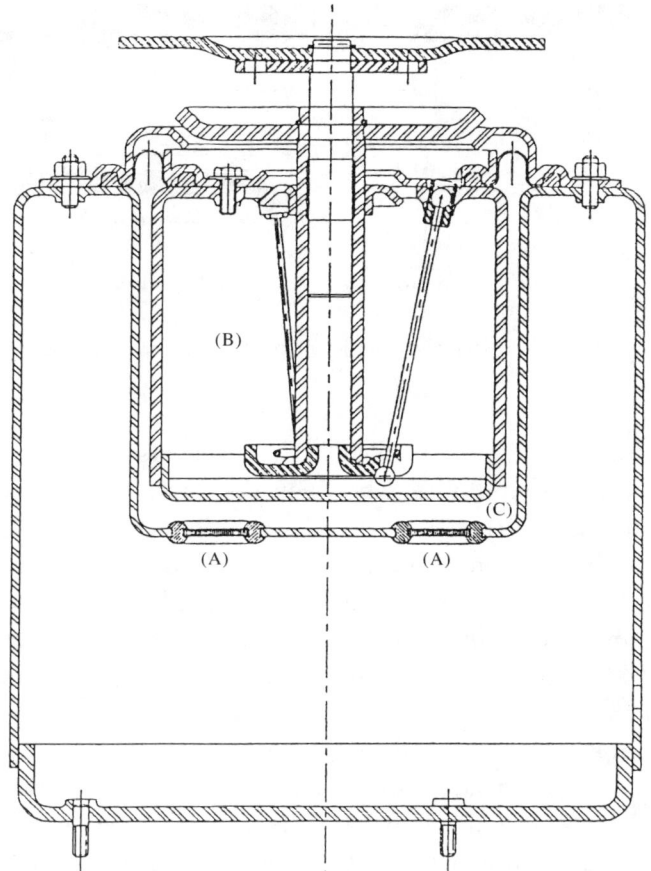

Fig. 4.9.6. Pneumatic Vibration Isolator with Three Direct Pendulums for Lateral Mobility.

employed by many manufacturers of vibration isolation means. Such systems can be called "semi-active" since they use external power sources, but only for leveling and not for actual vibration isolation, which is provided by stiffness and damping of the passive pneumatic system.

A typical design of the self-leveling vibration isolator is shown in Fig. 4.9.7 [4.28]. The main chamber of the isolator has a relatively small volume shaped by the "Isolator Assembly," guided by the "Rolling Diaphragm," and connected via the "Laminar Flow Damping Element" with the much larger damping "Hybrid Chamber". While the large volume of the latter results in high damping, see Section 3.3.5, it also slows down

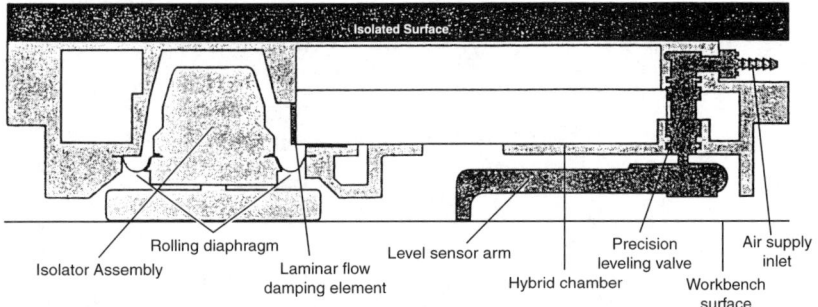

Fig. 4.9.7. Self-Leveling Pneumatic Vibration Isolator.

the pressure equalization between the chambers during the leveling process; the 8:1 volume ration between the damping chamber and the main chamber is often considered as the optimum. The damping chamber is continuously connected, via the "Air Supply Inlet" and the "Precision Leveling Valve," to a compressed air line that has higher air pressure than the maximum air pressure inside the isolator. When the object ("Isolated Surface") is properly leveled in relation to the supporting surface ("Workbench Surface"), the "Precision Leveling Valve" is closed and the isolator acts as a self-contained damped pneumatic vibration isolator. Any change in the weight load on the isolator or in the level (deformation) of the isolated surface would change the distance between the top surface of the isolator and the supporting surface, resulting in a motion of the spring-supported "Level Sensor Arm." If the height of the isolator increases (e.g., due to a reduction in the weight load on the isolator), the "Level Sensor Arm" activates opening of the "Precision Leveling Valve," thus bleeding air from the damping and, consequently, from the main chamber. The air pressure inside the main chamber decreases until the initial level of the isolator is restored, at which moment the valve closes. If the height of the isolator decreases, e.g., due to increasing weight load, the "Level Sensing Arm" activates the other side of the "Precision Leveling Valve," thus connecting the "Air Supply Inlet" with the damping and, consequently, the main chamber, resulting in increasing air pressure inside the isolator until the isolator height is restored, at which moment the valve closes. The possibility of maintaining the isolator level within one μm was reported. The level stabilization time depends, besides the volume ratio, on the time history of the height-changing event and on the vertical natural frequency of the isolator; usually, this time is on the order of magnitude of one second.

4.10 VIBRATION ISOLATED FOUNDATIONS AND INSTALLATION SYSTEMS

4.10.1 Installation Systems with Inertia Blocks

Vibration isolated foundations are usually constructed as massive blocks made from reinforced concrete and supported by flexible elements (steel springs as well as rubber blocks or mats). The foundation blocks, depending on the circumstances, are used for enhancing the structural stiffness of the object to be isolated, for increasing stability of the vibration isolating installation, for decoupling in the vibration isolation system, and for improved (reduced) transmissibility in the high-frequency range of isolation. In many cases, the foundation block is called for to satisfy more than one of the above requirements.

Since stiffness of vibration isolators is much lower than stiffness of conventional rigid mounting elements—such as wedge and screw mounts, steel shims with grouting, etc.,—the isolators do not provide as complete a connection between the object frame and the supporting structure (floor) as do the rigid mounts (Section 2.2.6). The properly selected type, number, and locations of the rigid mounts can assure tying of the frame and the floor/rigid foundation plate (supported by deformable soil) into a sufficiently rigid structure, e.g., see [4.44], [4.53], [4.54]. While a judicious approach to direct installation on vibration isolators of a precision object whose frame/bed stiffness is marginal can result in successful results (as illustrated in Section 2.2.6B), in many cases stiffening of the frame is necessary. It can be achieved by using self-leveling vibration isolators (as described in Section 4.9.3) in cases when the supporting structure (floor) is rigid, or by using a self-leveling installation system with a reference frame (like the one briefly described in Section 4.10.2) in cases when the stiffness of the supporting structure is inadequate. However, in many cases the stiffening effect is still being achieved by passive means, such as fastening of the object frame to a steel plate—thus transforming it into a box-like rigid structure—or, for larger objects, by attaching the frame/bed of the object to the concrete foundation block which, in turn, is installed on vibration isolators. If the foundation block (supported by vibration isolators) is used to enhance the frame/bed stiffness, its dimensions should be selected using the same approaches as those used for designing the soil-supported foundation. For high-precision objects, such as CMMs, machine tools, or photolithography tools, the inertia block may be integrated with the base frame of the object and made from a low-distortion polymer-concrete composition or from zero-distortion granite.

The robustness of objects supported by vibration isolators is their resistance to excessive movements while they are acted upon by external exertions (touching by operators, spurious hits by material handling devices, vibrations transmitted through the floor or through the air, etc.), by internal dynamic exertions (such as accelerations/decelerations of internal moving parts in the object), or by static action of the moving heavy masses causing undesirable tilting of the object. There are numerous approaches aimed to enhance robustness (some are discussed in Section 2.2.6B). However, low natural frequency installations featuring low stiffness vibration isolators may still have unacceptable high static and/or dynamic mobility. In such cases, use of foundation blocks between the object and the isolators (or the integrated frame blocks) may be required. Since requirements for vibration isolation are formulated in terms of magnitude of the natural frequency or the vibration isolation criterion (both in vertical direction parallel to the vector of the gravity force), increasing the effective mass of the object results in a proportional increase in the isolators' stiffness to realize the same natural frequency. Thus, use of this type of "inertia block" enhances robustness in two ways: The same external or internal exertion is applied to a larger mass, thus reducing its accelerating/decelerating effect; and the installation is more "sturdy" due to the higher stiffness of the isolators for the given vertical natural frequency f_z.

The importance of reducing inter-coordinate coupling in many realizations of vibration isolation systems for precision objects was demonstrated in Sections 1.3, 1.6, and 2.2.5. It is not less important for vibration isolation of the vibration producing objects addressed in Section 2.3. It was mentioned before that in some cases the modal coupling is beneficial, e.g. for a more uniform distribution of damping in a system with pneumatic isolators having very low damping in x, y directions. Most of the conditions for decoupling that were formulated in Section 1.3 are relatively easily satisfied by using CNF isolators with constant stiffness

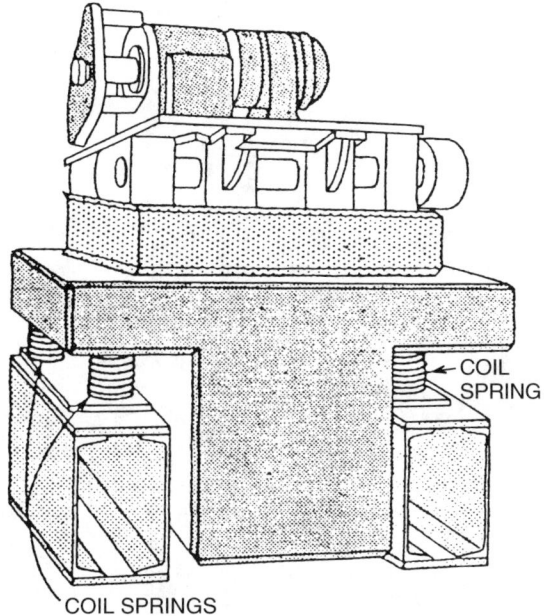

Fig. 4.10.1. Vibration Isolated Foundation with Elevated Isolators to Reduce Distance from the C.G.

ratios across the CNF load range. However, these conditions only reduce the modal coupling but leave two pairs of coupled translational-angular modes. All coordinate motions can be decoupled by locating the isolators in the plane containing the C.G. This can be easily located if an inertia block is used. The vertical position of the C.G. should be determined and isolators placed at this level (e.g., see Fig. 4.10.1). This arrangement can be recommended if the decision to use foundation/inertia block is made because of other considerations; otherwise it is better to achieve decoupling, if desired, by the other techniques.

In cases when an extreme degree of high-frequency isolation needs to be achieved, the foundation/inertia block can be modified and two sets of isolators used (one set between the object and the foundation block, the other set underneath the foundation block) to achieve the high-frequency suppression effect by using the foundation block as a large intermediate mass (see Fig. 1.5.8).

The foundation (inertia) block is usually supported by pneumatic low-stiffness isolators (for high-precision objects) or by steel spring isolators as shown in Figs. 4.6.2 and 4.6.3. Damping for the spring isolators is usually

398 ● **PASSIVE VIBRATION ISOLATION**

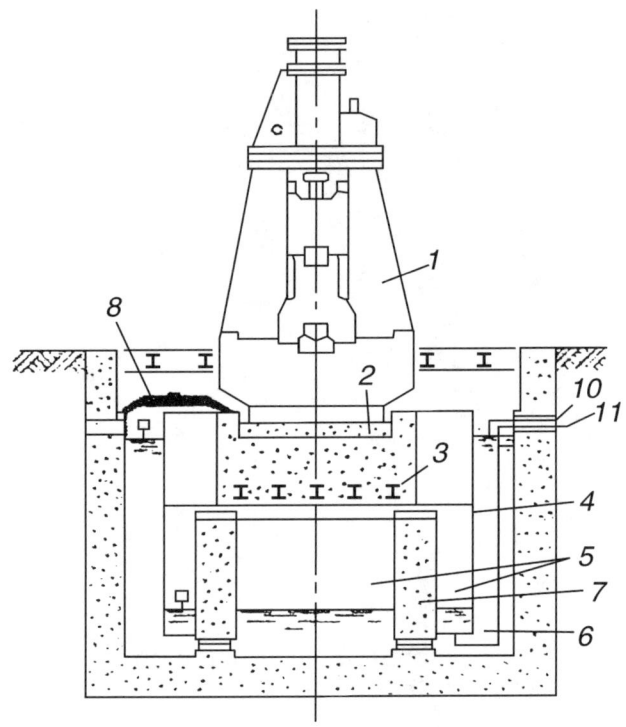

Fig. 4.10.2. Vibration Isolated Foundation for Forging Hammer with Combined Leaf Springs/Air Cushion Suspension.

introduced by using large rubber blocks that are parallel to the spring blocks. The rubber mats described in Section 4.3.1 have the advantage of combining stiffness and damping, and being less sensitive to moisture than the steel springs. By stacking the rubber mats, very low natural frequencies can be realized. An interesting approach to designing vibration foundation (for a forging hammer) was described in [4.55], Fig. 4.10.2. Hammer 1 is installed on foundation block 3 via stiff underanvil isolator 2. Weight of hammer 1 with foundation block 3 is supported by leaf springs 8 and by air pressure in chambers 5 generated by steel "bell" 4. Chamber 6 is filled with water. Water is supplied through inlet 10, and air is supplied through inlet 11. After filling chamber 6 and free space around block 3 with water, the air is let into chambers 5. The air squeezes out water and lifts block 3 with hammer 1 until it reaches the specified height. Losses of air and water are automatically compensated. The relatively low air pressure cushion provides (together with leaf springs 8) the required low-frequency isolation as well as stabilization of the foundation level. Simultaneous running of even five hammers installed on such foundations in one shop does not generate objectionable vibrations.

4.10.2 Installation System With Reference Frame

The self-leveling air mounts described in Section 4.9.2 maintain constant distance between the object's frame and the supporting structure, while providing a low natural frequency vibration isolation. However, the supporting structures, even large and expensive foundation blocks, may not represent ideal supporting surfaces. Due to sagging of the underlying soil, aging of concrete, deformations in the surrounding building structures, microseismic events, temperature gradients, etc., foundations deform and the geometry of supporting surfaces change. Let's assume that a long cast-iron bed (thermal expansion coefficient $\alpha = 10 \times 10^{-6}$ per 1 °C) is attached to a concrete foundation block (thermal expansion coefficient $\alpha = \sim 14 \times 10^{-6}$ per 1 °C). Then for the $l = 20$ m long bed and temperature change 1 °C, the difference in expansion between the bed, $\Delta l = 0.2$ mm, and the foundation block, $\Delta l = 0.28$ mm, is 0.08 mm. If a flat bi-material strip (modeling the system bed-foundation) is heated by t °C, its curvature radius becomes

$$\rho = \frac{\frac{(E_1 h_1^2 - E_2 h_2^2)^2}{E_1 E_2 h_1 h_2 (h_1 + h_2)} + 4(h_1 + h_2)}{6t(\alpha_1 - \alpha_2)}, \qquad \text{Eq. (4.10.1)}$$

where $E_1 \approx 1.15 \times 10^5$ MPa; $E_2 \approx 0.3 \times 10^5$ MPa are Young's moduli of cast iron and concrete; $h_1 = 1.0$ m; $h_2 = 2.0$ m are heights of the bed and of the foundation block. For $t = 1$ °C, $\rho = 0.5 \times 10^6$ m, which corresponds to deflection in the mid-span

$$\Delta = \frac{l^2}{4\rho} = 2 \times 10^{-4} \, m = 200 \, \mu m, \qquad \text{Eq. (4.10.2)}$$

or angular deformation at the ends 0.02/1000 per 1 °C. Such deformations are excessive for precision optical devices and other high-precision systems. Even in temperature-controlled facilities there might be changing temperature gradients between the object attached to the foundation and the lower levels of the foundations.

400 ● **PASSIVE VIBRATION ISOLATION**

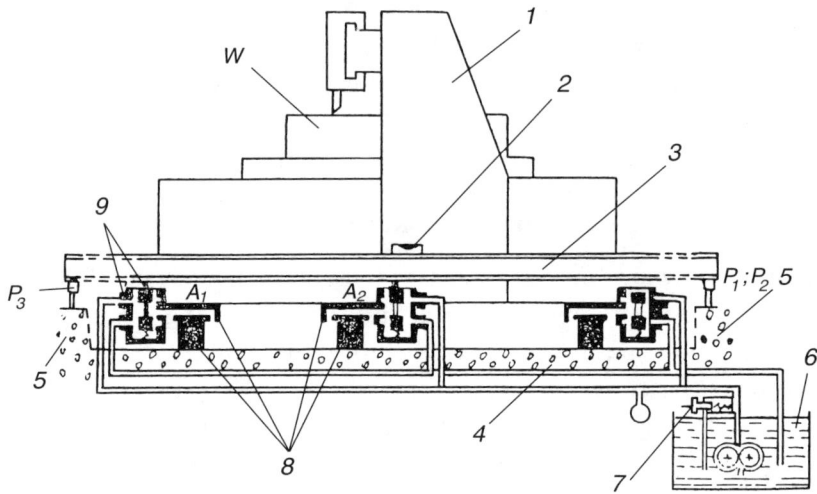

Fig. 4.10.3. Self-Leveling Installation with Separate Reference Frame.

To eliminate the negative influence of distortions in the supporting surfaces, [4.56] suggested developing a system combining vibration isolation with maintaining the attachment surface of a precision object parallel to a reference frame whose geometry is not subject to the deterioration influences listed in the above paragraph. Fig. 4.10.3 shows installation of a large precision machine tool (planer). Machine frame 1 is attached to supporting structure 4 through mounts 8 located in points A_1, A_2, ... placed so close to each other that deformations of machine frame 4 between the mounts are negligible. Reference frame 3 surrounds the machine and is kinematically supported at three points, P_1, P_2, P_3, located on the relatively low deforming part 5 of the supporting structure (foundation). Horizontal positioning of reference frame 3 can be checked using spirit or electronic level 2 and can be manually or automatically corrected if necessary. The machine level sensors are hydraulic valves 9 placed between reference frame 3 and machine frame 1. The pilot valves control hydraulic fluid flow from hydraulic pump 6 with pressure control valve 7 to mounts 8. If the machine frame deforms downward at a sensor location, upper passage 9 in the respective pilot valve increases and pressure inside the attached mount increases until the distance between the machine frame and the reference frame at the mount location returns to the initial setting. After this happens, the passage in the pilot valve closes. Similarly, at an upward deformation of the machine frame, the

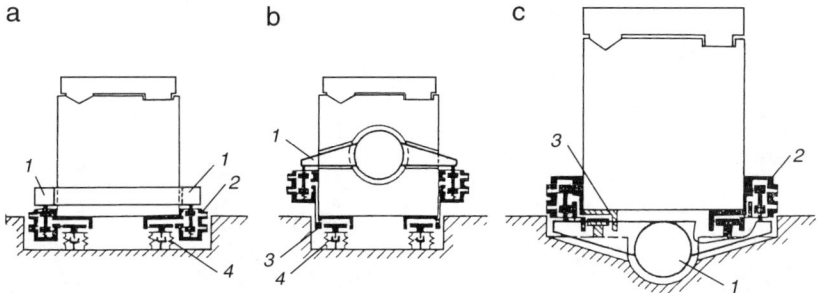

Fig. 4.10.4. Alternatives for Positioning the Reference Frame 1.

lower passage in the corresponding pilot valve opens, resulting in bleeding of the hydraulic fluid and reduction of pressure inside the mount attached to this pilot valve, until the distance between the machine frame and the reference frame returns to the initial setting.

While there might be many alternatives to sensor/actuating mount designs, the most important component of the system is the reference frame. Fig. 4.10.4 shows three variants of positioning of the reference frame 1. Fig. 4.10.4(a) (analogous to positioning of the frame in Fig. 4.10.3) presents the simplest approach since the pre-existing foundation structure does not have to be changed. However, in this configuration the reference frame is not protected from accidental damage and from temperature gradients; also, it is difficult to use this reference frame for objects whose beds have complex outlines in the plane view. Fig. 4.10.4(b) shows the reference frame made as a hollow tube placed inside the object and connected to the sensors and to the mounting points by linkages; it is supported on the foundation at three points. In this embodiment, the system can be installed on the floor without any required modifications. However, the object has to be designed together with the reference frame. The schematic in Fig. 4.10.4(c) has the reference frame located underneath the object. The reference frame is protected from mechanical damage, is less exposed to temperature gradients, while all critical units are easily accessible and the footprint of the protected object is not enlarged. This approach requires making a recess on the surface of the supporting structure.

The reference frame is loaded only by its own constant weight forces and by forces of interactions with the sensors, which can be made very low. However, due to usually large dimensions of the objects requiring

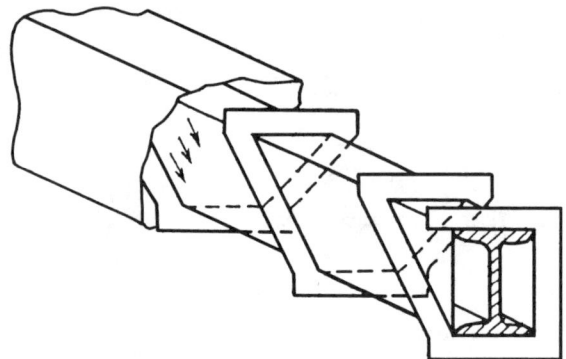

Fig. 4.10.5. Design Scheme for the Reference Frame.

self-leveling installation systems, the reference frame can be noticeably distorted even by small temperature gradients. The reference frame design suggested (and tested) in [4.56] is shown in Fig. 4.10.5. It consists of a rigid oversized I-beam surrounded by a spiral tubular conduit through which constant temperature air is continuously blown, and by a thermo-insulating duct. This design provides a very stable temperature field for the beam, with very small temperature gradients. It is much easier to maintain constant temperature of the reference frame than to maintain (to the same degree of precision) constant temperature in a big room housing the object. Since the largest effect of changes in temperature is structural distortion of a large object, the self-leveling systems in many cases may replace much more expensive constant temperature facilities.

4.11 ISOLATION OF NON-SUPPORTING CONNECTIONS OF ISOLATED OBJECTS

A significant vibratory energy, both in the relatively low-frequency range and especially in high (acoustic) frequency range can be transmitted to and from vibration isolated objects through so-called non-supporting connections, such as power transmission shafts, piping, cables, etc. If a *vibration producing object* is isolated in order to reduce vibration transmission to the environment, the same isolating properties should be designed for all the connections. It is especially important for the objects installed on relatively low stiffness structures, such as ships, surface vehicles, higher floors of buildings, etc.

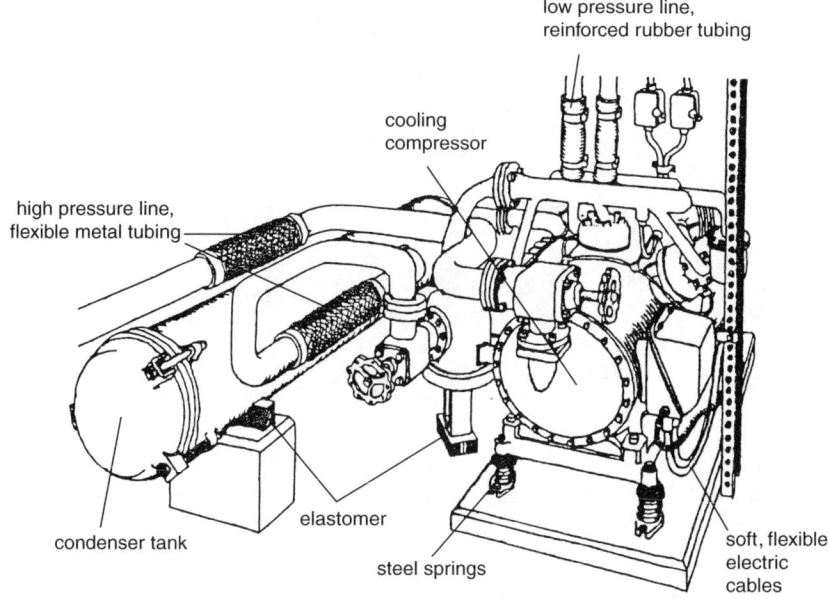

Fig. 4.11.1. High Air Pressure Compression Station.

There are several rules for designing vibration isolation for the non-supporting connections of vibration producing objects [4.57]:

- The connections should be isolated along their whole length, no matter how far from the isolated machine they go,
- Isolating means with high damping should be placed at the attachment points of the connections to the supporting structure,
- The number of the attachment points should be minimal,
- The impedance of the supporting structure at the attachment points should be such that the vibration isolation properties of the isolating means can be realized; essentially, it means that stiffness of the supporting structure at these points should be significantly higher than stiffness of the isolating means,
- The isolating means should be able to protect the isolated connections from outside static and dynamic exertions; for onboard ship applications, this means inertia forces caused by general vibration of the hull due to working mechanisms or rough seas, collisions, etc.,
- The isolating means should provide for necessary mobility of the isolated object on its isolators.

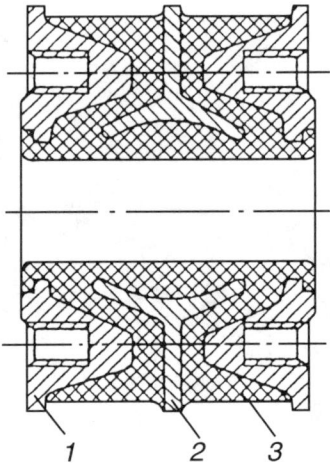

Fig. 4.11.2. Rubber-Metal Insert for High-Pressure Hydraulic Line.

If an isolated object is connected with other objects via a power transmission shaft (e.g., engine with reducer or with propeller, etc.), then the shaft should be equipped with power transmission coupling that has misalignment-compensating abilities and, in some cases, torsional flexibility and damping tuned for preventing intense torsional vibrations that can induce translational vibrations of the supporting structure (see Section 4.12).

Isolating means for hydraulic and compressed air piping include flexible inserts (couplings) and flexible brackets for attaching pipes and hoses to supporting structures. Fig. 4.11.1 [4.58] shows a typical installation of a high-pressure air compressor. The flexible insert in Fig. 4.11.2 [4.57] has a limited compliance in all directions; it is used for high-pressure hydraulic lines. In cases where contact of rubber with the fluid inside the pipe is undesirable, flexible inserts like those in Fig. 4.11.3 can be used. In this design the fluid has contact only with metal bellows 1, which is supported and protected by metal shield 2; ends 3 of the insert can be made to accommodate the accepted connection standards. Usually, installation of at least two inserts is recommended—one close to the isolated object and another close to the attachment point of the conduit to the supporting structure. Special attention should be given to reducing axial loading of the flexible inserts. To reduce the required deformations of the pipes and the inserts, the pipes should be attached to the isolated

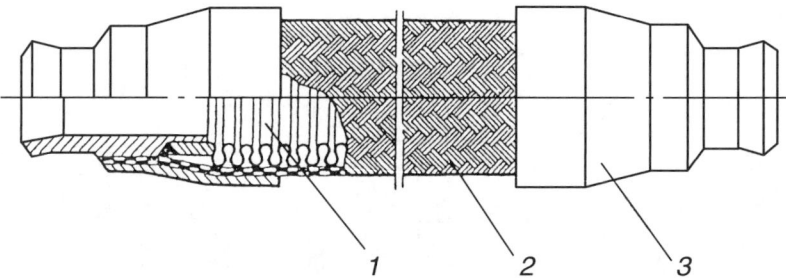

Fig. 4.11.3. Metal flexible Insert for High-Pressure Hydraulic Lines.

object at the points where its vibrations are minimal, and clearances between the pipes and adjacent objects should be adequate to prevent their contact. It is desirable that the combined stiffness of all connections does not exceed ~20% of stiffness of the main vibration isolators in a given direction.

A typical suspension bracket for a pipeline is shown in Fig. 4.11.4. It comprises collar 1 surrounding the pipe and housing rubber isolating insert 2, and attachment member 3. Sometimes a coil spring is used in series with the rubber insert in order to reduce stiffness.

Electric cables at the attachment points should be configured as loops or spirals to enhance their compliance.

Precision objects sensitive to vibration can get excited by an acoustic environment (air vibration generated by fans, air-conditioning units, etc.). Vibration also can be transmitted by energy and signal conducting conduits, such as electric cable, compressed air and vacuum tubing, hydraulic plumbing. There should be a clear distinction between vibration and jolts transmitted by movements of these conduits, and vibration induced by pressure fluctuations in pneumatic and, especially, hydraulic lines. Sometimes these excitations can be as significant as floor excitations [4.4]. The reduction techniques for alleviating harmful effects of the above sources of environmental vibration differ in accordance with the mechanisms of their action.

Since the acoustic energy is broadband, it may excite various structural modes of a precision object even more readily than the ground motion. Acoustic excitation harmful for precision objects can come not only from the adjacent noisy equipment, fans, air-conditioning units, etc., but from overhead airplanes, especially from sonic booms [4.59]. Acoustic isolation can be achieved by enclosures, e.g., see [4.59], by reducing air pressure

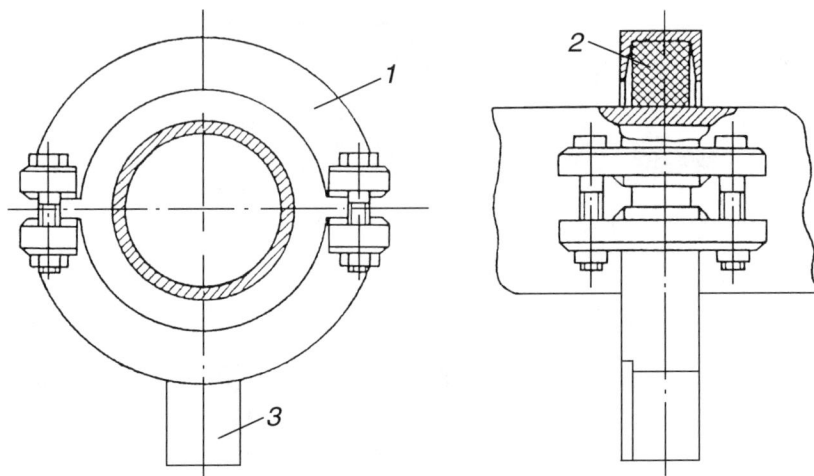

Fig. 4.11.4. Vibration Isolating Bracket for Pipelines.

around the object as well as by active sound cancellation. Combining enclosures with reducing air pressure inside them is a very effective technique that can be especially applied to precision devices that do not require the continuous presence of a human operator.

Pressure fluctuations in pneumatic and hydraulic piping are generated by pumps and can be transmitted as vibration excitation to the isolated object. Pumps that do not exhibit pressure fluctuations are usually rated for lower pressures. Pumps that service vibration-sensitive devices have to be designed for low vibration delivery of gases or fluids. Pressure fluctuations can be significantly reduced by using vibration absorbers (pulsation dampeners) and/or surge suppressors, e.g., see [4.60]. Practically "no ripple" flow of gases can be provided if, instead of pumps or compressors, the gas is supplied from a high-pressure bottle through a pressure regulator.

Mechanical attachment of various conduits, including electric power and signal cables, can lead to deterioration of the vibration condition of a precision object, especially for objects installed on soft vibration isolators. These conduits are very good transmitters of vibrations in the units from which they originate, from walls through which these conduits, from accidental contacts with moving vehicles or human operators, etc. All rigid connectors, such as tubes, must be dynamically uncoupled from the vibration isolated object by bellows or other flexible connectors.

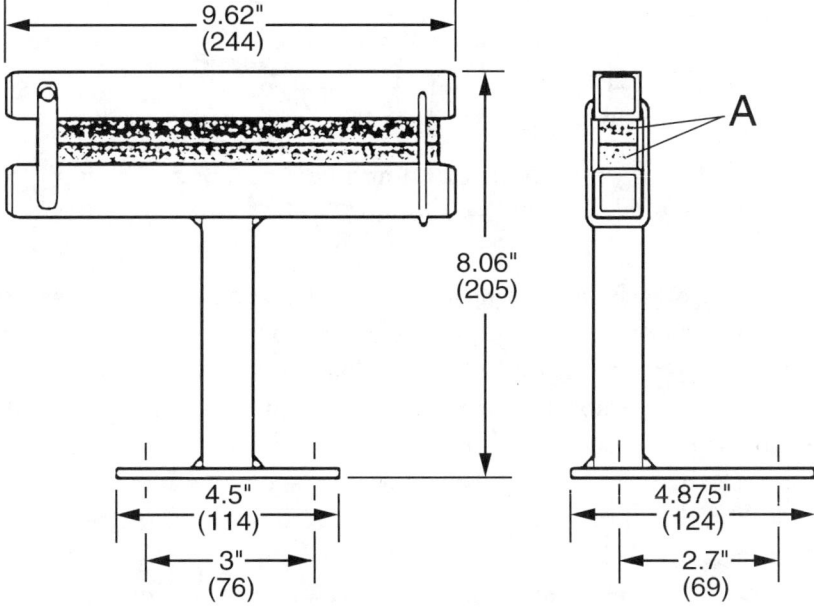

Fig. 4.11.5. Cable Management System; Dimensions in in.(mm).

In addition to this, the conduits must be in large loops or spirals that have low bending stiffness, thus isolating the objects.

Transmission of cable/hose vibrations to precision objects can be reduced by attaching relatively large masses to the cable/hose, with the loop- or cable-configured cable/hose between the mass and the object. A similar role is played by the so-called cable management system [4.28] shown in Fig. 4.11.5. The cable before the loop is clamped between two foam-coated faces A.

4.12 POWER TRANSMISSION COUPLINGS

If at least one of the connected power transforming devices (engine/motor, reducer, working organ of a machine) is vibration-isolated, the connection must provide for a relative mobility between the devices, otherwise the expected vibration-reducing effects would be negated. When a vibration isolated object is connected to another object and significant forces and/or torques are transmitted, then both angular and offset (radial) misalignments can be expected (Fig. 4.12.1). Some power transforming and power

408 • PASSIVE VIBRATION ISOLATION

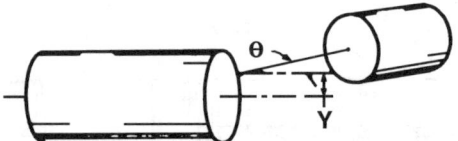

Fig. 4.12.1. Schematic of Combined Angular and Offset Shafts Misalignment.

transmission devices are prone to torsional resonant vibrations, which may be transmitted to the supporting structures. In both situations a properly selected connecting power transmission coupling may significantly improve the vibratory environment. Solution for the first of these tasks can be provided by misalignment-compensating couplings; for the second task, by torsionally flexible coupling—but the most widely used are combination purpose couplings combining both properties to a limited degree [4.61].

The purely misalignment-compensating couplings usually have high torsional stiffness and thus do not influence torsional vibrations. The popular designs comprise rigid and frictionally connected mechanical components. Oldham coupling has an intermediate member sliding relative to both connected hubs in mutually perpendicular directions. Such kinematics compensate only for radial (offset) misalignments. Due to high-pressure connections, friction between the sliding parts is quite high. Because of this, for small misalignments sliding does not commence at all and significant radial forces are transmitted between the connected shafts. The most widely used are gear couplings. These couplings have much lower contact pressures and thus are better lubricated. However, the friction forces transmitted to connected shafts are still significant, especially at small misalignments when lubrication is unreliable and static friction may prevail. These couplings can compensate only for angular misalignments. In order to compensate for offset misalignments, two gear couplings at the ends of a long extension shaft should be used. This approach is expensive, heavy, and requires an additional space.

A modification of Oldham coupling that is not based on sliding is shown in Fig. 4.12.2[4.62]. Hubs 101 and 102 have slots 106 and 107, respectively, whose axes are orthogonal. The intermediate disc can be assembled from two identical halves 103a and 103b. Slots 108a and 108b in the respective halves are also orthogonally orientated. Holders 105 are fastened to slots

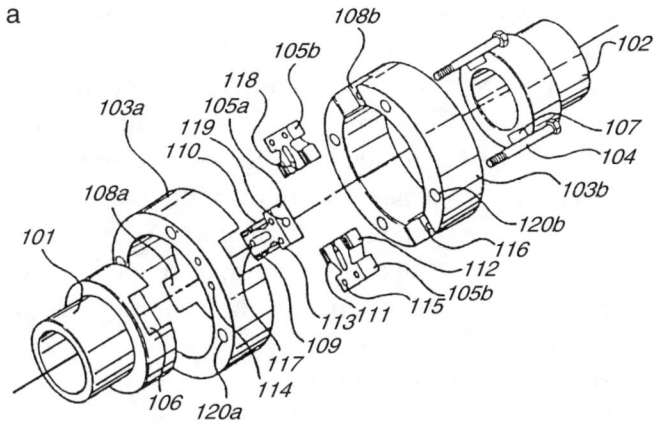
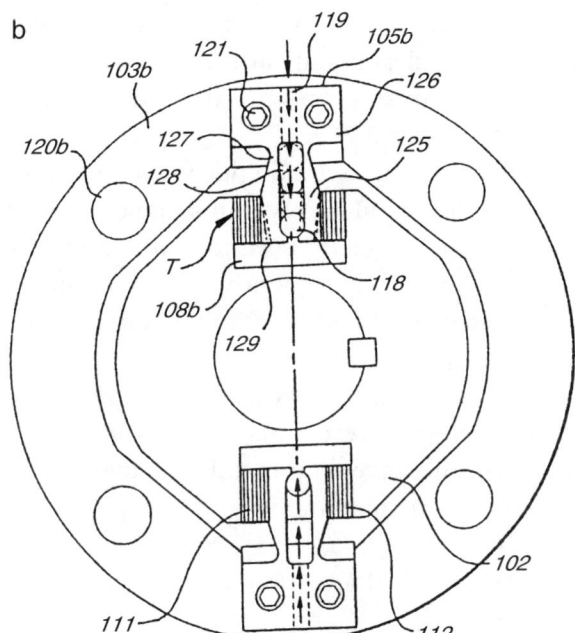

Fig. 4.12.2. Oldham Coupling with Thin-Layered Rubber-Metal Laminates.

108 in the intermediate disc and are connected to slots 106 and 107 via thin-layered rubber-metal-laminated elements 111 and 112, as detailed in Fig. 4.12.2(b). These elements are preloaded by sides 125 of holders 105, which spread out by moving preloading roller 118 radially toward the center.

This design provides for the kinematics advantages of the Oldham coupling without creating the above-listed problems associated with conventional Oldham couplings. The coupling is much smaller for the same rated torque than the conventional one, due to the high load-carrying capacity of the laminates and the absence of stress concentrations. It has very low stiffness in both radial directions and also compensates for some angular misalignments. The intermediate disc (usually the heaviest part of the coupling) can be made from a light strong material, such as aluminum. This makes the coupling suitable for high-speed applications.

The misalignment compensation stiffness, including ability to accommodate angular misalignments, and the rated torque can be varied by proportioning the laminates (their overall dimensions, thickness and number of rubber layers, etc.). The loads on the connected shafts are greatly reduced and are not dependent on the transmitted torque since the shear stiffness of the laminates does not depend significantly on the compression load (see Section 3.3.2C).

The torsionally flexible and combination purpose couplings are usually represented by the same designs. However, in some cases only torsional properties are required; in other cases both torsional and compensation properties are important, and most frequently these couplings are used as the cheapest available and users do not have much understanding of what is important in a specific application.

Torsionally flexible couplings are used in transmission systems when there is a danger of developing resonance conditions and/or transient dynamic overloads. Their influence on transmission dynamics can be due to one or more of the following factors:

- Reduction of torsional stiffness and consequent shifting of the natural frequencies,
- Increase of the effective damping capacity of the transmission system by using high-damping flexible elements or special dampers,
- Introduction of nonlinearity into the transmission system.

Torsional flexible couplings with elastomeric flexible elements always provide some compensation for radial and angular misalignments. The radial compensation stiffness of a typical coupling transmitting torque by compression of discrete elements uniformly distributed around the

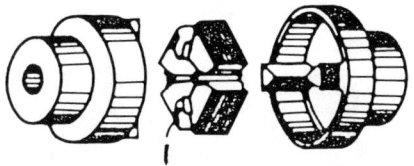

Fig. 4.12.3. Spider Coupling Giubomax (*Pirelli Rubber Co.*) (1—Rubber "Spider").

circumference (e.g., so-called spider or jaw coupling in Fig. 4.12.3) is [4.61]

$$k_{\text{com}} = \frac{F_{\text{rad}}}{e} = \frac{n}{2}(k_t + k_r), \qquad \text{Eq. (4.12.1)}$$

where k_t, k_r are stiffness constants of each torque-transmitting element in the circumferential (tangential) and radial direction, respectively; F_{rad} is radial force generated by radial misalignment e and exerted to the shafts and their bearings; and n is the number of torque-transmitting elements (only half of the elements in the coupling design transmit torque in each direction of rotation). If $k_t \ll k_r$, then the coupling can be called a purely torsionally flexible one. If k_t and k_r are commensurate, it can be called a combination purpose coupling.

It was shown in [4.61] that combination purpose couplings with $n = 2$ (like the one shown in Fig. 4.12.3) are characterized by undesirable fluctuations in the radial force they exert on the connected shafts—both in magnitude and in direction—even for a constant vector of the radial misalignment.

For a given design and for a certain value of the torsional stiffness k_{tor}, radial stiffness k_{com} of the coupling diminishes with increasing external radius R_{ex}. The ratio of radial (compensating) stiffness to torsional stiffness can be expressed as [4.61]

$$\frac{k_{\text{com}}}{k_{\text{tor}}} = \frac{A}{R_{ex}^2}, \qquad \text{Eq. (4.12.2)}$$

where the "coupling design index" A allows one to select a coupling design better suited for a specific application. If the main purpose is to reduce the misalignment-caused loading of the connected shafts and their bearings for a given value of the torsional stiffness, then the smallest value of index

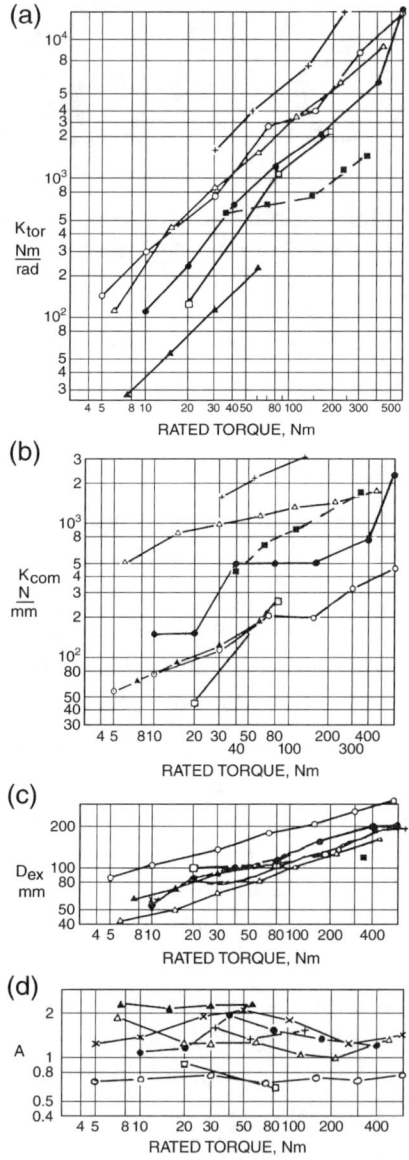

Fig. 4.12.4. Basic Characteristics of Some Combination Purpose Couplings: (a) Torsional Stiffness; (b) Radial Stiffness; (c) External Diameter; (d) Coupling Design Index ▲—Spider Coupling in Fig. 4.12.3; △—Spider Coupling with Straight Rectangular Legs; +—Finger Sleeve Coupling; o—Toroid Shell (tire) Coupling; □—Rubber Disc Coupling; ●—Centaflex Coupling; ■—Test Results for Coupling in Fig. 4.12.5 at Various Transmitted Torque for the Same Coupling.

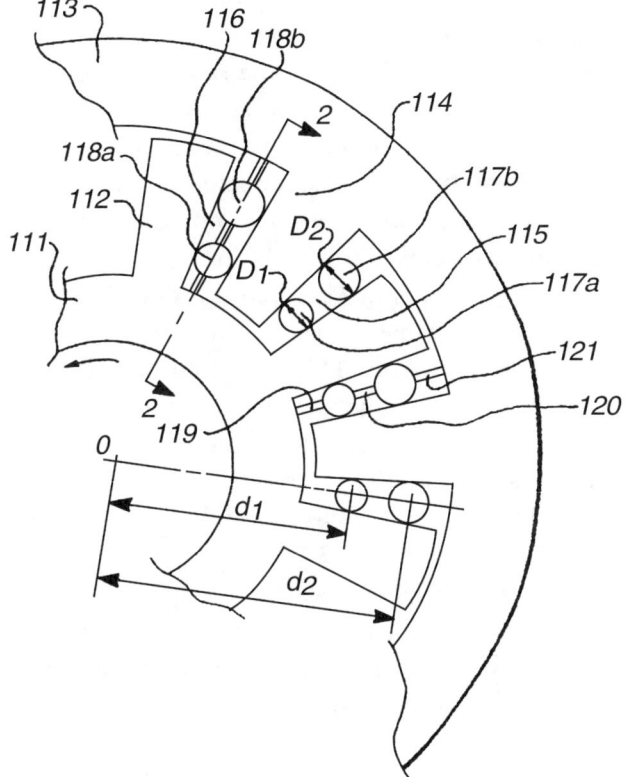

Fig. 4.12.5. Torsionally Flexible Coupling with Flexible Element Composed of Radially Compressed Rubber Cylinders.

A is the best, along with a large external radius. If the main purpose is to modify the dynamic characteristics of the transmission, then minimization of k_{tor} is important. Fig. 4.12.4 shows basic parameters of several popular coupling designs, as well as index A, as functions of the rated torque T_r.

The overwhelming majority of commercially available power transmission couplings with rubber flexible elements have both torsional and compensation load-deformation characteristics that are substantially linear in nature (more or less constant k_{tor} and k_{com}). Thus, for a given misalignment, forces acting on a vibration isolated object connected by such coupling do not depend on the transmitted torque. Of course, the same forces are acting on the shaft bearings. Since most of the installations are loaded with the rated torque magnitude only during a relatively small fraction of their operational time, the advantages of nonlinear couplings with a hardening load-deflection characteristic become apparent. While usually k_{tor} at the rated torque T_r of this type of a coupling is about the same as for the conventional (linear) coupling, most of the time k_{tor} and, consequently, k_{com} are much lower, thus reducing unwanted vibrations of the connected isolated object(s) and also the bearing loads. As noted above, torsional nonlinearity is also beneficial for torsional dynamics of power transmission systems.

The coupling in Fig. 4.12.5[4.63] employs radially compressed rubber cylinders 117 for torque transmission in one direction and 118 for the opposite direction, between hubs 111 and 113 attached to the connected shafts. This design combines the desirable nonlinearity with a significantly smaller size for a given T_r [due to the use of multiple cylindrical elements with the same relative compression in each space between protruding blades 112, 114, as well as due to high allowable compression of the rubber cylinders (Section 3.3.2B and 4.5.2), thus allowing use of smaller diameters and, consequently, many sets of cylinders around the circumference]. Test results for such coupling for $T_r = 350$ Nm are shown as ■ in Fig. 4.12.4; in this case the data refer to different T_r, but to the same coupling at different transmitted torques.

REFERENCES

[4.1] Rivin, E.I., 1999, "Shaped Elastomeric Components for Vibration Control Devices," *Sound and Vibration*, Vol. 33, No. 7, pp. 18–23.

[4.2] Rivin, E.I. and Lee, B.S., 1994, "Experimental Study of Load-Deflection and Creep Characteristics of Compressed Rubber Components for Vibration Control Devices," *ASME J. of Mechanical Design*, Vol. 116, No.2, pp. 539–549.

[4.3] Lee, B.S. and Rivin, E.I., 1996, "Finite Element Analysis of Load-Deflection and Creep Characteristics of Compressed Rubber Components for Vibration Control Devices," *ASME J. of Mechanical Design*, Vol. 118, pp. 328–336.

[4.4] DeBra, D.B., 1992, "Vibration Isolation of Precision Machine Tools and Instruments," *Annals of the CIRP*, Vol. 41/2, pp. 711–718.

[4.5] "Sorbothane Damping & Isolation Material," 1998, Catalog of *Sorbothane Inc*.

[4.6] EAR Corp. Catalog, 1999.

[4.7] Rivin, E.I., 1965, "Review of Vibration Insulation Mountings for Machine Tools," *Machines and Tooling*, Vol. 36, No. 8, pp. 37–46.

[4.8] Rivin, E.I., 1967, "New Designs of Vibroisolating Mountings and Carpets," *Russian Engrg. Journal*, No. 2, pp. 48–53.

[4.9] Rivin, E.I., 1979, "Principles and Criteria of Vibration Isolation of Machinery," *Trans. of ASME, J. of Mechanical Design*, Vol. 101, pp. 682–692.

[4.10] Rivin, E.I., 1986, "Design and Application Criteria for Connecting Couplings," *ASME J. of Mechanisms, Transmissions and Automation in Design*, Vol. 108, No. 1, pp. 96–105.

[4.11] NAVSEA 0900-LP-089-5010, 1977, Navy Resilient Mount Handbook. *A User Guide of Design, Installation and Inspection Information*.

[4.12] "Photo Briefs," 1966, *Mechanical Engineering*, Vol. 88, No. 2.

[4.13] Rivin, E.I. and Skvortzov, E.V., 1969, "Load-on-Each-Mount Measuring for Rubber-Metal Mountings," *Machines and Tooling*, No. 10, pp. 11–12.

[4.14] *Metalform '89 Catalog*, 1989, Vibro/Dynamics Corp.

[4.15] *Catalog ML-4RWF*, 1985, EFFBE Co.

[4.16] *Catalog 4 P X*, 1982, Paulstra Hutchinson Corp.

[4.17] *Catalog VP3*, 1976, Vibro/Dynamics Corp.

[4.18] "Pivotal Levelers," *Catalog*, TechProducts Corp.

[4.19] "Supports et pieds de machines antivibrations," 1986, *Catalog I.V.B.*, MarBett Plastc Engineering.

[4.20] *Industrial Products Catalog PC-2201g*, 1985, Lord Corp.

[4.21] Zavala, P.A.G., Pinto, M.G., Pavanello, R., and Vaqueiro, J., 2001, "Powertrain Mounting Development Based on Computational Simulation and Experimental Verification Method," *Proceedings of the 2001 SAE Noise and Vibration Conference (Noise 2001 CD), SAE Paper 2001-01-1509*.

[4.22] Stuecklschwaiger, W. and Ronacher, A., 1994, "Optimization of Engine Mount Parameters by Simulation and Statistic Techniques," *European ADAMS Users Conference*.

[4.23] Rivin, E.I., 1985, "Passive Engine Mounts—Directions for Future Development," *SAE Transactions*, pp. 3.582–3.592.

[4.24] "Vibrations. Constant Natural Frequency Rubber-Metal Isolators for Stationary Machinery. Parametric Series. Specifications," *USSR State Standard GOST 17712-72*.

[4.25] "Application Selection Guide," 1978, *Bulletin C5-178*, Barry Controls Corp.

[4.26] Rivin, E.I., "Constant Natural Frequency Vibration Isolator," *U.S. Patent 3,442,475*.

[4.27] Rivin, E.I., "Elastic Vibration Proof Support," *U.S. Patent 3,460,786*.

[4.28] "Vibration Control," 2000, *Newport Corp. Catalog*.

[4.29] "Affordable High Performance Reinforced Polyurethane Shock and Vibration Mounts," Report on U.S. Navy Contract N00167-98-C-0034, Globe Rubber Works Inc.

[4.30] Rivin, E.I., "Nonlinear Flexible Connections with Streamlined Resilient Elements," U.S. Patent 5,934,653.

[4.31] Huang, B. and Rivin, E.I., "Wheel Suspension," U.S. Patent 4,188,048.

[4.32] Huang, B. and Rivin, E., 1980, "Noise Abatement of In-Plant Trailers," SAE Paper 800494.

[4.33] Bulletin KDC/2, 2000, Korfund Dynamics Co.

[4.34] "Vibration and Shock Handbook," 1983, Bulletin VIB 8-12-83, Vibrachok USA.

[4.35] Frolov, K.V. (editor), 1981, "Vibratsii v tekhnike [Vibrations in Technology]," Vol. 6, Mashinostroenie Publ. House [in Russian].

[4.36] Kelly, J., 1984, "Concepts and Effects of Damping in Isolators," Proceedings of Vibration Damping 1984 Workshop, pp. L1–L15.

[4.37] "Aeroflex Isolators for Vibration Protection in All Environments," 2001, Aeroflex International, Bulletin 2M 1/2001.

[4.38] "Steelpaw® Wire Rope Isolators," 1998, Bulletin 3/98 CGWRI, Enidine Inc.

[4.39] Maltz, G., 1989, "Drahtseil-Federelemente im Vergleich zu Gummi-Metall-Schienen Maßstab: Verlustwinkel [Comparison of Cable and Rubber Flexible Elements]," KEM: Konstruction, Electronics, Maschinenbau, Vol. 26, No. 1, pp. 48, 50 [in German].

[4.40] Rivin, E.I., 1983, "Ultra-Thin-Layered Rubber-Metal Laminates for Limited Travel Bearings," Tribology International, Vol. 16, No. 1, pp. 17–25.

[4.41] Derham, C.J. and Thomas, A.G., 1980, "The Design and Use of Rubber Bearings for Vibration Isolation and Seismic Protection of Structures," Engineering Structures, Vol. 2, No. 3, pp. 171–175.

[4.42] Tikhonov, V.A. and Yakovlev, N.G., 1986, "Use of Thin-Layered

Rubber-Metal Elements for Vibration Protection Systems," *Kolebaniya i vibroakusticheskaya aktivnost mashin I konstruktsii*, Nauka Publish. House, Moscow, pp. 33–42 [in Russian].

[4.43] Timoshenko, S.P. and Gere, J.M., 1961, *Theory of Elastic Stability*, McGraw-Hill, N.Y.

[4.44] Rivin, E.I., 1999, *Stiffness and Damping in Mechanical Design*, Marcel Dekker Inc., N.Y.

[4.45] Platus, D.L., "Vibration Isolation System," *U.S. Patent 5,178,357*.

[4.46] Platus, D.L, 1993, "Smoothing Out Bad Vibes," *Machine Design*, February 23, pp. 123–130.

[4.47] Freakley, P.K. and Payne, A.R., 1978, *Theory and Practice of Engineering with Rubber*, Applied Science Publishers.

[4.48] Alabuzhev, P., Gritchin, A., Kim, L., Migirenko, G., Chon, V., and Stepanov, P., 1989, *Vibration Protecting and Measuring Systems with Quasi-Zero Stiffness*, Hemisphere Publishing, N.Y.

[4.49] Schubert, D.W., 1976, "Dynamic Characteristics of a Pneumatic-Elastomeric Shock and Vibration Isolator for Heavy Machinery," *Proceedings of Institute of Environmental Sciences*, pp. 244–251.

[4.50] Heiland, K.P., 1991, "Recent Advancements in Passive and Active Vibration Control Systems," *Vibration Control in Microelectronics, Optics and Metrology*, SPIE, Vol. 1619, pp. 22–30.

[4.51] Vukobratovich, D., 1987, "Principles of Vibration Isolation," *Vibration Control in Optics and Metrology*, SPIE, Vol. 732, pp. 27–44.

[4.52] Cellucci, T.A., 1991, "Matching Performance Requirements with Technology. An Applications Approach to Vibration Isolation," *Vibration Control in Microelectronics, Optics and Metrology*, SPIE, Vol. 1619, pp. 2–10.

[4.53] Kaminskaya, V.V. and Reshetov, D.N., 1975, *Fundamenty i ustanovka metallorezhuschikh stankov [Foundations and Installation of Machine Tools]*, Mashinostroenie Publish. House, Moscow, 208 pp. [in Russian].

[4.54] Burdekin, M., Huang, Z.J., and Hinduja, S., 1986, "Predicting the Influence of the Foundations on the Accuracy of a Large Machine tool,"

Proceedings of the 26th Intern. Machine Tool Design and Research Confer., edited by B.J. Davies, MacMillan Publishers Ltd., London, pp. 227–237.

[4.55] Coto, M., 1974, "Isolation des vibrations par l'utilization de foundations flottantes [Vibration Isolation by Floating Foundation]," *Formage et trait mitaux*, No. 58, pp. 65–67 [in French].

[4.56] Hailer, J., 1966, "Die selbstattige Ausrichtung von Werkzeugmaschinen—ein Beitrag zur Lösung des Aufstellungsproblems [Self-adjusting Straightness for Machine Tools–an Example of Solving the Installation Problem]," *Maschinenmarkt*, Vol. 72, No. 70 [in German].

[4.57] Kliukin, I.I. and Bogolepov, I.I., 1978, *Spravochnik po sudovoi akustike [Ship Acoustics Handbook]*, Sudostroenie Publish. House, Leningrad [in Russian].

[4.58] Nelson, F.C., and Nokes, D.S., 2001, "A Primer on Vibration Isolation," *Course Notes, 25th Meeting of the Vibration Institute*, Cincinnati, OH.

[4.59] Wilby, J.F., 2001, "Vibration of Structures Induced by Sound," Chapter 29/3 in *Harris's Shock and Vibration Handbook*, 5th Ed., McGraw-Hill, N.Y.

[4.60] "SENTRY," 2002, *Catalog*, Blacoh Fluid Control Co.

[4.61] Rivin, E.I., 1986, "Design and Application Criteria for Connecting Couplings," *Trans. of ASME, J. of Mechanisms, Transmissions, and Automation in Design*, Vol. 108, pp. 96–105.

[4.62] Rivin, E.I., "Torsionally Rigid Misalignment Compensating Coupling," *U.S. Patent 5,595,540*.

[4.63] Rivin, E.I., "Torsional Connection with Radially Spaced Multiple Flexible Elements," *U.S. Patent 5,630,758*.

INDEX

A

Absolute transmissibility 28, 37
Acceleration of gravity 42
Acrylonitrile butadiene 236
Active-hydraulic isolators 291
Active system 110, 141
Active vibration isolators 378
Adiabatic condition 276
Air conditioners 328, 405
Air-Loc™ isolator 271, 313
Ambient vibrations 110
Amplitude dependency 89
Anchoring 180, 315, 326
Angular acceleration 43
Angular natural frequency 33
Angular stiffness 9, 11, 13, 17, 19, 139, 147, 373
Anisotropic stiffness 251
Anti-aging agents 237
Anvil 174
Asymmetry factor 65
Axial compression 246

B

Ball screw 44
Bearings 185
Beca isolator 330
Belleville springs 214, 222, 303
Bessel's points 150, 151
Billet 174
Buckling 243, 255, 307, 370, 378
Buckling stability 239
Bulging 239, 308
Bulk modulus 238, 251
236
Butyl rubber (BR or IIR) 201, 236, 261, 264

C

Cable isolators 233, 368
Caisson 115
Capillary 227, 278
Capillary resistance coefficient 278
Captive design 199, 333, 342

Carbon black 36, 237, 263
CE isolator 364
Center of mass/gravity 5, 21
Center of percussion 70
Center of rocking 67
Centile spectra 112
Centrifugal force 184
Centrifugal moment of inertia 6
Characteristic equation 52
Chatter 106, 123, 180
Chiller 328
Chloroprene rubbers (CR) 236
Circuit cards 128
Circular arch isolator 372
Civil engineering structures 156
Coordinate Measuring Machine (CMM) 5, 23, 25, 106, 138, 146, 372
Coil spring 214, 303
Coil Spring Machined 216
Cold header 5, 170, 373
Complex dynamic stiffness 32, 259
Complex Young's modulus 259
Compressed air 275
Compression set 236
Conical coil spring 91, 340
Conical isolator 332
Conservative pulse 169
Constant natural frequency isolator (CNF) 20, 90, 144, 248, 304, 335
Constant stiffness isolator 20
Construction/agricultural machinery 344
Cork 272, 311
COTS xii
Coulomb friction 27, 228, 231, 263
Coulomb friction damper 27
Counterweights 148
Coupling 16, 64, 70
Coupling coefficient 51
Crank press 5
Creep 213, 231, 236, 258, 270, 301, 303, 313
Critical damping coefficient 29
Critical force 255

Critical strain 258
Cross stiffness coefficient 9
Cylindrical grinders 128, 155, 314

D
Damping 8, 26, 107
Damping coefficient 12, 13, 27, 29, 226, 229
Damping force 225
Damping Ratio 82
Decade of time 270
Decoupling 24, 70, 72, 194
Delta function pulse 80
Density 226
Die 174
Diesel engines 200, 330
Dimensional chains 120
Disc drive 128
Distributed parameter dynamic system 3
DK isolator 369
Double bell isolator 369
Drivetrain 192
Duhamel integral 79
Durometer 23, 197
Dynamic absorption 45
Dynamic coupling 20, 48, 131, 144, 174
Dynamic loads 180, 185
Dynamic model 26
Dynamic nonlinearities 89
Dynamic stability 180
Dynamic stability criterion 182
Dynamic stiffness 213
Dynamic stiffness coefficient 230
Dynamic-to-static stiffness ratio 301
Dynamic viscosity 226

E
EAR™ isolators 306
Earthquake protection 157, 372
Eddy-Current 229
EDM isolator 229
EES isolator 335
Effective compression modulus 252
Effective damping 190
Effective mass 28, 189, 191
Effective modulus of elasticity 242
Effective stiffness 146, 149
Elastic center 9, 70
Elastic coupling coefficients 50

Elastic modulus 32
Elasto-damping materials 31, 213, 229
Electromagnetic Damper 229
Elastomeric Materials 235
Electronic devices 106, 119
Electronic equipment 119
Electro-rheological fluid 226
Engine 25, 192
Engine isolation 26, 190, 193
Engine mounts 26, 266, 268, 284, 314
Epoxidized natural rubber (ENR) 236, 262
Ethylene-propylene-diene (EPDM) rubber 236
Euler force 255
Evidgom isolator 329

F
Fans 328, 405
Fast Fourier Transform 111
Fatigue resistance 265
Felt 232, 233
Fillers 237
Filters 78
Flexible production xi
Floor vibration 115, 131
Flow 279
Flow laminar 226
Flow turbulent 226
Fluoro-elastomers 236
Flywheel 42
Foam matrix 248, 360
Foam mats 305
Focalized systems xiii, 198
Force/moment matrix 14
Force transmissibility 29
Forging hammers 5, 88, 172
Foundation block 173, 189
Four-cylinder engines 284
Fourier integral 80
Fourier series 78
Fourier transform 80
Friction angle 43
Frictional joints 119
Frequency response 144

G
Gasoline engines 200
Generalized damping coefficients 50

Generalized excitation forces 50
Generalized masses 50
Generalized stiffness 50, 55
Glass transition 261
Grouting 315
Gyroscopes 373

H
Hammer crusher 8
Hardness 237, 239
High-damping metals 272
HVAC equipment 364
Hydraulic mount 201
Hydro/Level isolator 326
Hydromounts 285
Hydrostatic compressibility 255
Hydrostatic pressure 251
Hysteresis damping 36
Hysteretic damping 27, 134, 273

I
Ideal shape elastomeric elements 245, 353
Impact pulse 175
Impact testers 172
Impedance 26, 189
Impulse response function 80
Inclined mount 70, 76
Induction motor 167
Inertia block 19, 42, 59, 70, 81, 110, 147, 161, 169, 301, 366, 373
Inertia matrix 14
Inertias 55
Injection molding machine 172
In-plant trailers 361
Instrument mounts 314
Inter-coordinate coupling 16, 89, 95, 165, 323
Isodamp™ isolator 237
Isolating mats 304
Isolating pads 304
Isolation bushings 266
Isolation criterion 323
Isolation zone 30
Isoloss™ isolators 237
Isoprene rubber (IR) 236

J
Jolt tables/machines 172, 328

K
Kinematic excitation 29, 34
Kinematic mounting/support 22, 149, 152
Kinematic viscosity 226
Kinetic energy 54
Knee milling machine 321
Knife-type folding machine 202

L
Lagrange equation 55
Lagrange function 54
Laminar flow 226, 393
Lastosphere isolator 247, 354
Lathe 182, 188, 314
Lattice isolator 328
Leveling 141, 145, 154, 275, 314, 326, 351
Level sensor 400
Level stabilization time 394
LM isolator 321
Load-deflection characteristic 24
Load range 341
Lod/Sen isolator 321
Logarithmic decrement 27, 29, 230
Loom 166, 328, 373
Loss angle 32
Loss factor 230
Loss modulus 32

M
Machinery mounts 314
Machine tools 106
Magnetic resonance imaging 122
Magnetorheological fluids 226
Mass distribution 7
Mass ratio 98
Mean square value 78
Metal springs 213, 273
Metrology frame 148
Microcircuits 121
Microelectronics 112
Micro/Level isolator 323
Microscope 122, 146
Military standards 119
Misalignment-compensating coupling 109
Modal coefficient 53
Modeling 198
Mold conditioning machines 172
Moments of inertia 8, 76

Motion transformation xiii, 41, 162
Motion transformers 179
Motor generators 364

N
Natural angular frequency 29
Natural coordinate frame 5, 6, 13
Natural frequency 82
Natural period 85, 86
Natural Rubber (NR) 236
NBR Rubber 262, 351
ND isolators 353
Negative damping 180
Negative stiffness 381
Neoprene 321
Nitrile rubbers NBR 236
Nivofix isolator 323
Noise 180, 188, 316
Nonlinear coil springs 347
Nonlinearity 20, 89
Nonlinear metal spring 218
Normal coordinates 54, 68
NVH (Noise, Vibration, and Harshness) 192

O
Off-road vehicles 317
Oldham coupling 408
Optical components 121
Optimization 198, 338
Optimum damping 38
Organ effect 386
Overload 345
Ozone 236

P
Pad 360
Parallel connection 222
Partial natural frequencies 72, 96, 174, 181
Partial subsystem 20, 48, 126
Phase shift 29
Photolithography tools 106, 111, 146
Piles 115
Planar system 96
Planar Vibration Isolation System 61, 66, 69
Plane of symmetry 19
Planer 400
Plastic fibrous pad 310
Plasticizers 237

Plastics 271
Pneumatic damping 347
Pneumatic flexible element 275
Pneumatic isolator 12, 161
Poisson's Ratio 235, 239
Polyharmonic Excitation 166
Polyurethane 238, 271, 353
Polyvinyl Chloride (PVC) 238, 271, 313
Potential energy 29, 54
Power assist leveling 326
Power transmission coupling 404
Preload 268, 336
Principal axes of inertia 5, 6, 76, 193
Principal elastic axes 9
Principal moments of inertia 76
Printing shop 202
Probability 77
Product of inertia 6, 17
Projection Aligner 124
Proximity sensors 121
Psychological inertia 204, 272
Pulsation dampener 406
Pulse duration 86
Pulse excitation 88
Pumps 364
Punch Press 170

Q
Quasi-resonance 366
Quasi-zero stiffness 224, 233, 304, 381

R
Radial compression 246
Radially loaded cylinder 249
Radius of inertia 5, 7, 66, 68, 71, 127, 193, 242
Random excitation 77
Reaction frame 147
Reaction moments 169
Reciprocating compressor 330
Reciprocating mechanisms 106
Relative coordinates 48
Relative damping 29
Relative damping coefficient 29, 52
Relative energy dissipation 29
Relative transmissibility 28, 37
Relative velocity 27
Relative vibrations 82

Resonance xiv, 26, 29, 65, 98, 108
Resonance amplification factor 29, 37
Response spectrum 83
Reynolds number 226, 229
Rigidizing frame 154
RM isolator 375
Rocking 42, 63, 70, 138, 158, 171, 316, 364, 373
Rocking motions 17, 110, 145, 191
Rolling diaphragm 386
Root mean square 79
Rotor 5, 186
Rubber hardness 24
Rubber mat 310
Rubber-metal laminate 157, 376
Rubber mills 328
Rubber sphere 250

S

Scanner 146
Scanning electron microscopes 118
Screens 328
Seat 344
SEDU isolator 325
Self-leveling 387
Self-snubbing 91
Semi-active mounts 201, 393
Series connection 222
Servo-controlled ix, 378
Servo-controlled hydraulic isolators 291
Shaker 160, 173
Shape factor 242, 348
Shape memory alloys 213, 274
Shear Disc 361
Shear modulus 214, 238, 251, 259
Shim pad 313
Shock isolation criterion 179
Shock spectrum 83, 143, 170, 173, 174
Shore durometer 239
Sifters 328
Silicone fluid 226
Silicone rubber 236, 304
Six-degrees-of-freedom 16
Slotted springs 214, 220
Snubbers 89, 192, 199, 334, 344, 387
Softeners 237
Soils 174
Sonic boom 111, 405

Sorbothane™ isolator material 238, 306
Specific heats ratio 276
Spectral power density 78
Spectrum 80, 131
Stabiflex isolator 333
Stability conditions 217
Stages 146
Stamping press 169, 188
Standing waves 99
Start-stop 366
Static nonlinearity 89
Static stiffness 138, 213
Stay bolts 324
Steel springs 99
Stepper 125, 146
Stick-slip 106
Stiffness center 70
Stiffness coefficients 9, 13
Stiffness ratio 39, 68, 170, 288, 307, 331
Stochastic 77
Strain relaxation 154
Streamlined elastomeric elements 269, 353
Stress concentrations 243
Stress distribution 243
Sub-harmonic vibrations 91
Submarines 335
Surface grinder 5, 6, 25, 120, 146, 372
Surge 217, 278
Surge suppressor 406
Swaging 267
Symmetry 6, 63
Synchrotrons 122

T

Thin-layered rubber-metal laminates 147, 239
Thiocol RD rubber 36
Tico™ isolator material 271
Tilting 109, 148, 321
Torsional stiffness 193
Torque-reaction 199
Torus 249
Transfer function 80
Transient vibrations 25, 111
Transmissibility 38, 57
Truck 344
Tup 174
Two-Stage Isolator 56

U

Unbalanced rotor 65, 106, 160, 163
Uni-axial vibration isolation system 3, 26

V

Variable-stiffness isolators 202, 337
Vibra-Check™ isolator pad 271
Vibrachok isolators 121, 232
Vibration isolation criterion 131, 137
Vibration Isolation Synthesis 139, 140, 141, 142
Vibration level 180
Vibration level criterion 185
Vibration protection x
Vibration shaker 106
Vibratory conveyers 328
Vibratory velocity level criterion 185
Vibro-impact system 90
Virtual inertia 42
Viscous damper 27, 225
Viscous damping 134
Volume ratio 280
Vulcanizing agents 237

W

Wahl factor 215
Wave effects 4, 8, 96
Wear 316
Wedge mounts 121
Wedge-shaped isolator 331
Weight distribution 23, 144, 318, 320
Weight load distribution 343
Weight-sensing mounts 318
White noise 79
Wire-mesh 230, 263
Wire-mesh dampers 367
Wire rope isolators 370

Y

Young's modulus 239, 251, 381

Z

Zero mode 48
Zero stiffness 380